The Forgotten Man

ALSO BY AMITY SHLAES

The Greedy Hand

Germany: The Empire Within

The Forgotten Man

A New History of the Great Depression

Amity Shlaes

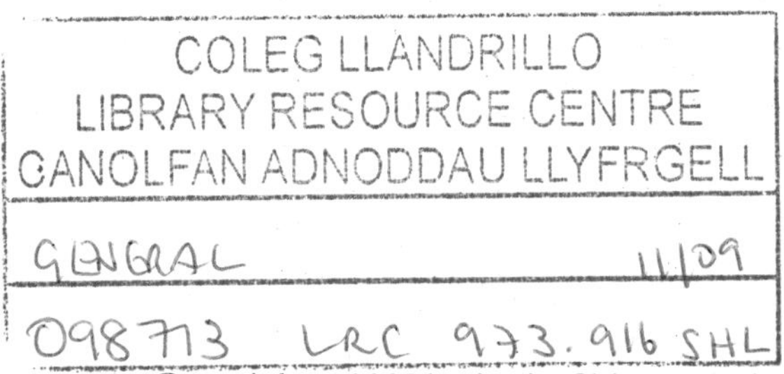

HarperCollins books may be purchased for educational, business, or sales promotional use. For information please write: Special Markets Department, HarperCollins Publishers, 10 East 53rd Street, New York, NY 10022.

FIRST HARPERLUXE EDITION

HarperLuxe™ is a trademark of HarperCollins Publishers.

Library of Congress Cataloging-in-Publication Data is available upon request.

ISBN: 978-0-06-128527-1

ISBN-10: 0-06-128527-7

09 10 11 NMSG/BVG 10 9 8

for my parents

These unhappy times call for the building of plans that rest upon the forgotten, the unorganized but the indispensable units of economic power, for plans like those of 1917 that build from the bottom up and not from the top down, that put their faith once more in the forgotten man at the bottom of the economic pyramid.

—Gov. Franklin Roosevelt of New York,
radio address in Albany, April 7, 1932

As soon as A observes something which seems to him to be wrong, from which X is suffering, A talks it over with B, and A and B then propose to get a law passed to remedy the evil and help X. Their law always proposes to determine what C shall do for X, or in the better case, what A, B, and C shall do for X. . . . What I want to do is to look up C. I want to show you what manner of man he is. I call him the Forgotten Man. Perhaps the appellation is not strictly correct. He is the man who never is thought of. . . .

He works, he votes, generally he prays—but he always pays. . . .

—William Graham Sumner,
Yale University, 1883

Contents

Introduction

One November evening long ago in Greenpoint, Brooklyn, a thirteen-year-old named William Troeller hanged himself from the transom in his bedroom. The boy had watched his family slide into an increasingly desperate situation. The gas for their five-room apartment on Driggs Avenue had been shut off since April. His father, Harold, had lost his job at Brooklyn Edison after suffering a "rupture"—a worker's hernia, probably. While the father waited for surgery in Kings County Hospital, Mrs. Troeller and six children waited too. Ruth, eighteen, wanted to work as a waitress, but she was unemployed. Harold Jr., twenty-one, had work in a government program. Harold told a newspaper reporter that his brother "was sensitive and always felt embarrassed" about asking for his share at mealtime. The Herbert Street police station near the Troeller home helped to arrange the funeral. Burial would be in a Catholic cemetery. "He Was Reluctant about Asking for Food," read the headline in the *New York Times.* New York that year had a Dickensian feel—an un-American feel.

William Troeller was just one story in a city of troubled stories. Across the East River, at an office desk at 20 Pine Street, a utilities executive named Wendell Willkie had spent the fall watching and pondering the downturn. Equally concerned was another executive who worked a block south of Willkie at 120 Wall—the vice president of American Molasses, an agriculture expert named Rexford Tugwell. Both men knew what the boy had probably only sensed: the despair was deep. Later in fact it would emerge that the birthrate that November, the month of William's suicide, was the second lowest on record for November for the city. A few weeks prior to William's act the Dow had dropped nearly 8 percent—the day had already come to be known as Black Tuesday.

It was a dark moment for the country as well. Willkie came from farm territory—Indiana. The next year two in ten Indiana families would collect some form of relief. Willkie had bought up a number of farms, in part to conduct a little experiment for himself: he wanted to see whether farming was still possible in America.

The Indiana story replicated itself across the states. A year before William Troeller's suicide, the Brookings Institution, a new think tank, had warned that the balance of the economy was "precarious." The economy was tipping. Nationally, durable manufacturing—the most important single meter on the dashboard of the economic engine—was plummeting, and more rapidly than anyone could remember. Unemployment was moving up by the millions. The next spring, the spring after William Troeller's death, one in five American men would be unemployed. Watching from Britain, the *Economist* would conclude in retrospect that the United States "seemed to have forgotten, for the moment, how to grow."

Yet Washington was doing all the wrong things. Officials in the capital seemed arrogant, obsessed with numbers, and oblivious to the pain the nation was suffering. People were angry that Congress and the president had recently raised taxes. With business so hard, why make it harder?

The very week of William Troeller's suicide, the treasury secretary went before an audience from the Academy of Political Science at the Hotel Astor. The secretary was so anxious about the speech that he had insisted that the president approve it. At his side stood Parker Gilbert, one of the old conservatives from the happy 1920s. There had been a national emergency in the past, the secretary told listeners. But now it no longer existed. The secretary then went on to conclude that the country must now "continue progress toward a balance of the federal budget."

A member of the audience laughed out loud in shock. The remark seemed so much at odds with the painful reality of that November. That week, there would be an article in the paper about Herbert Hoover. But it did not carry news of his views on the crisis. The article merely covered the fact that Hoover had traveled to Colby College in Waterville, Maine. Maine was one of the states that had voted for him, and now Colby would give Hoover yet another of his many honorary degrees.

The story sounds familiar. It is something like the descriptions we hear of the Great Crash of 1929. But in fact these events took place in the autumn of 1937. This was a depression within the Depression. It was occurring five years after Franklin Roosevelt was first elected, and four and a half years after Roosevelt introduced the New Deal. It was taking place eight years after President Herbert Hoover first made his own rescue plans following the 1929 stock market crash.

Washington had already made thousands of efforts to help the economy, yet those efforts had not brought prosperity. Rex Tugwell, the man at American Molasses, had led many campaigns, but now he had retreated to New York to reevaluate a record that counted too many failures.

The standard history of the Great Depression is one we know. The 1920s were a period of false growth and low morals. There was a certain godlessness—the *Great Gatsby* image—to the decade. The crash was the honest acknowledgment of the breakdown of capitalism—and the cause of the Depression. A dangerous inflation caused by speculating margin traders brought down the nation. There was a sense of a return to a sane, moral country with the crash. A sense that the economy of 1930 or 1931 could not revive without extensive intervention by Washington. Hoover, it was said, made matters worse through his obdurate refusal to take control, his risible commitment to what he called rugged individualism. Roosevelt, however, made things better by taking charge. His New Deal inspired and tided the country over. In this way, the country fended off revolution of the sort bringing down Europe. Without the New Deal, we would all have been lost.

The same history teaches that the New Deal was the period in which Americans learned that government spending was important to recoveries; and that the consumer alone can solve the problem of "excess capacity" on the producer's side. This explanation acknowledges that the New Deal did not bring the country to recovery fast, but emphasizes that the country got there eventually—especially with the boost of military spending in the late 1930s. The attitude is that the New Deal is the best model we have for what government must do for weak members of society, in both times of crisis and times of stability. And that the New Deal gave us splendid leaders and characters: Roosevelt himself, a crippled man who bravely willed us all back into prosperity and has been called the apostle of abundance. The brain trust, thoughtful men whose insights validated their experiments. The Hundred Days—that period at the start of his first term when Roosevelt legislated unprecedented reforms—was a thrilling period. From Adolf Berle, the expert on corporations, to Frances Perkins, the pioneering social reformer, to Tugwell, the New Dealers displayed a sort of dynamism from which today's moribund politicians might learn.

Without the New Deal, the country would have followed a demagogue, Huey Long, or worse, Father Coughlin. The rightness of Roosevelt's positions was only validated by what followed; FDR saved the country in peace, and then he saved it in war. Or so the story line goes.

The usual rebuttal to this from the right is that Hoover was a good man, albeit misunderstood, and Roosevelt a dangerous, even an evil one. The stock market of the 1920s was indeed immoral, too high, inflationary—and deserved to crash. Many critics on the right focus on monetary policy. Another set of critics focuses on Roosevelt's early social programs. They argue that New Deal programs such as the Works Progress Administration of the Civilian Conservation Corps spoiled the United States and accustomed Americans to the pernicious dole. Yet a third set of critics, an angry fringe, has argued that Roosevelt's brain trusters reported to Moscow. Stalin steered the New Deal and also pulled us into World War II, in their argument. For many years, now, these have been the parameters of the debate.

It is time to revisit the late 1920s and the 1930s. Then we see that neither the standard history nor the standard rebuttal entirely captures the realities of the period. The first reality was that the 1920s was a great decade of true economic gains, a period whose strong positive aspects have been obscured by the troubles that followed. Those who placed their faith in laissez-faire in that decade were not all godless. Indeed religious piety moved some, including President Calvin Coolidge, to hold back, to pause before intervening in private lives.

The fact that the stock market rose high at the end of the decade does not mean that all the growth of the preceding ten years was an illusion. American capitalism did not break in 1929. The crash did not cause the Depression. It was a necessary correction of a too-high stock market, but not a necessary disaster. The market players at the time of the crash were not villains, though some of them—Albert Wiggin of Chase, who shorted his own bank's stock—behaved reprehensibly. There was indeed an annihilating event that followed the crash, one that Hoover never understood and Roosevelt understood incompletely: deflation.

Hoover's priggish temperament, as much as any philosophy he held, caused him to both misjudge the crash and fail in his reaction to it. And his preference for Germany as a negotiating partner over Soviet Russia later blinded him to the dangers of Nazism. Roosevelt by contrast had a wonderful temperament, and could get along, when he felt like it, with even his worst opponent. His calls for courage, his Fireside Chats, all were intensely important. "The only thing we have to fear is fear itself"—in the darkness, Roosevelt's voice seemed to shine. He allowed Cordell Hull to write trade treaties that in the end would benefit the U.S. economy enormously. Roosevelt's dislike of Germany, which dated from childhood, helped him to understand the threat of Hitler—and, eventually, that the United States must come to Europe's side.

Still, Hoover and Roosevelt were alike in several regards. Both preferred to control events and people. Both underestimated the strength of the American economy. Both doubted its ability to right itself in a storm. Hoover mistrusted the stock market. Roosevelt mistrusted it more. Roosevelt offered rhetorical optimism, but pessimism underlay his policies. Though Americans associated Roosevelt with bounty, his insistent emphasis on sharing—rationing, almost—betrayed a conviction that the country had entered a permanent era of scarcity. Both presidents overestimated the value of government planning. Hoover, the Quaker, favored the community over the individual. Roosevelt, the Episcopalian, found laissez-faire economics immoral and disturbingly un-Christian.

And both men doctored the economy habitually. Hoover was a constitutionalist and took pains to intervene within the rules—but his interventions were substantial. Roosevelt cared little for constitutional niceties and believed they blocked progress. His remedies were on a greater scale and often inspired by socialist or fascist models abroad. A number of New Dealers, Tugwell included, had been profoundly shaped by Mussolini's Italy and, especially, Soviet Russia. That influence was not parenthetical. The hoarse-voiced opponents of the New Deal liked to focus on the connections between these men, the Communist Party, and authorities in Soviet Russia. And several important New Dealers did indeed have those connections, most notably Lauchlin Currie, Roosevelt's economics adviser in later years, and Harry Dexter White, at the Treasury. White's plan for the pastoralization of Germany takes on a new light when we know this. Lee Pressman and Alger Hiss duped colleagues in government repeatedly.

But few New Dealers were spies or even communists. The emphasis on that question is in any case misplaced. Overall, the problem of the New Dealers on the left was not their relationship with Moscow or the Communist Party in the United States, if indeed they had one. Senator McCarthy was wrong. The problem was their naïveté about the economic value of Soviet-style or European-style collectivism—and the fact that they forced such collectivism upon their own country. Fear of being labeled a red-baiter has too long prevented historians from looking into the Soviet influence upon American domestic policy in the 1930s.

What then caused the Depression? Part of the trouble was indeed the crash. There were monetary and credit challenges at the young Federal Reserve, and certainly at the banks. Deflation, not inflation, was a big problem, both early on and also later, in the mid-1930s. The loss of international trade played an enormous role—just as both Hoover and Roosevelt said at different points. If the United States had not raised tariffs at the beginning of the decade and Europe had not collapsed in the 1930s, the United States would have had a trading partner to help sustain it. Part of the problem was the challenge of the transition to industrialization from agriculture. Part was freakish weather: floods and the uncanny Dust Bowl seemed to validate the sense of apocalypse. With money and the weather breaking down, men and women in America felt extraordinarily helpless. They were willing to suspend disbelief.

But the deepest problem was the intervention, the lack of faith in the marketplace. Government management of the late 1920s and 1930s hurt the economy. Both Hoover and Roosevelt misstepped in a number of ways. Hoover ordered wages up when they wanted to go down. He allowed a disastrous tariff, Smoot-Hawley, to become law when he should have had the sense to block it. He raised taxes when neither citizens individually nor the economy as a whole could afford the change. After 1932, New Zealand, Japan, Greece, Romania, Chile, Denmark, Finland, and Sweden began seeing industrial production levels rise again—but not the United States.

Roosevelt's errors had a different quality but were equally devastating. He created regulatory, aid, and relief agencies based on the premise that recovery could be achieved only through a large military-style effort. Some of these were useful—the financial institutions he established upon entering office. Some were inspiring—the Civilian Conservation Corps, for example, which created parks, bridges, and roads we still enjoy today. From Wyoming, whose every county saw the introduction of projects, including the dramatic Guernsey State Park, to Greenville, Maine, whose CCC Road still bears the program's name today, the CCC heartened young Americans and found a place in national memory. CCC workers planted a total of three billion trees across the country. Establishing the Securities and Exchange Commission, enacting banking reform—as well as the reform of the Federal Reserve system—all had a stabilizing effect. Roosevelt's desire to control tariff law worked to the benefit of the economy, for, through Cordell Hull, he undid some of the damage of the Smoot-Hawley tariff.

Other new institutions, such as the National Recovery Administration, did damage. The NRA's mandate mistook macroeconomic problems for micro problems—it sought to solve the monetary challenge through price setting. NRA rules were so stringent they perversely hurt businesses. They frightened away capital, and they discouraged employers from hiring workers. Another problem was that laws like that which created the NRA—and Roosevelt signed a number of them—were so broad that no one knew how they would be interpreted. The resulting hesitation in itself arrested growth.

Where the private sector could help to bring the economy back—in the arena of utilities, for example—Roosevelt and his New Dealers often suppressed it. The creation of the Tennessee Valley Authority snuffed out a growing—and potentially successful—effort to light up the South. The company that would have delivered that electricity was Willkie's company, Commonwealth and Southern. The *New Yorker* magazine's cartoons of the plump, terrified Wall Streeter were accurate; business was terrified of the president. But the cartoons did not depict the consequences of that intimidation: that businesses decided to wait Roosevelt out, hold on to their cash, and invest in future years. Yet Roosevelt retaliated by introducing a tax—the undistributed profits tax—to press the money out of them.

Such forays prevented recovery and took the country into the depression within the Depression of 1937 and 1938, the one in which William Troeller died and Willkie worried. One of the most famous Roosevelt phrases in history, almost as famous as "fear itself," was Roosevelt's boast that he would promulgate "bold, persistent experimentation." But Roosevelt's commitment to experimentation itself created fear. And many Americans knew this at the time. In autumn 1937, the *New York Times* delivered its analysis of the economy's downturn: "The cause is attributed by some to taxation and alleged federal curbs on industry; by others, to the demoralization of production caused by strikes." Both the taxes and the strikes were the result of Roosevelt policy; the strikes had been made possible by the Wagner Act the year before. As scholars have long noted, the high wages generated by New Deal legislation helped those workers who earned them. But the inflexibility of those wages also prevented companies from hiring additional workers. Hence the persistent shortage of jobs in the latter part of the 1930s. New Deal laws themselves contributed to the sense of lost opportunity. This sense is what led to the famous description of the period that we have all heard—"the Depression was not so bad if you had a job." John Steinbeck described the same sense of futility more poetically in 1945 in *The Red Pony*: "No place to go, Jody. Every place is taken. But that's not the worst—no, not the worst. Westering had died out of the people. Westering isn't a hunger any more. It's all done." The trouble, however, was not merely the new policies that were implemented but also the threat of additional, unknown, policies. Fear froze the economy, but that uncertainty itself might have a cost was something the young experimenters simply did not consider.

The big question about the American depression is not whether war with Germany and Japan ended it. It is why the Depression lasted until that war. From 1929 to 1940, from Hoover to Roosevelt, government intervention helped to make the Depression Great. The period was not one of a moral battle between a force for good—the Roosevelt presidency—and forces for evil, those who opposed Roosevelt. It was a period of a power struggle between two sectors of the economy, both containing a mix of evil and virtue. The public sector and the private sector competed relentlessly for advantage. At the beginning, in the 1920s, the private sector ruled. By the end, when World War II began, it was the public sector that was dominant.

The contest was a brutal one, fought across the land, through famines and floods, and in a Washington that knew neither air-conditioning nor angiograms.

Roosevelt was clear about it. As he put in his second inaugural address, he sought "unimagined power." He, his advisers, and his congressional allies instinctively targeted monetary control, utilities, and taxation because they were the three sources of revenue whose control would enlarge the public sector the most. Since the private sector—even during the Great Depression—was the key to sustained recovery, such bids did enormous damage. Today we even have an economic theory, public choice economics, that sheds light on this. Public choice economics says that government is not higher than the private sector but rather a coequal combatant. Public choice theory tells us as much about the New Deal as the traditional economics Americans have been taught.

This particular school postdates the Depression, but the notion that something destructive was going on was evident, even to Roosevelt's allies. A number of them tried to articulate the problem. Ray Moley and Tugwell, two of Roosevelt's original brain trusters, dedicated years to grappling with the hypocrisy and damage of Roosevelt's actions. Wendell Willkie, at first a Democrat and enthusiastic reformer, would demonstrate that the contest between the TVA and his Commonwealth and Southern was not merely about electric power but also about control of the American future.

There remains a question. If so much of the New Deal hurt the economy, why did Roosevelt win reelection three times? Why, especially, the landslide of 1936? In the case of the third and fourth Roosevelt terms the answer is clear: the threat of war, and war itself. Roosevelt, unlike his narrow-minded Republican opponents, understood the dangers that Nazi Germany represented. In 1936, however, the reason for victory was different.

That year Roosevelt won because he created a new kind of interest-group politics. The idea that Americans might form a political group that demanded something from government was well known and thoroughly reported a century earlier by Alexis de Tocqueville. The idea that such groups might find mainstream parties to support them was not novel either: Republicans, including the Harding and Coolidge administrations, had long practiced interest-group politics on behalf of big business. But Roosevelt systematized interest-group politics more generally to include many constituencies—labor, senior citizens, farmers, union workers. The president made groups where only individual citizens or isolated cranks had stood before, ministered to those groups, and was rewarded with votes. It is no coincidence that the first peacetime year in American history in which federal spending outpaced the total spending of the states and towns was that election year of 1936. It can even be argued that one year—1936—created the modern entitlement challenge that so bedevils both parties only.

Roosevelt's move was so profound that it changed the English language. Before the 1930s, the word "liberal" stood for the individual; afterward, the phrase increasingly stood for groups. Roosevelt also changed economics forever. Roosevelt happened on an economic theory that validated his politics and his moral sense: what we now call Keynesianism. Keynesianism, named after John Maynard Keynes, emphasized consumers, who were also voters. The theory gave license for perpetual experimentation—at least as Roosevelt and his administration applied it.

Keynesianism also emphasized government spending. Yet focusing on consumers meant that Washington neglected the producer. Focusing on the fun of experiments neglected the question of whether unceasing experimentation might frighten business into terrified inaction. Admiring the short-term action of spending drew attention away from its longer-term limits—economies often go into recession when the spending disappears. Supplying generous capital to government made government into a competitor that the private sector could not match. Keynesianism provided the intellectual justification and the creation of constituencies.

Too much attention has been paid to what political polls said about the New Deal. Too little has been paid to two other measures, both also polls, in their way. One was the unemployment rate, which did not return to precrash levels until the war. The other was the stock market. It told a heartbreaking story. Uncertainty about what to expect from international events and Washington made the Dow Jones Industrial Average gyrate, both daily and over longer periods, in a fashion not repeated through the rest of the century: seven out of the ten biggest "up" days of the twentieth century took place in the 1930s. The uncertainty made Americans doubt themselves as investors. The Dow did not return to 1929 levels until nearly a decade after Roosevelt's death. The goodwill of the New Dealers, and there was enormous goodwill, could not excuse such consequences.

About half a century before the Depression, a Yale philosopher named William Graham Sumner penned a lecture against the progressives of his own day and in defense of classical liberalism. The lecture eventually become an essay, titled "The Forgotten Man." Applying his own elegant algebra of politics, Sumner warned that well-intentioned social progressives often coerced unwitting average citizens into funding dubious social projects. Sumner wrote:

"As soon as A observes something which seems to him to be wrong, from which X is suffering, A talks it over with B, and A and B then propose to get a law passed to remedy the evil and help X. Their law always proposes to determine . . . what A, B, and C shall do for X." But what about C? There was nothing wrong with A and B helping X. What was wrong was the law, and the indenturing of C to the cause. C was the forgotten man, the man who paid, "the man who never is thought of."

In 1932, a member of Roosevelt's brain trust, Ray Moley, recalled the phrase, although not its provenance. He inserted it into the candidate's first great speech. If elected, Roosevelt promised, he would act in the name of "the forgotten man at the bottom of the economic pyramid." Whereas C had been Sumner's forgotten man, the New Deal made X the forgotten man—the poor man, the old man, labor, or any other recipient of government help.

Roosevelt's work on behalf of his version of the forgotten man generated a new tradition. To justify giving to one forgotten man, the administration found, it had to make a scapegoat of another. Businessmen and businesses were the targets. Roosevelt's old mentor, the Democrat Al Smith, was furious. Even Keynes was concerned. In 1938 he wrote to Roosevelt advising him to nationalize utilities or leave them alone—but in any case cease his periodic and politicized attacks on them.

Keynes saw no point "in chasing utilities around the lot every other week." Roosevelt and his staff were becoming habitual bullies, pitting Americans against one another. The polarization made the Depression feel worse. Franklin Roosevelt's forgotten man, the constituent X, perpetually tangled with Sumner's original forgotten man, C.

This book is the story of A, the progressive of the 1920s and '30s whose good intentions inspired the country. But it is even more the story of C, the American who was not thought of. He was the Depression-era man who was not part of any political constituency and therefore lived the negatives of the period. He was the man who paid for the big projects, who got make-work instead of real work. He was the man who waited for economic growth that did not come. As an editorialist in Indiana wrote in 1936, "Who is the 'forgotten man' in Muncie? I know him as intimately as I know my own undershirt. He is the fellow that is trying to get along without public relief and has been attempting the same thing since the depression that cracked down on him."

Among the people whom the New Deal forgot and hurt were great and small names. The great casualties included the Alan Green-span figure of the era, Andrew Mellon, treasury secretary for the Harding, Coolidge, and Hoover administrations—a figure so towering it was said that "three presidents served under him." Another was Samuel Insull, a utilities magnate and innovator to whom the New Dealers assigned the blame for the crash. Yet another was James Warburg, a Roosevelt adviser who became so angry with the president that he penned book after book to express his rage. George Sutherland and James McReynolds, two of the four justices on the Supreme Court who fought back against Roosevelt, were also important. It was Willkie who spoke out most explicitly for the forgotten man on the national stage.

Others were of humbler background: those farmers who found themselves forced to kill off their piglets in a time of hunger because FDR's Agricultural Adjustment Administration ordained they must; a family of kosher butchers named Schechter who believed in Roosevelt but fought the New Deal all the way to the Supreme Court; a black cult leader named Father Divine; Bill W., the founder of Alcoholics Anonymous, who taught Americans that the solution to their troubles lay not with a federal program but within a new sort of entity—the self-help community.

Of course the Hoover and Roosevelt administrations may have had no choice but to pursue the policies that they did. They may indeed have spared the country something worse—an American version of

Stalin's communism, or Mussolini's fascism. That is the position that author Sinclair Lewis was taking when, in 1935, he published *It Can't Happen Here,* a fantasy version of the United States under fascist leaders remarkably similar to Roosevelt's opponents. The argument that democracy would have failed in the United States without the New Deal stood for seven decades, and has been made anew, by scholars of considerable quality, quite recently. But it is not right that we permit that argument—even if it is correct—to obscure some of the consequences of the two presidents' policies. Nor is it right that we overlook the failures of their philosophies. Glorifying the New Deal gets in the way of getting to know all the Cs, the bystanders, the third parties. They spoke frequently of the forgotten man at the time—the phrase "forgotten man" recurred throughout the decade—but eventually became forgotten men themselves. Going back to the Depression is worthwhile, if only to retrieve their lost story.

1
The Beneficent Hand

January 1927
Average unemployment (year): 3.3 percent
Dow Jones Industrial Average: 155

Floods change the course of history, and the Flood of 1927 was no exception. When the waters of the Mississippi broke through banks and levees that spring, the disaster was enormous. A wall of water pushed down the river, covering the area where nearly a million lived. Commerce Secretary Herbert Hoover raced to Memphis and took command. Hoover talked railroads into transporting the displaced for free and carrying freight at a discount. He commandeered private outboard motors and built motorboats of plywood. He urged the people who were not yet flooded out, such as the population around the Bayou des Glaises levee, to evacuate early, then rescued with the trains those tens of thousands who had ignored his warning. He helped the Red Cross launch a fund drive; within a month the charity had already collected promises of more than $8 million, an enormous figure for the time.

Several hundred thousand ended up in new refugee camps, many planned, right down to the latrines, by Hoover and his team. Hoover asked governors of each state to name a dictator of resources—he used the word "dictator"—and the governors complied. The dictators then managed the dysentery and the hunts for the missing along the floodwaters in their states hour by hour. He and the Red Cross sent the refugees to concentration camps—a phrase not so freighted then as it is today—at Vicksburg, Delta, and Natchez. One hundred thousand blankets from army warehouses were shipped to warm the refugees.

Things felt calmer on Hoover's watch. By mid-May, though the flooding was far from over, the anecdotes began to compete in the news with the reports of tragedy. Northerners read in *Time* magazine that a town called Waterproof, Louisiana, had not proven waterproof, and that its switchboard operators were still working—albeit from new posts, high up above the waters, on scaffolding. Not far from Memphis, Tennessee, bootleggers had also set up shop on high, in treetops. New babies were receiving flood names—Highwater Jones, Overflow Johnson. Now from Memphis, now from Little Rock, now from the Sugar Bowl, the itinerant flood manager, Hoover, wired or broadcast his analyses of the meaning of the disaster. Such flooding, he said, "is a national problem and must be solved nationally and vigorously." But the commerce secretary also spent a lot of time reassuring. The waters might hide the land, the crops might be lost, but the mood was now hopeful. More than any single figure, Hoover was succeeding in making Americans feel that the South would be all right again.

Hoover was already so famous that his name was a verb—to Hooverize, after the efforts in food rationing that he had led from a post as Washington's food administrator at the end of World War I. Americans recalled that he had led the humanitarian drive to feed occupied Belgium during the war. Now Hoover had outdone himself—and on a home territory whose geographic area covered more than Belgium's. What the public liked about Hoover was their sense of him as guardian, that he would protect them and what they had. If Hoover could win the presidential election the following year, then he might hold back whatever waters of adversity threatened. He was a Republican, like the sitting president,

Calvin Coolidge. He would pick up where Coolidge left off—though he might update things, for everyone knew that Hoover, a mining engineer, could do amazing things with technology. One of Hoover's neatest feats—and he pulled it off right around the time of the flood—was to acquaint the public with an early version of television. "Herbert Hoover made a speech in Washington yesterday afternoon. An audience in New York heard and saw him," the *New York Times* wrote in awe, adding that Hoover had "annihilated" geographic distance and commenting in a headline: "Like a Photo Come to Life." It was not yet modern television but wired images and the telephone combined. Still, the idea took hold in the minds of the reporters. Under Hoover, it was easy to believe that the 1920s were merely the American beginning.

The idea of philosophical continuity from Coolidge to Hoover seemed ironic to one man: Calvin Coolidge himself. The two were party allies. Hoover had loyally campaigned for Coolidge in 1924—indeed, had helped to defeat a Coolidge opponent in 1924 in California to clear the Republican presidential nomination for Coolidge. But Coolidge did not especially like Hoover. In the very period when the Mississippi waters were rushing, in fact, Coolidge's press spokesman had taken an explicit shot at Hoover, telling reporters that the commerce secretary would not be considered for the job opening if the secretary of state happened to retire.

The differences between the men had started with small things. Hoover was a fly fisherman. Coolidge fished with worms. Hoover liked the microphone. Coolidge shied away from it. After a landslide presidential victory in 1924 Coolidge had sent a clerk to read aloud his State of the Union address. Hoover ignored politics for the first thirty-five years of his life. Coolidge held his first office, that of city council member in Northampton, Massachusetts, at the age of twenty-eight, and had rarely been out of government since. Hoover was a mining engineer; Coolidge was a country lawyer. Hoover was a worldly American, a blend of regions and cities, the most successful in his field of his generation. He believed in the Anglo-American gold standard, not only because it had made him rich but because he had seen firsthand how it kept the world running, like a grandfather clock. Coolidge was a pure New Englander who seemed to re-create New England wherever he went. The very concept of "overseas" was a bit vague to Coolidge. The typical Republican of his day, he supported tariffs in the belief that they strengthened the United States. His failure to recognize the consequences of his policies, both abroad and for his country, was his greatest shortcoming.

Hoover believed that government might help business do better, functioning as a sort of beneficent hand. Coolidge liked Adam Smith's old invisible hand. The men were different breeds of Republican. Hoover believed that action was necessary to make the country live up to its potential. Coolidge had long ago determined that the world would do better if he involved himself less. Finally, there was a difference in temperament. Hoover strewed around phrases about individuality, but he could not control his own sense of agency. He was by personality an intervener; he liked to jump in, and find a moral justification for doing so later. People like Frederick Winslow Taylor, the great efficiency expert, and Herbert Hoover, the great engineer, had done so well in the private sector. Bringing some of them into government might allow some of that knowledge to rub off.

Coolidge by contrast believed that the work of life lay in holding back and shutting out. He conducted his official life according to his own version of the doctor's Hippocratic Oath—first, do no harm. It sounded easy, and many mocked Coolidge as being lazy in office—the same people who made fun of him by calling him Silent Cal. But Coolidge was not silent; he later estimated that each year as president he wrote or spoke 75,000 words, a share of those involving laying out his explanation for vetoing legislation. And Coolidge's "no harm" rule came out of strength of character. By holding back, Coolidge believed, he sustained stability, so that citizens knew what to expect from their government. If things were going well, he adhered to a stricter version of his rule: change less.

That, in fact, was how Hoover had come to be Coolidge's man at Commerce in the first place. When President Harding died suddenly one night in August 1923, Coolidge, then vice president, looked around and decided that stability was the most important factor in the Harding-Coolidge transition—important to the people, and important to the economy. Harding's last months had been clouded by scandal, and Coolidge wanted no more disruption. So he had kept on Harding's cabinet, including those he liked—Andy Mellon, at Treasury—as well as others, like Hoover. Coolidge's transition plan had been a success, at least insofar as the most precise measure of such things, the stock market, was concerned. The Dow Jones Industrial Average had stood at just below 88 the Friday before Harding died, and it was just below 89 a week later.

As a new president, Coolidge had been especially confident because the country had just seen a demonstration of his philosophy at work. During the Harding administration, recession had hit, and the downturn had been hard: one in ten men lost his job. But struggling firms had cut costs by reducing wages, and the country bounced back fast. By 1923, it was hard to find an unemployed man. Left alone, the majority's impression was, the economy would usually bounce back. That same year, a few months before Harding's death, Justice George Sutherland had led the Supreme Court in a sweeping rejection of the minimum wage in the District of Columbia. In his opinion—the case was called *Adkins*—Sutherland said that the minimum wage infringed on the individual's liberty to contract with his employer. The Sutherland opinion fit in with Coolidge's own general attitude, that the individual should have primacy—"All liberty is individual," he had said in a speech in 1924.

Coolidge's personal wager about the 1920s was that the private sector would and should take the lead, and that then the possibilities for progress would be boundless. His reserve did not mean that he was not interested in modern technology. He followed Charles Lindbergh's flights avidly, and the daughter of one of his closest friends, Dwight Morrow of Wall Street, would eventually marry Lindbergh. Citizens had proven their willingness to test Coolidge's propositions again by voting overwhelmingly for the refrainer in 1924, despite the fact that he had been vice president to Warren Harding, whose short time in office had been clouded by scandal.

At first, the differences between Coolidge and Hoover were nearly indistinguishable to the public eye. Both men, after all, deeply respected the Constitution and the gold standard. Both respected the independence of the Supreme Court—like Woodrow Wilson before them. Wilson had said that while it was within the power of government to overwhelm the Court on an issue, by, say, "increasing the number of justices and refusing to confirm any appointments," presidents recognized that this violated the spirit of the Constitution, and that the public would make "such outrages upon constitutional morality impossible by standing ready to curse them." Both believed in enterprise. In 1925 Coolidge summed up his philosophy, telling the American Society of Newspaper Editors that "the chief ideal of the American people is idealism." But he also offered a counterpart to that: "The chief business of the American people is business." It was the latter line that was remembered, and proved too moderate for some. They shortly altered it to the now better-known phrase "the business of America is business."

Finally, both men were humble about the position of the federal government relative to business. Compared to the private sector, after all, the federal government was a pygmy. Its size was less than 2 percent of the national economy, smaller even than that of state and city governments. Lawmakers of their generation constantly feared that the fast-growing private sector might further diminish their already questionable relevance. Back in 1910, word of the rise of the skyscraper in New York had panicked congressmen, who promptly zoned height limits for buildings in the District of Columbia, so that no private building could ever overshadow the Capitol.

Both men, too, shared an understanding of traditional economics, with its emphasis on the producer. "Supply creates its own demand," the classical economist Jean-Baptiste Say had written in France a century earlier, and in the case of many new industries, that seemed to be proving true again. Why focus on Coolidge, or Hoover? had been the attitude of the mid-1920s. The business leaders were the ones who would pull the country forward, and were therefore the ones worth watching.

Most dramatic was Henry Ford, who, the week after Harding's death, reported he was selling automobiles under a new trade agreement to Russia. Twenty years before, Ford had started with just a few employees and $27,000. By constructing the modern assembly line, he was creating modern Detroit and conquering the rest of the country, buying up its own coal and iron mines. Part of his success was due to his religiously plowing back profit into the business, forgoing dividends. In fact, as Benjamin Anderson, the chief economist at Chase Bank, would write later, the first quarter century had offered "case after case of Fords," all "showing the history of small businesses which, employing three or four laborers, had in relatively short periods of time (fifteen or twenty years) grown into very substantial businesses." In 1923, Ford plants were already producing 6,000 cars a day, a record.

During the war the government had begun work on a dam and power operation at Muscle Shoals, Alabama—the point being to make ammunition. The war ended before the plant was running, and recently Henry Ford, who had his own political ambitions, had put in a bid to take over Muscle Shoals. Only the private sector, some believed, had the wherewithal to develop America's most important new industry, and the key to its growth: electric power. Muscle Shoals could be "the Detroit of the South." Ford had tried to write a contract with the government to run Wilson Dam as a nitrate plant—indeed, Hoover hoped to broker the deal. But Congress had rejected it. Lawmakers like George Norris believed the government should control power.

Another hero on the horizon throughout the decade was Thomas Edison, the man who started the electrification boom. Across the country, people revered him; from time to time Edison would mount a contest to find young men of "all around ability" at his East Orange labs, and hundreds signed up. A young Vermonter named Bill Wilson who sat for and won one contest later recalled that seeing Edison in his lab coat, with the faint scars from a chemical explosion on his cheek, was to see the personification of American genius. In the summer of 1928, Congress would ask Mellon at the Treasury to strike a medal in Edison's honor. Mellon traveled to Llewellyn Park in New Jersey to give Edison the medal. From the base of a "small and illustrious company," Mellon told the crowd in his whispery voice, Edison had delivered what a businessman ought to. Edison had "not only changed the conditions under which men live" but also "helped to bring about a new social order."

Yet another hero was the British-born Sam Insull, who had started out in America keeping Edison's books. While on the East Coast with Edison, Insull had discovered that selling electric irons to Schenectady housewives was a good way to increase household use of power. His second insight had involved the economy of scale. Wall Street believed that each family or street needed a generator. There was even the idea that each gentleman should have his own generator, just like his own yacht; J. P. Morgan had one belching on his property in New York's Murray Hill. But Insull, feeling that the yacht was inefficient, had gone to Chicago. What Wall Street could not see, La Salle Street perceived, providing Insull capital to finance central power stations. In this fashion he achieved a miracle: he established power prices that were acceptable to the small consumer.

Insull's Chicago was a rough place. Before the war a college student from Indiana named David Lilienthal described his experience on a visit in 1917: he came across a crowd surrounding a puddle, and stuck his head among the others to see "what these busy-men-of-the-world were watching with such evident enjoyment." It was "but a tiny mouse, swimming about in the pool." Lilienthal was disgusted to see that "whenever he would struggle to a place of safety—someone would stick out his mahogany cane and throw the poor quivering thing back to his death. When this would happen," Lilienthal noted, "some portly comfortable looking son-of-a-gun would shift his cigar and chuckle." The young man commented on the Chicagoans in his diary: "And such creatures expect mercy for themselves from some higher authority, as they are to mice!"

Where others saw lawlessness, though, Insull saw opportunity. He was brave, he was aggressive—a frontier man—and few laws stood in his way. Insull wired the city, then the state, and then other parts of the country—like Ford, always plowing cash back into projects. When banks could not provide cash, he had used equity vehicles to raise the money, repeatedly creating holding companies, parent companies that owned operating utilities.

Critics said that he was watering down stock. But all Insull saw was the need for cash—the industry was the world's most promising and would grow only if it got capital. And it did: the line on graphs of the American utilities industry in the 1910s and '20s moved up in an incontrovertible diagonal, consumption increasing each year, even in the early 1920s recession, seemingly independent of the overall economy. At his high point Insull provided a full eighth of America's electrical power. To reward Chicago for the prosperity it had given him, Insull would spend the late 1920s building an opera house as ambitious as his business empire: a forty-five-story giant equipped with electric elevators designed so that every seat in the house, including the high gallery seats, would offer as good a view as the dress circle. Insull's opera house had no boxes for aristocracy; he wanted to prove that the world of electricity was a democratic one. The plans revealed a building in the shape of an armchair, a symbolic throne for Insull.

The armchair faced west, the ultimate gesture of defiance toward New York.

On Wall Street, there were other, different figures to watch. Henry Morgenthau Sr., Felix Warburg, and Bernard Baruch were joining the Morgans as leaders in the financial district in the early part of the century. Some were continuing the success of a dynasty—Warburg. Others were bent on establishing new dynasties—Morgenthau especially. In the 1920s a young man named Alfred Lee Loomis had taken a firm that was nearing bankruptcy, Bonbright, to heights of profitability with innovative investments in Insull's industry, public utilities. Half of the nation's homes were electrified; together with his friend and partner Landon Thorne, Loomis wanted to electrify the rest. The industry began recruiting talent wildly. One of its finds was a corporate lawyer who had worked for Firestone, one of the tire companies in Akron, Wendell Willkie.

Like Insull, Alfred Lee Loomis and Thorne had seen that older investment houses were not sure they wanted to pour cash into the new utilities industry. And like Insull, they had seen the efficiency of holding companies: little local companies could save cash if they banded together into "superpowers." One of the most promising markets, they had recognized, was the South, a laggard in modernization. Electrifying the

South, they had realized, would also do enormous social good. As it was often said, the South was tired of living in the dark. Lone state companies could not do the work; they needed to hook up a network and share resources. Georgia Power Company had provided indifferent service to some customers, in part because it lacked the advantages of a larger holding company. One customer who wrote to complain was a polio patient from Warm Springs, Georgia—Franklin Roosevelt, the future New York governor.

In the 1920s two other big figures loomed large. The first was the treasury secretary, Andrew Mellon of Pittsburgh. Mellon's father, whose family had come from county Tyrone in Ireland, had been a faithful reader of Benjamin Franklin: "The way to wealth, if you desire it, is as plain as the way to market. It depends chiefly on two words, industry and frugality. That is, waste neither time nor money, but make the best use of both." Thomas Mellon, a merchant banker, believed in investing in commodities, but also, like Insull, in investing in ideas. Among the new bank's first visitors—while Andrew was still a student—had been a twenty-one-year-old bookkeeper with a scheme to build fifty coke ovens. He thought he could serve the growing steel industry, but he needed $10,000. An agent whom Mellon sent to evaluate Frick wrote:

"Lands good, ovens well built, manager on job all day, keeps books evenings, may be a little too enthusiastic about pictures but not enough to hurt." Judge Mellon struck a deal with him.

That man, Henry Clay Frick, remained affiliated with the Mellons from that point on. He and Andrew, good friends, traveled to Europe together; on such a trip Andrew started to build an art collection, buying his first picture.

But what was more important at the time was that Andrew also built a great business empire. He and his brother Richard created first a national bank, then a steel concern, and then an empire. Young Mellon cornered the bauxite market. He shared in the profits of Carnegie Steel, of which Frick was president. The Mellons together established the enormous Aluminum Company of America; later they picked up Bethlehem Steel. They invested in Spindletop, the Texas gusher that opened the Gulf Coast oil industry. By the time Hoover reached adulthood, the Mellons also were players in the steel, railway, construction, and insurance industries. Succeeding Andrew Carnegie and Henry Clay Frick, Andrew Mellon ruled Pittsburgh in a way that not even the president ruled Washington.

Mellon had stayed close to his father's original formula: thrift that emphasized the accumulation of capital. But he had also created value—and not merely by cornering a market as a robber baron would, though he had done this with bauxite. Mellon invested in new innovations, functioning as an early version of the modern venture capitalist. The magazine *World's Work* described the Mellon formula thus: "Find a man who can run a business and needs capital to start or expand. Furnish the capital and take shares in the business, leaving the other man to run it except when he is in trouble. When the business has growth sufficiently to pay back the money, take the money and find another man running a business and in need of money and give it to him, on the same basis."

In the late 1880s an inventor named Charles M. Hall had showed up in Mellon's offices. Hall had developed a new way to smelt aluminum, but he lacked capital to sell his product. Mellon, for his part, agreed to lend $250,000 in return for a controlling share of a new firm, the Pittsburgh Reduction Company. Pittsburgh Reduction, iterations later, became the Aluminum Company of America.

Hoover himself, impressed, later recounted an anecdote an official from the same company would tell him about working with Mellon. The Mellon Bank had refused to lend to an inventor. But Mellon told the man, "I sometimes personally loan money on the security of character." Mellon gave the man $10,000 and then invested yet more, and then yet more when the pilot showed prospects. For his cash efforts, Mellon eventually agreed to a fifty-fifty ownership. The inventor wondered aloud why Mellon was settling only for half. Now that his project had value, Hall reminded Mellon, Mellon might foreclose and own all. But the answer came back: "The Mellons never did business that way."

To Mellon the formula was obvious—invest in the private sector, do not intervene too much, wait silently, and the returns would be all the greater. Eventually, he was so successful at producing innovations that he created the Mellon Institute in Pittsburgh to make the resources of his empire available to other, less innovative companies. The institute came under the general Mellon rubric "self-improvement," though whether that "self" was Mellon or his business or the U.S. economy generally even he left unclear. In the same spirit, Mellon also undertook to improve his knowledge of the French language at around the same point. It was an early version of the modern science think tank. Companies brought their cash to the Mellon Institute; this funded scientists who could solve their problems through efficient applied research. Hundreds of books and papers, as well as more than 600 patents, resulted.

By the time he entered public life, Mellon would serve on the board of more than 150 corporations, presiding over hundreds of millions.

When it came to the press, Mellon was suspicious—Victorian, in fact. More than Coolidge, he hid from newspapermen, even when he had good news. To show off was, to his mind, as to Coolidge's, unseemly. Besides, it brought on bad luck. As a result, he lost out on his share of puff pieces. And he was often surprised to find that the press did not side with him when he needed it. He once found himself calling a newspaper editor he had never met to tell the man he had a story wrong—Mellon's son was not sick, as the paper had alleged. Mellon lore was so scarce that people ended up trafficking in the stereotype of Mellon as miser. They noted that Mellon rarely spoke, that he seemed "half-frightened" (columnist Drew Pearson), and that he was quick to dismiss ideas that he considered shallow. He was reported to be an incredible penny-pincher. His wife left him. Mellon's son Paul, who bore great affection for his father, also found much to criticize. He later compared Andrew to Soames Forsyte, the cold husband and "Man of Property" in John Galsworthy's novels, who also collected pictures. Paul wrote later of Andrew's "ice water smile."

The last of the giants was Herbert Hoover himself. On the surface, again, he seemed much like the others, with the same story of success. A Quaker, Hoover had been born in 1874 in West Branch, Iowa. He had been orphaned as a child and, like the Mellon men, demonstrated early a love of thrift, enterprise, and new technology. Before she died, his mother led Quaker meetings, and, observing her, Hoover acquired a sense of righteousness: if one was truly a leader, the rest would be silent and follow. There was no reason to criticize a leader. He was the sort of boy who seemed years older than the rest, the one who always jingled a ring of keys in his pocket. An uncle, Henry John Minthorn, later recalled that he had trouble when he asked the boy to ride a horse; the youth persisted in preferring bicycles.

As an undergraduate at a new university, Stanford, Hoover made the brilliant move of studying geology and engineering. By reason of geography the university had advanced instructors, themselves pioneering mining even as they taught it. When the professors were not in the classroom, they were surveying the frontier. In the space of several years, therefore, Hoover acquired a grounding that the best miners across the world could only hope for, studying the newest techniques for finding minerals, paleontology, engineering, and chemistry. In those years the great nations were on a gold standard. Economies depended upon gold in the way that, for example, they depend on microchips today. Yet only a few engineers knew how to get that gold out of the ground. As Hoover would later note in his memoirs, snobbery made Britain and the Continent reluctant to elevate engineering above the status of a trade: "the European universities did not acknowledge engineering as a profession until long after America had done so."

Europe's loss was the gain of the gentlemen at Stanford—especially Herbert Hoover. The discovery of gold deep below the surface in western Australia coincided with Hoover's Stanford graduation. A London firm, Bewick Moreing, offered the graduate a starting salary of $6,000, the equivalent of $124,000 today. The firm would send him to that corner of the world, where mining was the most challenging but also, potentially, the most rewarding. The place was rough; at a mine called the Sons of Gwalia, he endured blackflies, white heat, and seemingly inferior Australian labor—"noddle heads," he called the workers. Still the principles were the same: thorough, serious work, U.S. engineering technique.

By the time he was twenty-five, Hoover had brought a failing mine to fabulous profitability. Next he oversaw the rehabilitation of mining throughout the region. He acknowledged that his work in Australia was "practically a new science" but was not content with his gains there, raking in thousands in stock market winnings on the side. By the age of twenty-seven, Hoover had turned around the production and the books of mines in the United States, Australia, and China. He and Lou, his wife, lived in Tientsin, China, where they helped rescue others in the Boxer Rebellion. Then it was on to London, where the Hoovers lived in style—with a chauffeur, even. The prewar London of that time was a truly international city, a city that showed off the gold standard at its best. The gold that Hoover dug out of the mines helped the standard to function, in turn making global trade possible.

In foreign places, without peers to challenge him, Hoover became accustomed to solo management. Ignoring the noddles seemed the fastest route to success and the best way to avoid painful criticism. Hoover's pride grew with each project. It was said of Hoover that there was "no cleverer engineer in the two hemispheres." Luck makes talent look like genius, and every era has its own kind of luck. In Hoover's era, luck blessed mining engineers. It was an era of commodities. In 1901, the first item on the Dow Jones Industrial Average was Amalgamated Copper, the last U.S. Steel

Preferred. In between were also mostly commodity-driven companies: American Sugar, National Lead.

Hoover, who was willing to travel and not afraid of bankers, knew better how to exploit that luck than his geologist teachers. Equally at home in a mine and a bank boardroom, he moved forward fast. From improving mines themselves he went to improving mining projects and mining finance. Before he turned thirty, the papers reported that he was the best-paid man of his generation.

One of the insights that continued to serve Hoover in this period was that mines were cats—they had nine lives, if only one found the technology to sustain them. In 1905 and 1906, for example, scientists knew both that zinc was a valuable mineral and that it was hard to isolate. Frustratingly, it could be found in chunks of other rocks mined—so-called tailings. In those years they were also discovering that if unusable chunks of rock—the tailings—containing zinc ore were crushed and mixed with sulfuric acid, the zinc in them would float to the surface. Hoover traveled to Broken Hill, hundreds of miles northeast of Adelaide, to observe the ore deposits; then he led Bewick Moreing in creating a company, the Zinc Corporation, to exploit the new process. After a few dicey months in 1907 and 1908—"if we failed, our reputation would be gone," a colleague on the project later recalled—the Zinc Corp. proved intensely profitable.

By 1908, Hoover had outgrown Bewick Moreing and established himself as an enormously successful freelance engineer. There was more work in Belgium, Germany, France, China, Japan, and Burma. Hoover got to know Russia through his work at a baron's estate in the Ural Mountains, not far from Ekaterinburg, where the Romanovs would later be killed. The Kyshtim estate provided a livelihood for 100,000 and was rich in copper and other metals. Its engineers, as Hoover discovered when he came to Kyshtim, were working from the wrong metallurgical concept. They were using blast furnaces. Hoover and his engineers thought an unusual process, pyritic smelting on a large scale, might work better. They found additional financial resources both to reorganize the company's finances and to bring men from Butte, Montana, where a similar quality of ore had been treated. There were also new furnaces. And to his satisfaction, Hoover was later able to report that the property after his rationalization earned a net of about $2 million annually. Following the doctrines of Taylor, the efficiency guru, the new company paid its workers 25 percent above the going wage in the area.

Hoover had seen great promise for Russia but also recognized the potential for great trouble. While working at Kyshtim he found himself one day at a train station, observing as "a long line of intelligent, decent people brutally chained together were marched aboard a freight car bound for Siberia. Some were the faces of despair itself, some of despondency itself, some of defiance itself." The whole dark scene, Hoover wrote, robbed him of sleep. But Hoover's own life had been nearly all sunshine and had seemed to move only forward.

The events in Russia had strengthened Hoover's conviction about the need for firm leadership in Europe and even the United States. In 1916 Bolsheviks began agitations at his own Kyshtim plants. In 1917 the Communists took power, throwing out the ownership and management at Kyshtim and giving themselves a 100 percent raise. The Americans on the project were sent off on trains to Vladivostok, but the Russian experts were brutalized or even killed. What made it worse was that without the experts the delicate Kyshtim furnaces broke down within a week; the Communists could not read the blueprints left behind that would have told them how to do repair work. "In a week the works were shut down, and 100,000 people were destitute," Hoover recalled, rightfully disgusted, in his memoir. After that

Hoover led a great relief action in "Bololand," the nickname the relief staff used for the Bolshevik state. But unlike Henry Ford, he did not nurse much hope that doing business with Russia would bring Bolsheviks back to market ways.

By age thirty-five or forty Hoover still feared criticism, but less than before—he encountered it so infrequently. Luck and talent had done their work, and he began to feel his greatness was unlimited. Not only Americans but also foreigners had the same impression. Getting to know Hoover at the peace conference after World War I, the economist John Maynard Keynes was deeply impressed. For Hoover grasped what others had missed: the crushing blow that the reparations payments demanded by France would deal to the future of Germany. Hoover, Keynes said, operated in "an atmosphere of reality, knowledge, magnanimity and disinterestedness, which, if they had been found in other quarters also, would have given us the Good Peace." Others might live lives of periodic setbacks; Hoover seemed immune. Sherwood Anderson, the novelist who chronicled such setbacks in *Winesburg, Ohio,* would write with astonishment that Hoover's was the face of a man who "had never known failure."

All the while, four characteristics made Hoover subtly different from some of the others, especially from Mellon. The first was that he was younger, born in 1874, which made him a full generation younger than Mellon. His world was more Edwardian than frontier or Victorian; the experience of the war had made a big impression on him. Hoover came to believe that life was like wartime, and that government, therefore, ought to plan more, as if in a war. The second was his love of publicity. The third difference was that Hoover, who had profited so much from commodities, tended to distrust Wall Street, whose wealth he viewed as ephemeral. The fourth was that Hoover loved to jump in where, say Mellon, had stayed back. That was why he had been so successful as a consultant.

Under Harding, the differences between Hoover on the one hand and Mellon and Coolidge on the other became more visible, at least for those who looked carefully. Mellon, true to form, had focused on allowing businesses to work on their own, which to him meant reducing taxes. Income taxes in those days applied only to the rich—and those rich paid extremely high rates, the top being 73 percent. Mellon believed that rate was prohibitive—it killed investment, and therefore jobs. He did not see taxation as a moral matter. Taxes were a practical thing: a tax was a price. And one could only charge "what the traffic will bear," as he put it, drawing on a metaphor from his own railroad freight days.

When a government overtaxed, it hurt itself, for it got less revenue. Taxes that were too high, Mellon noted, simply were not paid. In the end lawmakers wrote loopholes that enabled high earners to escape. Instantly, with the aid of his staff, Mellon began to work on plans to cut taxes. Overtaxation was the very sort of intervention he had abhorred as a venture capitalist, and he would do what he could to reduce it from Washington.

Mellon's other preoccupation was with the efficient use of family money—his own, and that of other people. Money, he believed, must stay in the private sector, with the family, either for further investment, or for one's children, or for charity. He had watched in admiration—and perhaps envy—as his friend Henry Clay Frick built up his art collection. Every once in a while a headline trumpeted "Another Vermeer Bought by Frick." In the autumn of 1919, though, Frick died, and Mellon saw originally enormous bequests to charity, as well as Frick's splendid collection of paintings, become embroiled for years in fights over estate taxes. Institutions such as Princeton University to whom Frick had promised gifts found themselves with far less than expected, a shameful outcome from the philanthropist's point of view. Mellon resolved to find a way to change the way estates were taxed.

At the Commerce Department, meanwhile, Secretary Hoover had begun to apply *his* notions. The department pre-Hoover had been relatively tame, almost a sinecure. One of its tasks was coordinating the nation's lighthouses. It was said of the post of commerce secretary that it consisted of "lighting the lamps along the coast and putting the fishes to bed." Early on Hoover invited a promising young man from the Warburg banking dynasty, James Warburg, to work as his assistant secretary. "But what do you do?" Warburg asked Hoover impudently—and did not take the job.

Hoover saw the post as grander: an opportunity to show the country what it could do if it had a national engineer. When the recession hit, he urged President Warren Harding to host a conference with business and labor on creating employment. The idea was that Washington could encourage business, and perhaps the states and labor, to work together to make the economy grow. Among those attending was John L. Lewis of the United Mine Workers. Later, in 1922, Commerce Secretary Hoover argued that union wages were too high, and that there was "an overplus of mines and miners" in general. The miners struck, and Hoover intervened; this time the innocent in need of protection had been consumers, not Belgian children. "The administration is not injecting itself into the strike," he wrote Lewis; "it is trying to protect the general public from the results of the strike." Meanwhile, Hoover was also working on the problem of the Colorado River and the parched Southwest. The river was a monster, violent and unpredictable. In one season it swelled and destroyed farmlands. In the summer its flow dropped to a trickle. It had knocked out a dam built earlier in the century as if the dam were nothing. Hoover agreed with the conservatives that it was wrong for Washington to intervene, but he also firmly agreed with Franklin Roosevelt of New York, who believed that state governments must involve themselves in hydropower. Hoover therefore assigned himself the task of playing go-between among the seven states through which the Colorado flowed. In the early 1920s he visited each of the states, convincing them to make various key concessions.

In this period, the first half of the 1920s, both Mellon and Hoover published books codifying their philosophies. The austere Mellon gave his an unexpectedly populist title: *Taxation: The People's Business.* In it he laid out the theories of his fellow Scot Adam Smith to justify his program of continued tax-cutting. Reaching to find an image from the day—it may have been hard—he talked about Henry Ford's company. "Does anyone question that Mr. Ford has made more money by reducing the price of his car and increasing his sales than he would have made by maintaining a high price and a greater profit per car, but selling less cars? The government is just a business." The lesson of the book was simple: people responded to tax rates, and lower rates might promote growth in the 1920s and pull in higher revenues for government. The whole idea of overtaxation was to Mellon un-American. "Any man of energy and initiative in this country can get what he wants out of life. But when initiative is crippled by legislation or by a tax system which denies him the right to receive a reasonable share of his earnings, then he will no longer exert himself and the country will be deprived of the energy on which its continued greatness depends." When failure attended business, after all, noted Mellon, "the loss is borne by the adventurer."

Mellon took pains to be consistent, even when it was against his own personal interest. In the same volume he attacked tax-exempt bonds, the very sort of bond rich men like himself tended to purchase. His action here was the proprietor's: municipal bonds were bad because they deprived the Treasury of revenue. The better philosophy was to lower rates over all. Mellon also disliked other tax loopholes, and did not mind mentioning that fact, even though his own empire availed itself of the same loopholes in the preparation of the Mellon returns. After all, what he did as a private citizen was his own affair, as long as it followed the law books.

Hoover gave his book a title similar to Mellon's—*American Individualism.* But the text, like Hoover's work, was distinct. Hoover rejected the old brand of absolute individualism and disdained laissez-faire economics as "theoretic and emotional." Private property, he also said, was "not a *fetich*" for Americans (using the spelling acceptable in that period). He also made clear that he believed America must move toward regulation: "Our mass of regulation of public utilities and our legislation against restraint of trade is the monument to our intent to preserve an equality of opportunity," he wrote—and, describing Mellon's Adam Smith, put the word "capitalism" in quotation marks, to signal that he wanted to keep his distance. The message was clear: as far as he was concerned, men like Mellon could have Adam Smith's invisible hand. In this view Hoover was in line with two academics popular in the 1920s, William Trufant Foster and Waddill Catchings. The pair were the ones who had popularized the phrase "beneficent hand." Catchings and Foster introduced other novel theories. They deplored the traditional emphasis on supply; economic policy rather should pay attention to the consumer. Spending, they argued, including spending by the consumer, could promote and smooth over economic growth. Many readers, especially those trained in classic political economy, found Catchings and Foster hard to take. One was Franklin Roosevelt of New York, who in 1928 inscribed the front pages of his copy of the authors' *Road to Plenty* thusly: "Franklin D. Roosevelt, Hyde Park 1928. Too good to be true—You can't get something for nothing." Hoover did not like the writers' emphasis on increasing debt and deficits, but his book was, essentially, a version of the beneficent-hand argument.

To continue growing, Hoover argued, the economy had to be better organized. Standards and precision were important now. He had faith in the concept of economy of scale—the bigger a thing was, the more efficient. He hated Soviet Russia. Still, in *American Individualism* and elsewhere Hoover made it clear that he viewed as retrograde the old battles of left versus right. What mattered was efficiency and being right. The common craftsman, even the guild member—these small traders were too idiosyncratic and individual for a modern economy. Though he had lived in London, the world's financial capital, for more than a decade, he still thought of America as a commodities country, whose wealth was something one mined (gold) or planted in the ground (wheat). As for government leaders, they ought not to intervene unconstitutionally but must play the broker. More important, they could be a beacon of reform. If the Commerce Secretary was to "light the lights," as the old quote said, then let him have a substantial lighthouse: Hoover envisioned a giant new department building.

In Coolidge's presidency both members of the cabinet continued to advance their agendas. Mellon warned that for prosperity to continue, taxes must come down: "High taxation, even if levied upon an economic basis, affects the prosperity of the country because in its ultimate analysis the burden of all taxes rests only in part upon the individual or property taxed. It is largely borne by the ultimate consumer." He had established the charitable deduction he talked about. Tax revenues promptly behaved just as he had predicted, increasing after rate cuts, and the country moved into surplus. He too had a few standardizing projects—he made uniform the size of the various denominations of bills, a step, he estimated correctly, that would save the Treasury money. He brought the budget into balance and supervised buildings for Coolidge. And he battled for the repeal of the estate tax, perhaps thinking of the various conflicts over Frick's grave.

As for Hoover, he had decided that, at least for a while, he could live with John L. Lewis's idea of higher wages. He and Lewis determined that long periods of labor peace would bring economic benefits sufficient to offset the higher wages that Lewis sought. After an agreement with labor in 1924 at Jacksonville, Florida, the coal companies found themselves cornered into committing to paying higher wages.

Hoover and Lewis now had nothing but praise for each other, with Lewis, to date a movement radical, now declaring himself a free-marketeer. "It is the survival of the fittest. Many are going to be hurt, but the rule must be the greatest good for the greatest number." Hoover openly flattered Lewis: "Mr. Lewis is more than a successful battle leader. He has a sound conception of statesmanship." But by the mid-1920s it became clear that they had been wrong: nonunion mines were driving the companies that had gone along with the Jacksonville agreement out of business. The union men were not better paid; they were out of work. *Cushing's Survey,* a newsletter, reported that Lewis assailed Hoover, saying, "You got me into this mess; it's up to you to get me out." He even wrote his own book, *The Miner's Fight for American Standards,* a plea to Hoover. But this time Hoover did not back him up.

There had also been far smaller projects, which Hoover relished. One was strengthening the Bureau of Foreign and Domestic Commerce, whose job it was to promote trade. He expanded the number of its offices across the country from six to twenty-three; a businessman could now go to any good-sized city and learn about the possibility of business abroad. As one of his biographers, Will Irwin, noted, this expansion generated an increase in inquiries about trade from 200,000 to 2.4 million. Irwin, a supporter of Hoover's, reported that American businessmen would often tell Commerce that their resources obviated a tiresome research trip abroad: "You fellows have more than I could get in a dozen trips."

Hoover had also created a Division of Simplified Practices, whose job it was to standardize and harmonize the distressingly fractious and unresponsive manufacturing and construction sectors. In those days roads were often still paved in brick, and brick was a typical example: sixty-six different sizes were being produced by manufacturers when Hoover ordered research on the topic. This was sheer waste, as far as the utilitarian Hoover was concerned. He therefore pulled the nation's paving-brick firms into a room and settled the matter; the range of sizes dropped from sixty-six to eleven. Emboldened, Hoover also looked into brick for homes; here he claimed victory outright, for the number of sizes went "from forty-four to one," the praiseful Irvin reported. Then there were beds. Seventy-four different sizes were available; as a result of encouragement from

Hoover, the figure went down to four. If these latter accomplishments had a comical aspect of "putting the fishes to bed," Hoover did not notice.

Whether the public was ever conscious of the contrasts among all these probusiness leaders is hard to discern. In 1924 they gave Coolidge 382 electoral votes, far more than his Democratic and Progressive opponents, whose total together was 149. But Hoover was also popular. What was clear, however, was that Coolidge's initial impressions of his two cabinet members had only strengthened. The supposedly cold Coolidge heartily approved of Mellon's tax policy, saying that "the wise and correct course to follow in taxation and all other economic legislation is not to destroy those who have already secured success but to create conditions under which every one will have a better chance to be successful." Mellon, with Coolidge's support, reduced the national debt from $24 billion to $16 billion. He did away with the excess-profits tax—it was wrong to say that profits were excessive anyhow, when they created the work. Negotiating past the progressive George Norris, he put through the Revenue Act of 1926, a dramatic series of rate cuts, repealing gift taxes, slashing estate taxes, and taking one-third of those who had paid in the preceding year entirely off the tax rolls. Norris commented of one of the drafts of the act, "Mr. Mellon himself gets a larger personal reduction than the aggregate of practically all the taxpayers in the state of Nebraska"—Norris's state. Still, Mellon also paid more taxes than the people of such states.

Mellon's goal had been to get the top tax rate down to 25 percent, a full 50 percent below where it had stood at Frick's death, and he achieved this. Mellon had also lowered the base or "normal" rate at the bottom of the schedule. Growth was up, but so, more importantly, was the average real wage, solid evidence that a tax cut for the rich was also good for Henry Ford's worker. The after-inflation earnings of employees grew 16 percent from 1923 to 1929. Revenues continued to flow in just as the treasury secretary had so pointedly predicted. Mellon was managing to balance the budget and to reduce the staff of tax officials at the Bureau of Internal Revenue. The policy was so well regarded that even Democrats, the party of the income tax, argued for lower taxes. John Nance Garner, a representative from Texas and a leader in the House, had a plan for a corporate tax lower than Mellon's, at least in some brackets. Between Coolidge and Mellon there was also a personal bond. Later it would be said of them that they conversed entirely in pauses.

Matters were different when it came to Hoover. Coolidge understood the political success of the beneficent hand, but he did not believe in it. Man himself, he would write toward the end of the 1920s, was after all "but an instrument in the hands of God." More and more Coolidge was thinking of God—in 1924, his son Calvin got a blister on his toe playing tennis on the South Lawn of the White House, and in those prepenicillin days, the blister brought on an infection that killed him. This tragedy made Coolidge brittle, impatient, and irritable, and one of the people who irritated him was the persistent Hoover, so different from Mellon. Where the president eschewed technology, Hoover was always playing with it. Coolidge also hated Hoover's tendency to react to news with grand, intrusive plans. Could not Hoover see where some of his rescues had led? At one point later on, the minimalist president Calvin Coolidge concluded quite simply that "that man has offered me unsolicited advice for six years, all of it bad." He had a nickname for Hoover: "Wonder Boy."

Beyond grief lay Coolidge's accurate perception that in the 1920s Mellon's and his own policies were yielding the good that the men had predicted. Today we estimate that the highest level of unemployment under President Coolidge had been 5 percent in the year he was elected. From there it dropped to 3.2 percent in

1925 and then into the twos and ones. Citizens could afford all the new products. There was nothing bubbly about the potential for productivity gains. By the end of 1925 Ford's peak production was 8,500 a day, up substantially from the 6,000 from a few years before. Overall in the years from 1923 to 1929 car production would double. Another emblem of the new progress was the price of Henry Ford's cars. The Model T, $600 before the war, sold for $240 in the mid-1920s. Right after the war it seemed that the United States had become the greatest power through might. With the growth of the 1920s, the country was showing that it deserved to be that power. Coolidge began his December 6, 1927, yearly message to Congress by announcing that "It is gratifying to report that for the fourth consecutive year the state of the union in general is good." There was a sense among political leaders that industrial America would do even better than agricultural America had done.

The progressives, who once had hoped to take the country, now seemed to be fading. The name of Bob La Follette, the great reformer of Wisconsin, simply did not mean as much as it had even as recently as 1924, when La Follette had won 16.6 percent of the vote in the presidential election, promising to end "an orgy of corruption" and quick and easy access to the White House.

Among citizens, it was increases in the standard of living that seemed to be raising hope. In Indiana two social scientists, Robert and Helen Lynd, undertook a study of a single town, Muncie, which they called Middletown. In Middletown, the Lynds found that new inventions were remaking home, work, and leisure: the radio, the automobile, and the telephone, a new medium that created a "semi-private, partly depersonalized means of approach to a person of the other sex." This was the decade in which Americans began traveling the country in automobiles, and the decade they began to prefer the telephone to the postchurch visit.

The stock market rose faster toward the end of the decade. A new generation of Americans turned to it. One was Bill Wilson, the Vermonter who had won Edison's contest. Wilson, like so many men, came to Wall Street after serving in Europe to see if he could play the equities game. An uneven fellow, Wilson drank too much and found himself subsidized by his wife's family in Brooklyn Heights. But he was a tinkerer, an Edison if only in temperament, and when it came to the financial world, he spotted a genuine flaw. Everyone traded stocks, but even the experts knew relatively little about the companies they came from. He himself, the Edison enthusiast, had bought a few shares of General Electric at $180 a share; now they were worth $4,000 or more.

But why? Wilson bought a Harley-Davidson with a sidecar for his wife, Lois; they would ride around the country investigating companies. It was an early version of modern stock analysis. Companies shouldn't merely be reported on, as in the papers; they ought to be studied. He began to make money.

Wilson's attempts to shed light on the mystery of markets were the humblest version of what his colleagues on Wall Street and at universities were trying to do. One was Irving Fisher, a professor at Yale who had been a student of the great philosopher William Graham Sumner.

Sumner and Fisher had disagreed on many things. Sumner had told Fisher that if life came down to a choice between socialism and anarchy, he would take anarchy. Fisher, reporting this to the Yale Socialist Club, told his audience that given the same choice, he would take socialism. But Sumner had imparted to Fisher, and generations of other students, a deep understanding of the degree to which tariffs could slow the economy. And he also started Fisher off on his career by suggesting that he look into the mathematical side of the economy, then a new approach. Fisher, an erratic but brilliant man, proceeded to create indexes to measure the rise of commodity prices, then a new concept. He also looked at prices and money generally in a new way. He believed that the gold standard slowed growth. One of its problems was that gold was too unpredictable, and that a stabler currency might be arrived at by linking the dollar to a group of commodities, instead of to gold. Fisher, like Wilson, was investing even as he philosophized. Wilson bought a used Dodge. Fisher's son noted that he moved up from Dodge (1916–1920) to Buick to the luxury of a Lincoln and a chauffeur.

As the productivity gains sank in, the Dow marched upward still more aggressively, from 155 in February of 1927 to 200 by the end of that year. Many investors were now wilder than Wilson. New investors had discovered that they could buy shares without the cash to pay for those shares—they simply borrowed on margin and hoped that the rise in the stock prices would cover their loans. The margin rule was not new, but the investors were. Whole households with very small resources speculated, the women as well as the men, though the women's wagers were usually smaller. Stock pickers were borrowing so much from brokers to buy stocks that the amount—quantified later by the economist David Hale—was equivalent to a full 18 percent of gross domestic product (GDP). The excitement over stocks was so high that stock exchange management—the children of Victorians, after all—decided, like Hoover, that it was time to forestall inappropriate behavior. Stock exchanges therefore created separate rooms for men and women investors.

There was even hope for farms, which had done much less well than the rest of the economy over the decade. Perhaps the same productivity gains that Henry Ford had achieved in industry might also be achieved in agriculture. In Montana a giant farming experiment, funded by J. P. Morgan, was under way. The idea was to bring to the farm all the economies of a Ford assembly line. The "farming factory," as it was called, covered 95,000 acres; its head farmer, Thomas Campbell, was written up in *Time.*

The Gilded Age was generally proving to be gilded for the average, even the poor, man. Groups hoping to rise out of poverty also did well. Their general conviction was that individual effort was the key to advancement. In Muncie, a Russian Jew gave a speech that explained the immigrants' hope and excitement: there was not gold paving the streets in the United States, he found, but the gold of opportunity in the small towns, and gold "in the hearts of your citizens, the gold which, too, makes each of us able to go all over the world with respect and safety as American citizens."

In New York, Italian Americans became symbols of success; one of these, the half-Jewish Fiorello LaGuardia, represented the state as a Republican in Congress. Another proud group were his cousins, the Jews, both the older German Jews and the newer East European Jews. Jews at the time had a general belief in charity and taking care of one another: "All Israel is responsible for one another." In addition, they were aware of a specific history in New York; Peter Stuyvesant had asked the Dutch West India Company to ban Jewish settlement, but the company had allowed Jews to stay as long as the Jewish poor "be supported by their own nation." The colonial Jews had pledged that they would, and the commitment was still alive. As late as the 1910s, philanthropist Jacob Schiff said that "a Jew would rather cut his hand off than apply for relief from non-Jewish sources."

The paramount symbol of such immigrant independence was the Bank of United States, which served immigrants and within a few years was establishing sixty offices spread out around New York. The bank's very name—Bank of United States, not Bank of *the* United States or Bank of America—was awkward. Its position was also awkward—while it was large, because immigrants were arriving fast and saving aggressively, it was not a member of the New York Clearing House, and therefore outside the established network of banks. Indeed, one likely reason for the bank's official-sounding name was to signal that the bank was part of the American dream, and as close as a private bank could come to being as trustworthy as government.

The Bank of United States served the textile and clothing businesses—the rag trade and many others, for depositors would soon number 400,000. From the jewel trade to the wholesale meat business, immigrants were integrating into the New York economy. Among the Jewish families in the city throngs were kosher butchers named Schechter, in Brooklyn. In the late 1920s several banded together to open the Schechter Brothers wholesale poultry slaughterhouse.

Blacks were part of this story. Some were coming up in the great migration from the South—leaving the flood zone, giving up homes and, sometimes, land—to establish themselves in an entirely new place, northern cities. At the time, the step seemed wise. There was now a small black upper class in cities like New York, the old middle class in Harlem's Strivers' Row being joined by wealthy blacks. In the North, unlike the South, black children could attend school, and their parents had some choice of work. Black illiteracy decreased to 16.4 percent in 1930, from 45 percent in 1900. Fewer black babies died at birth—by half. Black life expectancy was rising. Most important, blacks were able to find work at about the same rates whites did. Data from the 1930 census would show black unemployment nationally standing slightly below white unemployment.

Coolidge of the party of Lincoln was not content with this. He wanted to see an end to lynchings in the South but was not clear whether Congress had the authority to reach over the states and do so. In December 1923 he said that "the congress ought to exercise all its powers of prevention and punishment against the hideous crimes of lynching." But Congress was not inclined to act, and here his federalism seemed disingenuous.

The black impulse to strive, and black impatience at presidential hesitation, now found a strange expression: the cult of a self-taught preacher whom his followers called Father Divine. Father Divine had little education and, like other Baptist preachers, his own style of evangelizing. He did not speak his ideas as much as speak his way *toward* them. "God," he would say, "is not only personified and materialized. He is repersonified and rematerializes. He rematerializes and he rematerialates." Unlike the black nationalist Marcus Garvey, Father Divine taught that the salvation of blacks was through the Gospel of Plenty—through making it in Middletown, as it were. Here he was in line with Coolidge. Coolidge emphasized the economic progress of blacks when he spoke, arguing that equality would come after.

Father Divine also believed that it was destructive for blacks to think of themselves in racial terms—indeed, he himself refused to recognize racial differences and did not allow his followers to do so. In other words, heaven was not a black Middletown or a white one but simply a heaven of the middle class. Though Father Divine preached in churches like others, his most famous center in the 1920s and early 1930s would be his Sayville, New York, home. Seventy-two Macon Street was a sprawling house, in a town that was a model of conservative and white suburbanism. Father Divine called it his "heaven." Contemporary journalists mocked Father Divine's movement as a parody of the general culture of aspiration in the 1920s; certainly he provided a contrast to Garvey. But Father Divine's followers, like the followers of Booker T. Washington before him, had a respectable purpose: to improve themselves as individuals. Father Divine used movement money to acquire as much property as possible, mostly in all-white neighborhoods. Through him, he believed, poor blacks would become part of the American people, great and anonymous.

Overseas, all these American successes were being noted. In 1927 *Literarische Welt,* a German periodical, quizzed eleven Berlin citizens chosen as typical to see if they recognized names on a list. All eleven knew who Thomas Edison was; ten knew Henry Ford. Only four knew who Joseph Stalin was. Edison himself knew whom he should thank for his own and America's standing in the world. That same year the *New York Times* published an interview with Edison, now just on the brink of retirement. It contained the following exchange: *NYT:* "Do you think President Coolidge will be renominated and reelected?" Edison: "He ought to be."

Still, evaluating the specific worth of Mellon's contribution or Coolidge's reticence remained hard for most. Only a few favored Mellon over Hoover as Coolidge did. To the rest of the country Mellon was a distant figure. To the farmers, he was even the enemy; his gold standard kept grain prices low. As another election approached in 1928, it was Hoover who knew how to put on a political show, and Hoover who was becoming ever the greater figure. His plan for the Colorado River was coming together as the states agreed to the project. Under the Compact Clause of the Constitution, Congress would approve the dam agreement and allocate funds for it—but the project would also be the states'. Later in the decade, the project to dam Black Canyon was put up for bid. Six Companies, a group of West Coast titans specially put together for the project, won the job. There had long already been a Roosevelt Dam, named after Teddy Roosevelt. Even Coolidge, who so detested displays of public power, would get a dam: in 1924 Congress authorized participation in construction of the Coolidge Dam on the Gila River in Arizona. A town in the area, reclaimed with the erection of the dam, would also be called Coolidge. It seemed obvious that one day the Colorado River dam would be the Hoover Dam.

Hoover's meticulousness about the legal process for the Colorado dam reflected the tensions of the times. The Muscle Shoals project had certainly employed people, as many as eighteen thousand workers at peak. But the nitrates that had arrived too late were a sort of national joke about the inefficiency of government. Coolidge stood for privatization—he had said that "if anything were needed to demonstrate the almost utter incapacity of the national government with an industrial and commercial property, it has been provided by this experience." Hoover too opposed a government-owned Muscle Shoals.

The Colorado dam was becoming Hoover's demonstration that the problem of power could be solved another way than by nationalization. Of his 1927 flood work he summed up: "We saved Main Street with Main Street." This statement reflected his own nuanced positions on such projects; he had not said, "Main Street saved Main Street by itself." Washington's task was to referee.

As the floodwaters receded, the commerce secretary himself withdrew to Bohemian Grove, a retreat for the western elite in the redwood country of northern California. At an all-male powwow, he conferred with political leaders. The papers reported that his following in the South was now so strong that he did not even need the administration's support to get a share of that region's electoral votes.

Coolidge was increasingly perplexed. As Hoover later recorded, the two had discovered that there was no getting around the essential difference in their philosophy: "One of his sayings was, 'If you see ten troubles coming down the road, you can be sure that nine will run into the ditch before they reach you and you have to battle with only one of them.' . . . The trouble with this philosophy was that when the tenth trouble reached him he was wholly unprepared, and it had by that time acquired such momentum that it spelled disaster."

Coolidge could see that now the spotlight was on Hoover, and that, looking closer, the country liked what it saw. More than ever, the fact that Hoover was a businessman had an appeal. That he was a westerner—

after Coolidge of Vermont and Massachusetts, Harding of Ohio, Wilson of New Jersey, and Theodore Roosevelt of Oyster Bay, New York—was also a plus. If Hoover was more active than Silent Cal, that might not weigh against him. The Engineer, as he was called, seemed to the public to represent a useful update of the old laissez-faire man in these more vigorous times. After his flood feat, citizens were ready to give him a wider mandate.

Coolidge now had a problem. If he didn't want Hoover to supplant him, he didn't necessarily want to stay either. "It is difficult for men in high office to avoid the malady of self-delusion. They are always surrounded by worshipers. They are constantly and for the most part sincerely assured of their greatness," he would write shortly after leaving the presidency. In the same volume, a short autobiography, Coolidge made a thoughtful argument against long service in the job, noting that "the presidential office is of such a nature that it is difficult to conceive how one man can successfully serve the country for a term of more than eight years." Too often, the man became the office. He did not want to be such a man.

That same summer, the summer of 1927, Coolidge issued a short statement: "I do not choose to run for president in 1928." There was coyness there—what if there were no choice, and candidacy were foisted upon him? But with each month it became clearer that he would indeed leave the presidency after his five and a half years. It was another of Coolidge's acts of refraining, his last and greatest. And again, it opened a door for Hoover.

2
The Junket

July 1927
Average unemployment (year): 3.3 percent
Dow Jones Industrial Average: 168

One late July day in 1927, in the same week that Hoover would present a ten-year flood-prevention plan to Congress, a steamship left the New York piers and headed out into the Atlantic. Its name was the *President Roosevelt,* after America's first progressive president, Theodore Roosevelt. The *New York Times* reported that the ship would dock at Plymouth, Cherbourg, and Bremen.

Among the travelers on the roster was "Prof. R. G. Tugwell"—Rexford Guy Tugwell, a dark-haired Columbia economist who had just turned thirty-six.

Joining Tugwell was another professor, John Bartlet Brebner. The pair were part of a larger junket—academics, magazine writers, union men. Stuart Chase, a certified public accountant and economic commentator, had come with the party. Chase that year had with partner F. J. Schlink laid the ground for a new movement, the consumers' movement, by establishing a group called Consumers' Research—it would lead to the Consumers' Union and the magazine *Consumer Reports.* Together the pair had also published a manifesto for consumers, *Your Money's Worth.* Their idea was that the average consumer deserved more recognition than the Sumners or the Coolidges allowed. Indeed, Chase tended to break down the world into simple categories—engineers were good, consumers were good, businessmen were suspect.

Another traveler was Carlos Israels, a Columbia law student and the son of one of New York governor Al Smith's advisers, Belle Moskowitz. Silas Axtell, an attorney from the Seamen's Union, was on the trip. The senior figure on the junket was the sixty-three-year-old socialist James Hudson Maurer, president of the Pennsylvania Federation of Labor. John Brophy of the United Mine Workers District 2 in Pennsylvania was on the ship, getting over his recent defeat in the contest for UMW leadership by John L. Lewis. Another voyager on the *President Roosevelt* was Albert Coyle, the editor of the *Brotherhood of Locomotive Engineers Journal*, and a conservative in the union world.

In Europe, the men would meet up with yet others: Paul Douglas, a labor scholar from the University of Chicago, whose rough, honest face and independent demeanor would later remind observers of the actor Spencer Tracy. Traveling too was George Counts, a professor at Columbia Teachers' College and a disciple of the legendary educator John Dewey. There was a school superintendent from Winnetka, Illinois, Carleton Washburne, and a professor from the Divinity School at Yale. Alzada Comstock of Mount Holyoke, an expert in government finance, also came along. Robert Dunn, a labor researcher and early staffer at the American Civil Liberties Union, joined them; he would serve as secretary. In Russia around the same time was Roger Baldwin, a peace activist who had founded the ACLU. Baldwin was already a radical hero; he had been convicted in the United States of instigating unlawful assembly at a textile strike in Paterson, New Jersey, several years before. He had also spent nearly a year in prison for refusing to serve in World War I. The others respected him as a player in the world of ideas, a friend of the anarchist Emma Goldman.

At the time Benito Mussolini was popular across the world. Conservatives admired him for his efficiency; he made the cover of *Time* and filled the pages of *Forbes.* Mrs. William Randolph Hearst had embarked just a few months earlier from the same piers on the liner *Duilio,* on a trip to visit Il Duce. Some in this group wanted to stop in Italy, too, but Mussolini turned them down. The travelers spent time in Britain, some of it sampling clotted cream, strawberries, and meat pies, as Tugwell, a bit of a gourmet, later recalled. They also visited Germany and Belgium. But the ultimate destination of these travelers was a country that had been through a revolution just a decade earlier—the Soviet Union.

Soviet Russia was controversial in a way that Italy was not. The United States did not recognize the Soviet Union diplomatically. Even labor leaders like William Green of the American Federation of Labor opposed recognition, arguing that there was little in common between the cause of American labor and Russia's revolutionaries. Green explicitly rejected the idea of an official delegation. John L. Lewis, the same man who had beaten Brophy and been Hoover's partner, also made it clear he would not negotiate with Communists.

So the travelers carefully gave themselves a label: they were the first non-Communist, unofficial American trade union delegation. They saw themselves as objective and wanted to make sure others saw them that way as well. The trip was not truly independent, however, for Stalin's regime controlled the itinerary inside Soviet borders, selecting the factories and farms they visited. What's more, some of the less-known people on the trip, support staff included, were affiliated with the U.S. Communist Party. Either Tugwell, Chase, and the other scholars did not know this, or they were confident of their ability to protect themselves, to judge accurately regardless of whose company they were in.

"I want to study conditions for myself," Maurer had retorted when he'd heard of the ban imposed by Green. If conditions were as awful as labor leadership said, then "those gentlemen should encourage American labor to go and get the data to bear them out." The junket was funded by a nonprofit regarded as respectable, the Garland Fund, though Paul Douglas took the extra precaution of paying for his own tickets. In any case, the travelers planned to demonstrate their good faith through their work. They would write out extensive notes and reports—at least one, maybe two volumes—to document the experience for U.S. readers. Interviews with Bolshevik leaders seemed possible.

Joseph Stalin had two reasons to host the guests. The first involved winning over American labor. The official policy of the Soviet Union might now be "Socialism in One Country," but, if Stalin could draw the smaller fry to his side, he might split the U.S. movement and, eventually, capture it.

The second motive was more immediate. The Soviet revolution was failing. Communism needed Western cash to survive. The situation in the Soviet Union of 1927 was grave. As Comstock would note in a later report on the trip, the weakness of foreign trade represented "an ever-present source of danger." This was why, just a bit earlier, Stalin had dropped the original Soviet goal of worldwide revolution. To get his money, Stalin needed legitimization. If he could win acceptance, or even better, formal diplomatic recognition, he would receive loans, perhaps enough to make him invincible at home.

The American side of the story was subtler. Diplomatic details aside, the concept of such a great land in the process of such great change was exciting, so exciting that the world seemed to split into those who had been there and those who had not. In this regard, the travelers on the *President Roosevelt* felt like the young Herbert and Lou Hoover, who had headed off to China or London for the excitement of the foreign adventure. The American frontier had been tamed; here was a new frontier, where projects might be tried that even

Americans had never dreamed of. Already, a year earlier, Henry Ford had invited Russian factory workers to Detroit to train them to make Fordson tractors. It was Washington that seemed retrograde when it balked at granting visas. Moscow for its part had ordered ten thousand Fordsons. In the years after the *President Roosevelt* trip, thousands of Americans, especially engineers, would come to the Soviet Union to work, also in the spirit of adventure. Moscow in this period heard the accomplishments of Thomas Campbell, the manager of the farm-factory in Montana. Stalin would counter with his own bid: a million acres for Campbell to work with if he started a Soviet farm. And Campbell would sail back to the States with praise for the Soviet experiment. The Soviet Union also appeared to be playing catch-up in other areas. Stalin, like the czars before him, had a flair for developing show projects that drew attention away from economic flaws.

The day that the *President Roosevelt* left Manhattan's piers, the New York office of Amtorg, the Soviet trade representative in the United States, had announced that the Soviet Union had begun work on what would be Europe's largest dam, Dnieprostroi, a giant in the Ukraine whose turbines would each deliver 50,000 horsepower. The engineer on the project was an American, Colonel Hugh L. Cooper of New

York. Soviet boldness and American know-how might achieve what American know-how at home could not.

The same scale of ambition showed up in the Bolsheviks' plans for the cities. Enormous buildings went up. Hoover might standardize brick size for paving roads. The Soviets went a step further, as Alzada Comstock would note in an article several years later, and simply commanded the cobblestone's removal. Within several hours, volunteers removed 1,062 meters of cobblestones before the state opera house. The Soviet Union was fresh, and therefore interesting. Three years earlier, the progressive Lincoln Steffens had penned his famous summary of the achievement of revolutionary Russia, a summary that rang in the ears of the 1920s travelers. That summary did not say, "I have seen socialism and it works," although Lincoln Steffens was a man of the left. It said: "I have been over to the future and it works." Steffens also said, "I would like to spend the evening of my life watching the morning of a new world."

Bob La Follette, the Wisconsin reformer, was one who had been over in 1923 with Lincoln Steffens; the Bolsheviks had put him up in a palace expropriated from a sugar merchant, across from the Kremlin. Another traveler who had preceded them was Roger Baldwin's friend the anarchist debater Emma Goldman.

Goldman's lover, Alexander Berkman, had shot Mellon's friend Henry Frick three times and stabbed him twice in a rage over Frick's treatment of workers. Washington had deported both Goldman and Berkman, sending them back to Russia on the U.S. Army transport *Buford,* nicknamed the Soviet Ark. Lillian Wald, a New York social worker, had also been to Russia and spoken of "the vast promise of the Soviet government and the strength and wisdom and social passion of Lenin." The big question in the travelers' minds was not therefore whether they ought to be going to a nation unrecognized by Washington—it was whether they were going there too late. Like Hoover, they knew it was advantageous to get somewhere early.

But there was a fundamental difference between men like Hoover and these travelers. Hoover belonged to a group that exported American ideas. His whole career had been about exporting American know-how and values—the know-how to mine zinc, the Quaker vision of doing fieldwork in occupied nations. The same held for, say, Mellon, who also saw little to learn from in the Soviet model. Mellon treated the old land of the czars as a salvage project: a few years after the travelers headed for Russia he would, through his art brokers, buy up Raphael's *Madonna of the House of Alba* and several other masterpieces from the Hermitage. Samuel

Insull was also an American exporter, albeit a naturalized one; after developing his power grid in Chicago, he went back to Britain to share his insight with Westminster, which planned construction of a similar grid for Britain. To such men, places like the Soviet Union were not the future, unless one could call a nightmare-come-true the future. "Inherent in Communist destruction," Hoover at one point concluded, was a "shift from intelligence to ignorance." Men like Hoover and Mellon believed in the primacy of the American idea. They might want to modify America, but they did not doubt her.

Men like the travelers, by contrast, were importers. They did doubt. Their progressivism went beyond the progressivism of, say, Theodore Roosevelt, after whom their ship was named, for they also believed specifically that ideas should be collected abroad and then used at home to improve the country. They despaired of purely domestic solutions, and that despair often came out of long experience in those areas of national life that did still need improvement.

Agriculture, Tugwell's area, was a good example. Tugwell's father farmed fruits in Sinclairville in western New York. His mother, Dessie, had taught school. The pair were the leaders in their community, and people with a sense of style, a fact reflected in their baby's audaciously long name. Tugwell had been a mischievous child, so much loved that when he ran away, his family served him lemonade and cookies upon his return. His family spread butter on their bread while others had bacon drippings.

Still, Tugwell concluded in his childhood that the American game was rigged against farmers. They had so many poor neighbors, and so much chance of again becoming poor themselves; Tugwell's mother had been born in a place called No God Hollow. The first big downturn in the Tugwells' lives came in 1893, when Rex was two, and others followed with disconcerting regularity. The boom-and-bust cycle of New England agriculture was too rough.

Farther north, in Maine, Douglas as a young man came to the same conclusion about life on the land. Douglas observed the struggles of lumberjacks sending timber south down Sebec Lake in the icy spring; each year, several drowned. What was the point of stingy New England, with the Great Plains turning out to be so easily fruitful? The New England of the Tugwells and young Paul Douglas was not quite the lawyer's New England of Coolidge. It was more the New England that the poet Robert Frost wrote about—a rocky, small-scale place where the lawyer's rigid insistence on tiny bits of private property and endless stone walls seemed to constrain the future. "Something there is that doesn't love a wall," the poet would write about two farms. "Before I built a wall I'd ask to know what I was walling in or walling out?"

Tugwell and Douglas learned early that the plains that the New Englanders so envied had their own problems. The gold standard drove prices up and then down in a seemingly random fashion that sent punctilious farmers into despair. In the 1890s at Chautauqua, New York—a summer lecture resort not far from the Tugwells—the politician William Jennings Bryan captured the national rage by speaking of the "Cross of Gold" that burdened farmers by keeping prices down. *Wallace's Farmer,* a periodical edited by Henry Cantwell Wallace, chronicled farming troubles in every issue. Harding had made Wallace agriculture secretary.

Out in Elwood, Indiana, a German American schoolteacher named Herman Willkie had invited Bryan to teach his Sunday Bible class. Herman Willkie, like the Tugwells, had been through numerous downturns and had seen firsthand how the farms suffered. He watched Elwood rise and fall on the discovery of natural gas in the region. In 1892, a year after Tugwell's birth, the Willkies had a son to whom they gave a name as ambitious as Rexford Guy Tugwell: Lewis Wendell Willkie.

Tugwell as an adolescent extrapolated from the farm experience to a generalization: the world needed changing, and he would participate in making those changes. He made it to the University of Pennsylvania and the Wharton School. There, he discovered a magazine—the *New Republic*—and read it with increasing excitement; "as nearly as preachment could," he would recall later, "it gave us a text to live by." Unlike the doctrinaire Communist periodicals, the *New Republic* devoted itself to discussion and reform. It was truly liberal, liberal in the way Tugwell hoped he was himself. In his senior year, Tugwell taught sections for a Pennsylvania professor, Scott Nearing. He was horrified when Nearing was fired by the trustees for being a leftist who pointed out large disparities of wealth between Philadelphia's rich and poor. When Tugwell married his high school sweetheart, Florence Arnold, he took her to visit Nearing on their honeymoon. And in 1915, the year the Russian Revolution was building, Tugwell penned his own poem for the student periodical *Intercollegiate:* "I am sick of a nation's stenches, I am sick of propertied czars . . . I have dreamed my great dream of their passing . . . I shall roll up my sleeves—make America over!"

Rural life had also made a radical of Douglas, who was a year younger than Tugwell. On the debate team at Bowdoin College, Douglas argued through questions such as the popular election of senators—then not yet law—the tariff, the income tax, and workmen's compensation, also still just an ideal. Disgusted with the fraternities on campus, he also started a club for that minority in the class who had not made a fraternity. Douglas, who had vowed to serve the freezing logger, studied labor economics in graduate school before coming to the University of Chicago to teach. He was now in the process of splitting up with his wife, Dorothy Wolff Douglas, but the two shared a commitment to collectivism. He was greatly interested in pension plans—the image of the logger staying with him.

At Indiana, Wendell Willkie, born the same year as Douglas, had for his part donned a red sweater in solidarity with Red Europe. He also asked the faculty in Bloomington to introduce a course on socialism and then pulled together the students to fill up the class. At graduation, Willkie gave "the most radical speech you ever heard," the IU president later recalled.

Another disillusioned Hoosier, David Lilienthal, from Michigan City, to the north of Elwood, was younger than both Tugwell and Willkie. As a child, David had seen his father, Leo, struggle in the dry-goods business, from Morton, Illinois, to Valparaiso, Indiana, and then, yet again, to Michigan City. While

David was attending high school, relatives had taken him to a Chicago labor meeting where he heard Emma Goldman speak "with a voice deep and strong, with a quick turn and a lash to it which was more of the type possessed by some sharp-witted man." Lilienthal witnessed the men in Chicago torturing the drowning mouse around this time. After attending DePauw in Greencastle, he headed east to Harvard for law school. While there he learned that his father had lost $5,000 in business, the family fortune. Lilienthal felt moved to read John Reed, whose *Ten Days That Shook the World* had recently been published. He also contemplated foreign travel. But his father blocked it, warning that a trip to Russia might "make a goat" out of David. Lilienthal decided that he would stick to his old resolution of being "a *student* of the problem"—the italics were his. But he did not give up hope that in America there might still be a "happy orderly revolution." He too began writing in the *New Republic.*

Bryan's heyday had passed by the time these youths became men. But farms, especially in the late 1920s, once again had the same problem—falling grain prices. Tugwell found himself making a career of economics and reform, even spending a semester at Amherst College in Massachusetts, the academic home of Robert Frost. Surely, he thought as he moved through his twenties and thirties, the solution had to be collective. The individual was an ungenerous figure from the past.

Utilities were another area that concerned Tugwell and others. They noted that the prosperity seen in the cities was not as visible out in the country, and the rapid advance of utilities in the United States did not seem rapid enough to the farming crowd, for even in the 1920s, many farms did not have power. Wall Street's capital might serve Campbell out in Montana, but could it reach the rest?

One man who thought about the utility question continually was Bob La Follette. Another was George Norris, the senator from Nebraska, who vowed not to give up on Muscle Shoals. But yet another was an aristocratic young politician who lived south of the Tugwells, in Hyde Park, New York: Franklin Roosevelt, a cousin of the president. At Harvard as a young man, Franklin Roosevelt had taken History 10-B with a visiting professor who was an expert on the American frontier, Frederick Jackson Turner. Turner was developing a thesis that stuck with Roosevelt all his life: as the American frontier closed, the United States was entering into a period where the old rules did not hold. As an undergraduate Roosevelt had also studied with a monetary expert, Oliver M. W. Sprague, and at least thought about the money problem and the farms.

Franklin Roosevelt too had observed that rural America was not always keeping pace with the city, and written a college thesis on the need for electricity. Roosevelt would later tell his speechwriter and friend Samuel Rosenman of a family who lived not ten miles from Hyde Park yet were without power. "Now they have been watching electricity come along in Poughkeepsie and Rhinebeck and even in Hyde Park, more and more each year and they have tried to get electricity out where they are and they just can't. They and other farmers are willing to pay but the damned old electric corporation won't listen." Roosevelt had a goal: "I want to get cheap electricity out to that farm."

The answer for utilities was wider government involvement in the power industry, nearly all progressives believed—both those who liked foreign models and those who did not. In the 1920s, Roosevelt had thought a lot about Georgia, for he traveled there so often for his cures at Warm Springs. To many southerners, the back-and-forth debates between Hoover and Coolidge on the one hand and the progressives on the other were not inspiring; they were frustrating. Wilson Dam, the original dam at Muscle Shoals, produced power in the 1920s. But not much of that power had made it to the locals around the dam. For years the town of Muscle Shoals tried to buy power for itself from Wilson Dam.

Finally they would be allowed to, but at a higher rate than that charged to the Alabama Power Company. Frustrated town leaders would send a wire to their senator, Hugo Black: "Telegram you received from Muscle Shoals this morning framed by city fathers, in City Hall by light of kerosene lamps, though within 2 miles of tremendous power tumbling to waste over Wilson Dam with administration's consent."

Factories were another area that might be improved by foreign study. From the Triangle Shirtwaist fire of 1911, in which close to 150 perished, to the sweatshops of the West Coast, American industrialization seemed to them not progress but proof that Thomas Hardy was right: factories debased. Stuart Chase, for one, was hoping Soviet industry might provide a model to solve some of these problems. Chase did not think that the American factory would continue to improve by itself.

Chase believed it was the responsibility of the idealist to act. He had graduated from Harvard in the same class as the revolutionary John Reed, the class of 1910. Sitting in the economics section of the Boston Public Library in 1911, in the autumn after the fire, he wrote a note to himself: "Now I must choose my own path . . . from among the many and follow it in all faith and trust until experience bids me seek another." Chase did not dislike big projects; he was something of a

Taylorite, and in fact had worked at Hoover's Food Administration. But he believed that government ought to involve itself with something stronger than a beneficent hand. In the mid-twenties he had published a book expressing his disgust at all the leavings industry seemed so thoughtlessly to produce; its title was *The Tragedy of Waste.* He did not buy some economists' argument that increases in productivity were spontaneous. They had to be made to happen.

And he ignored the overall improvement in wages to focus on groups that hadn't seen great gains. Two years before Chase had written in the *New Republic* that "garment trades, the machinists, many miners and railroad workers have learned there can be no increase in wages until industrial waste is checked." Differences among political regimes seemed to him irrelevant. What mattered to him, as to many Taylorites, was that all countries confronted the same problem, taking advantage of the economy of scale to be more efficient and less wasteful.

The factory and city also obsessed James Hudson Maurer. Born in 1864, he had started life with a childhood so poor it made Tugwell's look smooth. In the panic of 1873 his mother handed him a letter and sent him across the county in search of a relative who might take him in—his own grandfather handed the letter back, saying, "Don't see how I can use you." School when he attended it had been more about "whipping than teaching." Maurer learned to read only just before age sixteen, when a fellow worker and member of the Knights of Labor handed him a political speech, and he saw a point to the whole exercise. Maurer was a socialist, but when he saw that socialism was not catching on in America, his reaction was to conquer his own territory for the socialists, however small: he had become the socialist mayor of Reading, Pennsylvania.

John Brophy, for his part, was angry at the moderate progress of unionism. He was from Lancashire, in Britain, and wondered why the United States had not kept pace with the United Kingdom or Europe when it came to developing social models. Mellon's Pennsylvania was not glorious; it was a Pennsylvania of brute exploitation.

To these men the heroes of the factory were people like Goldman, and her banishment a sign that the country was unwilling to concede the truths she highlighted. Goldman was especially beloved because she had been so articulate in her speeches about America's failings—Roger Baldwin would later remark on her fluency in English, rare in someone who had first come to the United States as a seventeen-year-old, as Goldman had. In her lectures she had pointed out that hunger was still a problem in the United States, that men still beat women, that Frick had his paintings while newcomers were without toilets. She accurately noted that the new industrial America had not yet found a way to take care of old people in the city, where no farmstead welcomed grandparents.

The rights of citizens were a third area of focus. To progressives, the anti-Red actions of both Democratic and Republican administrations were evidence of a trend toward repression. Recently two anarchists, Nicola Sacco and Bartolomeo Vanzetti, had been sentenced to death for killing a company paymaster in South Braintree, Massachusetts. Many progressives argued that the men were innocent of the crime and had really been convicted because of their radical political beliefs. A Harvard law professor, Felix Frankfurter, bravely fought for the pair, even though the president of his university, A. Lawrence Lowell, had taken the position that justice was being served sufficiently.

A number of the *President Roosevelt* travelers had also involved themselves in the Sacco and Vanzetti cause—Douglas, for example, had sent money for the defense. But the pleas failed, and the execution of Sacco and Vanzetti was expected to happen while the travelers were in Europe. This act seemed to confirm American barbarism. Roger Baldwin, especially, wondered whether Soviet Russia might have found a higher sort of freedom.

There were no African Americans on this junket, but the travelers were aware that the year before, William E. B. Du Bois, a black leader a magnitude more prominent than an eccentric like Father Divine, had been inspired by a visit to Moscow: "I stand in astonishment and wonder at the revelation of Russia that has come to me," he wrote. "I may be partially deceived and half informed. But if what I have seen with my eyes and heard with my ears in Russia is Bolshevism, then I am a Bolshevik."

Others too thought that Russia might have the answer to the race problem. There was a legitimate sense among them that legal change was too slow in coming, and that segregation was breaking down too slowly. The Ku Klux Klan was not going away; on the contrary, it flourished. Father Divine's house-by-house gradualism seemed almost a parody. Coolidge's general statements about equality and tolerance were no substitute for an explicit law to halt lynching.

The social barriers of the 1920s were all the more insupportable because of blacks' loyal service in World War I, and because so much time had now passed since slavery. The young Paul Robeson, for instance, in the 1910s received the third scholarship ever given to a black student at Rutgers, but other students trod on his hands when he was a football player. And now that he was out of Columbia Law School, in the 1920s, he could not practice law: a white secretary refused to take dictation from him. Robeson was turning to another of his many gifts—music—but it was a bitter turn because he had been pushed. He would later travel to Britain and meet with the miners; in the early 1930s, he would also start studying Russian, and travel to the Soviet Union in 1934. Of the travelers, Baldwin and Douglas were probably the most eager to find answers to problems like Robeson's.

Still, the expanding boom at home made these causes seem far weaker than they had even a decade or five years ago. When only two in a hundred men were out of work, it seemed absurd to preach radical reform. After college, many radicals gave up their causes, convinced that those causes were fading anyhow. That was what had happened to Wendell Willkie after the war. His solution to the problems of rural America had been to escape them. Following college and the military, he taught a bit, then turned to law school at Indiana University and life as an Akron, Ohio, lawyer for Harvey Firestone, who headed up one of the new tire companies. Like Tugwell a jovial man, Willkie had married a librarian named Edith Wilk—"How do you feel about adding a couple of letters to your name?" he had asked her. For Willkie, there was no more time to think about European socialism. Akron was growing so fast that the Willkies and their baby Philip had to share an apartment with another couple. Willkie was shortly earning thousands more than most people he knew back in the agricultural towns. Some of his work was in the utilities area, for power companies.

With each year that passed, the radicals fell more out of step with the country. Tugwell, more honest than many, acknowledged that. "Life in the 1920s was often frustrating for those people of my political persuasion—political progressives or radicals," he would later write. "We were, in fact, all but regarded as social misfits."

Aboard the *President Roosevelt,* the travelers settled in. Tugwell would later write of the ship that to him it always seemed "homelike," and the others likely had the same feeling. For the boat was a version of something that had become increasingly important to the American Left in the 1920s: a refuge. Rather than give up or bemoan their isolation, people like Tugwell simply redoubled their efforts to create a parallel reality of their own. The world shut them out, so they shut out the world—with a salon, a club, a ship, a trip. Or even a new identity: the 1920s intellectual.

Just a few years before, in 1922, the author Sinclair Lewis had indirectly done much to establish that identity when he wrote *Babbitt.* Babbitt, Lewis's protagonist, practiced a vague urban trade: he was a real estate broker. He had no high moral purpose, he "just got along" and lived a tedious life in the fictional yet typical neighborhood, Floral Heights, of a fictional yet typical American town, Zenith. He was annihilatingly provincial. The thinkers of the 1920s loved Babbitt because he supplied them with a goal. They would be everything that he was not: urban, bohemian, anti-money, idealist.

Inspiration also came from abroad, from the intellectuals of Leningrad or Paris. After all, the thinkers of those cities understood about the Babbitts and the sometimes equally tedious classes below him. Back in the 1910s a traveler from Russia had written in the magazine *Novy Mir* about his own reaction to the dull expressions of New Yorkers chewing gum on the subway: "The color of their face is grayish, their hands are hanging down weakly, their eyes are dim. . . . Only their jaws are moving, submissively, evenly, without joy or animation. . . . What are they trying to find in this miserable degrading chewing?" The name of the disgusted travel writer was Lev Bronstein, but he would later call himself Leon Trotsky.

The intellectual exclusivity of the Left would also be captured many years later in a novel about Vassar College at the start of the 1930s by the author Mary McCarthy, titled *The Group.* Alone together, dreamers reinforced one another.

Vassar, one hour and forty minutes north of Manhattan by train in Poughkeepsie, was one of the more important refuges. There a young theater director named Hallie Flanagan created a sense of utopian experiment. Flanagan had herself visited Russia to learn from Soviet theater, years earlier, and would do so again. In May 1930, at term's end, she would load her students onto another ocean liner, Holland-America's *Volendam,* and take them to Kiev, Moscow, and Leningrad to performances of the Moscow Art Theater, Proletcult, the Blue Blouse, and other revolutionary theater groups. A student influenced by Flanagan, a young woman from the Seattle area named Mary McCarthy, would later achieve a name of her own.

"The Vassar Girl is thought of as carrying a banner," McCarthy would write—the banner of the rebel and the reformer. Now that suffrage had been gained, in 1920, women of ideas were looking for new causes. One of Vassar's trustees, Franklin Roosevelt, would shortly run for governor of the state of New York. His wife, Eleanor, was the co-owner of a tiny furniture factory that made colonial reproductions, Val-Kill, and which counted Vassar College among its clients.

But Vassar was not the only refuge. Right in New York City, two progressive educators, Nancy Cook and Marion Dickerman, had joined Eleanor Roosevelt, Franklin's wife, in purchasing the Todhunter School; in 1928, Eleanor would teach girls there Mondays, Tuesdays, and Wednesdays. Todhunter was rigorous—Eleanor liked rigor—but also progressive. The exam questions Eleanor and her colleagues wrote up were challenges to the America of Coolidge and the doctrines of Mellon: "What is the object today of inheritance, income, and similar taxes?" "Why is there a struggle between capital and labor?" "What is the World Court?" "Who is the dominant political figure in Soviet Russia?"

Then there was the University of Pennsylvania, where Tugwell had studied with the radical Scott Nearing, or Douglas's university, Chicago. At Chicago John Dewey, the great philosopher of education, had established his own laboratory school on the Midway where children of professors worked in labs, shops, and kitchens. This followed Dewey's belief that learning by doing was the best way. Shortly progressives would also come to the University of Wisconsin, where the next year educator Alexander Meiklejohn would establish an experimental college in which undergraduates looked at only one topic at a time.

At all such places, the progressive intellectuals might make up a minority, but they still were a presence, reassuring one another. And then there was Harvard. There a few star professors encouraged their students to push for radical change—both in the law, and in government. The brightest star was Felix Frankfurter. Frankfurter was an immigrant himself; he had come over as a child from Austria. His rise was the result of merit, not birth. This fact alone made him exciting, especially to his students, many of a class that had enjoyed great advantage early. Frankfurter embraced his new country and the law with passion. His respect for American law was almost like respect for a church—he would describe his feeling for Harvard Law School as "quasi-religious." One of his heroes was Louis Brandeis, now on the Supreme Court. Brandeis might have seemed exotic to some observers—he was the first Jew to sit on the Supreme Court—but his philosophy was straight Thomas Jefferson. Another Frankfurter hero was the jurist Oliver Wendell Holmes, again, hardly a European radical; Holmes, the son of the wordsmith Oliver Wendell Holmes, was one of the most American of Americans, buried in Boston's

Mount Auburn Cemetery along with Amy Lowell. Still, Frankfurter did have a European model—Britain, where labor reforms were taking place. And when thinking politically, or as an advocate, Frankfurter viewed American law as a vehicle for European-scale reform.

Frankfurter's views on economics were near the opposite of Coolidge's, even though they both spent much of their careers in Massachusetts. Frankfurter liked the idea of an active American government very much, and he tended to dislike, or disapprove of, business. An observer, Raymond Moley, would much later sum up Frankfurter's worldview: "The problems of economic life were litigious, controversial, not broadly constructive or evolutionary. The government was the protagonist. Its agents were its lawyers and commissioners. The antagonists were big corporation lawyers. In the background were misty principals whom Frankfurter never really knew first hand and who were chiefly envisaged as concepts in legalistic fencing. Those background figures were owners of the corporations, managers, workers and consumers."

There was another element to Frankfurter's personality that impressed his fellow intellectuals: he knew how to get along politically, no matter how unpopular radical thought was. Many at Harvard, including

President Lowell, disagreed with him, yet Frankfurter managed to survive, even thrive at the university. Writing to Holmes, Frankfurter flattered the Supreme Court justice and won his friendship. He was close to Brandeis, who even subsidized him in the 1920s. Frankfurter, among all law professors, probably best knew Brandeis's aversion to the large, whether it be the large company or a large country. Brandeis would later publish a book titled *The Curse of Bigness,* which argued there was danger in large corporations. Brandeis was an early Zionist, liking the idea of a small Jewish state, but he also was fond of the modest Scandinavian countries, especially Denmark. (In the next decade, as the Soviet Union became more popular as a destination, Brandeis would tell young travelers to go to Denmark instead.)

In 1924 Frankfurter had supported La Follette, writing in the *New Republic* that both mainstream parties "have an identical record of economic imperialism," and describing his foreign model—Labour in Britain. As early as 1906, when he first encountered the young Franklin Roosevelt, the professor began to work to influence him. The Frankfurter touch reached to the smallest detail. One of Frankfurter's biographers reports that the pair talked about reading, and Frankfurter suggested to FDR that he indicate a significant passage with a line in the margin, like the great

English historian Thomas Babington Macaulay, rather than underscoring line after line.

Frankfurter could influence young Roosevelt and others because, even in *Babbitt* times, he was able to transmit a wonderful sense of possibility to those around him. Student after student after student came to him and stayed, sometimes simply for the pleasure of going mind against mind. One of those students was Adolf Berle, who, like Frankfurter, was not accustomed to coming in second. Berle attended Frankfurter's course two years in a row. "What, back again?" Frankfurter had asked. "I wanted to see if you'd learned anything," Berle replied. Another Frankfurter student was David Lilienthal of Indiana, the young man who had disliked adventurous Chicago. Lilienthal found Frankfurter so enthralling he would later describe him as a man "who could read the dictionary and make it exciting." Others were Benjamin Cohen and Thomas Corcoran, who later would be called his "hot dogs."

Frankfurter's next skill inhered in this: Better than any law professor in the nation, he knew how to place his students in important jobs. From Harvard, Frankfurter sent his students to Brandeis and fellow justice Oliver Wendell Holmes to clerk. Many of these students would later have leading roles in government or universities, among them Dean Acheson, David Riesman, and Alger and Donald Hiss. Frankfurter had a virtual monopoly when it came to clerk appointments at the Supreme Court. When Justice McReynolds, a few years later, selected a Harvard alumnus who was not a Frankfurter protégé as his clerk, Brandeis bluntly asked the clerk how he came to get his job. "There isn't one chance in a thousand for any graduate of Harvard Law School to come to the Court these days without Professor Frankfurter's approval."

Frankfurter still believed that the era of social legislation had only begun, that the country could make numerous changes, like introducing a minimum wage. He regarded Justice George Sutherland, the author of *Adkins v. Children's Hospital,* as especially retrograde. Frankfurter argued for state minimum wages before the Court, and his students had the satisfaction of knowing that at least one justice—Holmes—seemed to follow his line of reasoning. Frankfurter's impression on his students was especially profound in the area of utilities. The rest of the country might look to Commonwealth and Southern or Insull for progress, but in Frankfurter's course Public Utilities, the emphasis was on the "public." Frankfurter and other progressives felt strongly that governments should not miss this opportunity to regulate, the way they had failed to regulate industry in the preceding century.

In Frankfurter's classroom it mattered little that in the 1920s the constitutional obstacles to a grand federal program for power generation seemed greater than the boulders on the Colorado River. Removing them, Frankfurter suggested, might even be easy. A student, Frances Plimpton, wrote a skeptical rhyme about Frankfurter's crusades:

You learn no law in Public U
That is its fascination
But Felix gives a point of view
And pleasant conversation.

Lilienthal and others saw that Frankfurter and the *New Republic* were creating a new kind of liberalism, different from what Sutherland or Coolidge meant when they used the word. Maybe Frankfurter's liberalism was that of the future.

Another place with the feel of radical fraternity was Columbia, to whose economics department Tugwell had come in the 1920s. In 1925, together with two colleagues, Thomas Munro and Roy Stryker, Tugwell had produced an innovative economics textbook, *American Economic Life*. Tugwell saw potential in Stryker, a farm boy like himself, from Kansas. The three used photos in new ways to dramatize their arguments: a picture of the tall buildings rising in Manhattan was accompanied by a didactic caption: "Collective effort built this; the inference is inescapable; but we sometimes attempt to avoid the logical further inference that more collective effort is needed. Sometimes we say that what we need is more individual enterprise. No individual," the men concluded pointedly, "ever built a skyscraper." Tugwell and Stryker were proud of the book; in another time, they told themselves, it could become a model.

Raymond Moley, an acquaintance of Tugwell's, taught a course at Columbia's sister college, Barnard, on how reformers in Britain and Bismarck in Germany had solved some of the social problems of industrialization. Moley, and Douglas as well, had become interested in the British export of the settlement house, a community center for the urban poor.

Over at the Teachers College, Columbia University, another man on the Soviet trip, George Counts, was working in a different field: education. Counts had also researched and experimented with John Dewey at his progressive Laboratory School at Douglas's university, trying to establish a new form of child-centered education. Farther north up Lake Michigan, another man on the trip, Carleton Washburne of the progressive Winnetka school system, had also been experimenting with new methods of teaching. The educational progressives believed that competition among individuals in school—just as in Tugwell's economy—was wrong. Instead it was time to look for a model of the collective school for the new society. Families mattered less in such a model—the family was an old agricultural unit, after all. And the factory mattered more. Long before this trip, Dewey and Counts had argued that the best models might be found abroad. Dewey had also argued that in a new mass society, the school must promulgate social change, not respond to it.

There were other refuges. Settlement houses had spread across the United States, and these were homes for intellectuals too. One that achieved the most was Hull House on Halsted Avenue in Chicago. For the poor of that area its resourceful founder, Jane Addams, provided everything, from piano lessons to drilling in English to health care. At Addams's center Douglas met Europeans—"British journalists and politicians and fiery Indian nationalists"—who reinforced his sense that the United States must learn from examples abroad. At one point Addams called the Soviet revolution "the greatest social experiment in history." For generations, progressives had gathered here.

On Halsted Street Douglas found other reformers. At least one who had paid a visit before him was an apprentice social worker—Frances "Fanny" Perkins, a young alumna of Comstock's college, Mount Holyoke. Labor reform was deeply important, Douglas believed. He spent time familiarizing himself with the highly compelling case of the Brotherhood of Sleeping Car Porters, a black union.

In New York, Lillian Wald had established the Henry Street Settlement, which by 1916 was sending nurses out to see 1,300 patients a day. Henry Street helped the poor, but it also served as a safe haven for some of the children of the wealthy, who turned to progressive life as much out of desire to escape as out of any dedication to social change. One of those to work there—he spent time helping out as early as the summer of 1911—was Henry Morgenthau Jr., the son of the Morgenthau who had made a fortune on Wall Street and real estate. The younger Henry was not sure which field of work to choose, and was actually attracted to the idea of farming—the rural retreat. Another to come to Henry Street was a Vassar girl, Beatrice Bishop. Bishop, an heiress from an old Dutch family like Roosevelt's, had suffered at the hand of a vindictive mother, herself the daughter of a president of the New York Stock Exchange. Her mother, Amy Bend Bishop, had wanted to stop Beatrice attending college, predicting, "You will become a bluestocking and no one will look at you." Beatrice had a striking, wedge-shaped face; her mother had been right only about the first part. It was Adolf Berle who brought her to Lillian Wald's table. He described Henry Street to her as "a lay convent of sorts." Dining there with him and other intellectuals at a dark oak table, Bishop found a refuge of idealism.

At Christodora House on Tompkins Square, yet another young idealist, a Grinnell graduate named Harry Hopkins, worked with boys' clubs. Hopkins found himself intrigued with Europe in a more personal sense, secretly courting and marrying a bright young social worker who had emigrated from Hungary, Ethel Gross. Nor was he the only social reformer to take his romance with the European East to a personal level. Nearing the age of sixty, John Dewey in 1917 fell in love with Anzia Yezierska, a redheaded social activist from Russia who was also known as the "sweatshop Cinderella."

And of course there were the magazines where the intellectuals found one another: the *New Republic*, the *Nation*. The audiences were small but the editors were friendly. Earlier in 1927, Tugwell had written a desperate article in the *Nation* titled "What Will Become of the Farms?" His conclusions were gloomy, but having an outlet and collegial editors consoled him. Two brothers, Carl and Mark van Doren, held spots on the masthead; so did Mark's wife, Dorothy. Carl's wife, Irita, had already gone over to the *Herald Tribune* as an editor. But they all knew one another.

To make landfall in Europe was a relief for the travelers. Here at least the economic facts did not contradict their reform concepts so profoundly. There were some bumps along the road. In Warsaw they felt a jolt when their guide, Albert Coyle, acknowledged that he had misplaced the trip funds. In Dortmund, Tugwell got bored and skipped a meeting with trade union people at a steel plant to go to an art gallery. But from the time they met Soviet trade union leaders at the Polish-Soviet border, the travelers felt their spirits rise. This would indeed prove the junket of all junkets. It didn't hurt that their hosts gave them first-class treatment—free transportation, cheap or free hotel service, and so on. And there were to be meetings with leaders—Mikhail Ivanovich Kalinin, the Russian prime minister, Leon Trotsky, already out of the Kremlin's inner circle but not yet exiled, and others of high rank. For the travelers, who were at best respectable but not themselves of national rank at home, these introductions in and of themselves made for a high. There is nothing headier than finding one is more recognized abroad than at home. And that was not all:

rumor had it there would be meetings at the highest level, perhaps even with Stalin himself.

The travelers' enthusiasm was only strengthened by what they saw in the first few days. The failures of the economy were not all visible. Indeed, if one squinted, things looked almost reasonable in Soviet Russia. Lenin, before dying, had instituted his New Economic Policy (NEP), which allowed the survival of small artisans. The economy had finally begun to regain pre–World War I levels. The brutal collectivization of agriculture and the famines of the 1930s were still to come. The Soviets for their part tried to burnish their own reputation with unfavorable references to America. Everywhere the travelers went, Brophy would later note, they heard about Sacco and Vanzetti, who had been executed while the travelers were in Russia, just as predicted. For days after the execution, the towns the travelers arrived in were draped with banners hung in honor of Sacco and Vanzetti, "victims of 'American capitalism.' " To the travelers it seemed that Russia understood what the land of Babbitt did not.

Roger Baldwin, who had corresponded with Vanzetti until his death, was deeply impressed with what he saw. Baldwin understood that Stalin's Russia had a dark side. He didn't enjoy his time in what he called this "irritating place." But Russia still seemed somehow farther along than the narrow Massachusetts that could put the anarchist pair to death. Baldwin gave Leo Tolstoy's son letters he had received from Vanzetti so that Tolstoy might post them for Russians to see at a state bank. As he wrote of Russia, Baldwin's own conclusions were hopeful. "Everybody is poor together," he wrote to his mother. "There is much discontent, much regulation of life, but not much terrorism or repression except of the old upper classes."

For the high-spirited Tugwell, part of the trip was about having a good time. Half a century later, Stuart Chase would write Tugwell, asking whether he recalled when "you, Bart Brebner and I were the 'Three Musketeers' in Moscow in 1927." At one point the group split up, and Tugwell traveled down the Volga on a barge, insisting that his interpreter and captain teach him a folksong about a Russian Robin Hood, "Stenka Rasin." In exchange Tugwell taught the Russians "Beulah Land." He rode in private railway cars—"ancient but gaudy" first-class wagons-lits from the days of the Romanovs—through Cossack country. Tugwell kept notes; he dined out. He wondered, as he always did when he was abroad, whether his life was on the correct path: after another preceding period overseas he had taken leave from academia for a year to farm beside his father before deciding the move was a mistake. The more earnest Douglas, himself considerably distracted by his own dying marriage, at one point reproached Tugwell for his lack of gravity.

But when it came to their work, Tugwell, like the other travelers, was serious enough. Committed to researching agriculture reform, he fought off offers to see factories and demanded visits to farms instead. He noted, first of all, that while conditions were still terrible within the Soviet Union, they were probably improving: "The manor houses are gone; only the drab villages remain," he wrote, concluding that "here is a bit more to eat of a little better quality. There is a radio in the village hall. There is more wood for warmth," he would later write. New England might be slowly dying; the Soviet Union to his mind represented "a stirring of new life hardly yet come to birth." He loved the idea of economics being made subservient, itself like a serf, to the good of the rural village: "with us, prices are a result; in Russia they are agents of social purpose." Tugwell insisted on more visits and was duly granted them.

Tugwell found himself admiring the active role of the Soviet government toward farming. He liked the idea of the *agronom,* the farm manager or bureaucrat, who oversaw a set of farms or a region. The Russian farmer, he noted, "suffers from price-disadvantage, it is true; but so also do farmers all over the world." Tugwell pointed out a difference from the United States: in Russia, the farmer's challenges were the subject of genuine government controversy. "There is a disposition to do something about it. Can this be said of the U.S. government?"

Most of all, however, it was the villages that impressed Tugwell. Many had not yet been collectivized, but they were still relatively cooperative compared to rigidly fenced New England. This cooperation he perceived to be natural, indeed, inevitable—"cooperation is forced in the nature of things." In his own childhood, there had been similar cooperation. He remembered traveling over New York's Ellery hills to a friend's house with his father, only to find the family not at home. The pair had fixed a meal from what they found in the buttery nonetheless, a fact which did not bother their hosts, when they returned, in the slightest. That was the way things were, in the old agricultural community. Under the czars, Tugwell noted, village farmers too had shared—"Russia was communal in this sense long before it was persuaded to Communism in the Marxian sense."

Tugwell believed that what remained of private arrangements also needed to be ended; it was time for "abandoning the old one-man, one-plow method."

After all, in a big communal field "a tractor can go as far and fast as it is capable of doing without the bother of fence corner turnings. Socially, the village has great advantages if it is not too closely built or too big." Further rationalization might work if only the stubborn peasant would cooperate. And even though he disliked the Soviet dictatorship from the start, he was struck by the authority of Russian propaganda and its enormous success. Always, the Russians they met up with "told *us* what our country was like." This simultaneously horrified the progressive in Tugwell and pleased the efficiency expert in him: "I knew from then on how determined dictators come to manage a people."

Meanwhile Chase was looking into industry, his area. The official goals of the Russian state planning commission impressed him deeply. This was "the attempt to do away with wastes and frictions that do such dreadful damage in Western countries." The scale of the management took his breath away: "Sixteen men in Moscow today are attempting one of the most audacious economic experiments in history . . . they are laying down the industrial future of 146 million people and of one-sixth of the land area of the world for fifteen years." Chase continued, "These sixteen men salt down the whole economic life of 146 million people for a year in advance as calmly as a Gloucester man salts down his fish." And, Chase noted with enormous admiration, "the actual performance for the year 1928 will not be so very far from the prophecies and commandments so calmly made. . . . One suspects that even Henry Ford would quail before the order." Perhaps the United States could organize its economy in similar fashion. Chase, like Steffens, believed he saw something that worked. All this went far beyond the planned efforts to stimulate the consumer advocated by William Trufant Foster and Waddill Catchings, or Herbert Hoover's careful constitutional constructs.

Chase paid a call to the offices of Gosplan, the state planning commission now charged with running the economy. Here he found that phenomenon Tugwell had longed for in a *Nation* article: a nation unified as if at war—but during peacetime. "Its atmosphere," he recalled after the trip, "reminded me strongly of the Food Administration Barracks in which I worked at Washington—the temporary partitions, the hurrying messengers, the calculating machines, the telephones, the cleared desks."

George Counts, the education man, was even more excited than Chase. In Russia, he saw, schools had already moved beyond being John Dewey's Lab School or Eleanor Roosevelt's Todhunter: they were indeed, just as he had hoped, vehicles of "the collectivist social ideal." Professional education, denied the common man under the czars, was officially available to all: "All academic standards were abolished and the doors of the higher schools were opened to the members of the working class regardless of their qualifications." There were adult education courses for workers, schools for political literacy so that all manner of adults might familiarize themselves with Marx and Engels. Physical education, another emphasis of American progressive educators, was moving forward here at a pace they could not dream of at home. "Basketball and volleyball are good Russian words today," wrote Counts. In revolutionary Russia even women ran hurdles. The whole country seemed to be hurdling ahead of the United States.

When it came to the question of labor, James Hudson Maurer, the senior leader, was thrilled by the data: 92 percent of the eligible workers had enrolled in unions. "There were no anti-strike laws and nothing resembling our curbs on them," he would later note in awe. He conducted his own tests of union independence: "Everywhere I went I asked the workers: 'Are your unions controlled by the government?' " The reply? "It is our government and they are our unions." A woman at an electric supply plant told Maurer: "Now we are free, free!"

The Soviet literacy programs inspired Maurer, who had learned to read so late himself: "Under tsarism 85 per cent of the masses were illiterate. New schools were everywhere in evidence and compulsory attendance laws were strictly enforced." This sounded better than what he himself had grown up with as the son of a shoemaker. In Moscow, Maurer got a chance to meet an exiled hero of the left, Big Bill Haywood of the Wobblies, the Industrial Workers of the World. There was the wonderful feeling of political friends meeting in a new setting. As it turned out, Haywood would die a few months later, and Maurer never forgot the meeting.

Douglas for his part was less enthusiastic than the others. He was interested in so much: the trade union movement, wages, pensions for senior citizens, the consumers' cooperative—he would write up essays about four of these topics after his departure. Big innovations grabbed his attention, but so did little ones—fourteen million Russians, he noted with wonder, had created 60,100 cooperative stores, all since the time of the Russian Revolution. But he also found on his tours, to his shock, that differences of opinion "were not tolerated." Chase and he were asked to give a speech to workers on the night shift at an airplane factory. When Douglas completed his remarks, the workers began shouting,

"Sacco and Vanzetti." This meeting was taking place, after all, around the time of the execution. The cries touched Douglas—Sacco and Vanzetti were important to him, too. But Sacco and Vanzetti had enjoyed "the full defense of the law," Douglas told the workers. Then he launched a counterattack, repeating an ugly story he had heard about the factory. " 'But what about yourselves? Two months ago a group of bank clerks were arrested at two o'clock in the morning.' Here the interpreter stopped and refused to go on. . . . 'They were tried at four o'clock and executed at six. Where was their right to assemble witnesses, to engage counsel, to argue their case, and, if convicted, to appeal?' "

The workers shouted back, but what Douglas would remember for decades was a young woman who approached him with a countering argument. "You talked only about individual justice. This is a bourgeois idea." Douglas was taken with her, and talked for an hour. Leaving, she told him, "History will prove us right and you wrong" and wrote her name down in his notebook: Betty Glan.

Still, even Douglas set aside his hesitations when big interviews materialized. Trotsky, one of the Soviet Union's original ruling troika, was already on his way out that summer. But he still had a small post, commissar of foreign concessions, and of course found time to meet with this group. The group arrived with a long list of questions, and was kept waiting for half an hour. Trotsky entered the room wearing, Douglas the diarist would later note, "an immaculate white linen suit." He picked up their prepared questions, and pronounced it "a very nasty list of questions." Then he answered the questions rapidly—to Douglas, he seemed like a showman. The interpreter made it all seem elegant by delivering replies in Oxford English.

The end of the trip approached, and the group was still angling for an appointment with Stalin. The plan was an on-again, off-again one, a typical mid-junket arrangement that seemed unlikely to be followed by the reward of a meeting with the Soviet Union's leader. Preoccupied with his own plans and likely feeling tired of the tour, Tugwell opted to play hooky and headed off on September 9 again with Chase to see some modernist paintings. As Tugwell would later write, with appealing honesty, "We had been good the day before and gone to see Trotsky and thought we had done our duty by the high command." It was easy to understand why Tugwell ran the risk—the rebel in him probably found a lot more in common with Trotsky, the intellectual's Communist, than with Stalin.

But Tugwell missed his rendezvous with history. For this time, the appointment hour, 1:00 p.m., came without further vacillation by the Kremlin. Robert Dunn, John Brophy, and Paul Douglas all went for the interview. So did Louis Fischer, an American who was writing pro-Soviet articles for left-wing American periodicals out of Moscow at the time. So, as it turned out, did a journalist who was visiting Moscow for the *New York Times,* Anne O'Hare McCormick.

Those who did attend kept notes. Douglas: "Recalling the deeds of terror that had been committed there throughout its history, I shivered as we entered Red Square and then went through the gates of the Kremlin." A small pockmarked man met them in a cloakroom; Douglas assumed it was an attendant. But the man took the head place at the table. It was Stalin. "His low brow was clear under a square-ish brush of black hair that made his head look oddly cubist," wrote Anne O'Hare McCormick. "He looked like any of a million Soviet workingmen," commented Fischer. "Deep pockmarks over his face," read Fischer's notes; "low forehead"; "ugly, short, black and gold teeth when smiles." Whereas Trotsky had worn white, Stalin wore khaki. Douglas thought he saw a private's uniform, Fischer a civilian suit. The pants legs he stuck into high black boots. Fischer sought to capture the moment in every medium possible. In his notebook, next to the words, he made pencil sketches of the leader's head.

The group expected an hour with the leader. They got six and a quarter. One thing struck them even before the meeting started: Stalin's charm. He was not dashing like Trotsky, but he seemed in a way more genuine. What came through was that Stalin had done his homework and touched on the issues that interested them—workers' insurance, for example, Douglas's pet research area since the days of the loggers. Stalin knew all about La Follette's strong 1924 showing. A questioner asked how Stalin knew that the Russian people were behind him. He answered that the Bolsheviks would never have come to power if they were not popular; today heads of unions were all Communists, again a fact that reflected grassroots support.

Stalin also took time to emphasize that his government was an ethnically diverse one, with a Ukrainian, a Byelorussian, an Azerbaijani, and an Uzbek in the central executive committee of the Soviets. There were also, Fischer would later write, questions about religion: must a Communist be an atheist? Yes, Stalin answered, and even as he answered, church bells across the street rang. The guests laughed, and Stalin smiled—as if to signal the tolerance he could not articulate officially.

Stalin also rejected the notion that U.S. Communists worked "under orders" from Moscow as "absolutely false"—itself a lie. As the group drank lemon tea from a samovar, Stalin made his case: the Soviet Union and the United States might trade together even if they had different systems—the new doctrine of Socialism in One Country.

Fischer reported that no one but a serving woman entered the room during the course of the meeting; she brought cheese, sausage, and caviar sandwiches. (Brophy reported tea and cookies.) There must have been an interpreter and stenographer present. After several hours the guests made an attempt to go; Stalin would not permit it. Instead he turned the tables and asked questions of the delegates. The transcript of these questions, published within a week in *Pravda,* give as clear a snapshot as any document of the tactical and strategic goals of Soviet foreign policy. Stalin wanted to make the point that he had a genuine labor following in the United States, and he wanted to sideline those organizations that had sidelined him—with the aid of his interlocutors. He had already skewered the anti-Communist American Federation of Labor. Now he set about doing so again: "How do you explain the fact that on the question of recognizing the USSR, the leaders of the American Federation of Labor are more reactionary than many bourgeois?"

Brophy allowed that the AFL had a "peculiar philosophy." Dunn took time to point out that the AFL was too close to capitalists—especially Matthew Woll, AFL vice president. Brophy was the one who spoke the last formal words of the visitors to Stalin before they departed. In Stalin's official transcript, the travelers gave the Soviet leader what he sought, a form of U.S. blessing: "The presence of the American delegation in the USSR is the best reply and is evidence of the sympathy of a section of the American workers to the workers of the Soviet Union." As the group left, Douglas spied a bust of Karl Marx, with full beard, in the corner. Contemplating it, he was startled to feel a heavy hand on his shoulder. It was Stalin. They joked about whether Marx had worn a necktie.

Several of the travelers sensed that they had been used to an extent they had not foreseen: "we realized that in his speeches he was talking over our heads to the newspapers, in answer to Trotsky," Brophy would write. Anne O'Hare McCormick, confused, retreated to racialist imagery for her report: Stalin, she said, was a hybrid of east and west, almost "Occidorient in person."

The vessel that returned the group home to America was not the *President Roosevelt* this time but the *Leviathan.* The irony of that name may not have escaped some of them. On shipboard, Silas Axtell, the lawyer, bitterly objected that some of the other labor people on the trip were producing a report far too positive. As he later recalled, "The whole report was written with such a solicitous and affectionate regard for the welfare of the dominating group in Russia, whose guests we had been, and the impression from reading the report was so different from the one I had received, I could not possibly subscribe to it." Douglas likewise quarreled with Robert Dunn over the content of their joint essay. Dunn was painting the picture too rosily, Douglas maintained. Later, he discovered that Coyle had diluted his discussion of civil rights in the published report.

Axtell and Douglas may have been thinking of another intellectual pilgrim who had met Stalin before them: Emma Goldman. Goldman had had every reason to accept what she saw in Russia; the United States of Wilson, Harding, and Coolidge was unlikely to welcome her back. Yet when she learned that Stalin was imprisoning her beloved fellow anarchists, she had grown skeptical. And when the Bolsheviks—led by the same Trotsky of the white suit—bloodily put down their fellow Communists in Kronstadt in 1921, she had turned against the Soviet Union entirely. "I found reality in Russia grotesque, totally unlike the great ideal that had borne me upon the crest of high hope to the land of promise," Goldman wrote. Though she really had nowhere to go, she left Communist Russia and shortly published a monograph on the false freedoms of the Soviet Union, *My Disillusionment with Russia.*

A decade after Emma Goldman's experience, and five years after the 1927 delegation, Arthur Koestler, a young Communist, would also be repulsed. He found that the Soviet Union had developed a neat trick for bribing young intellectuals. Through its State Publishing Trusts it would buy the rights to a book or article—with a different payment for an edition in each one of the Soviet Union's multiple languages. Koestler reported selling the same short story to as many as ten different literary magazines, from Armenian to Ukrainian. The place really was, he would note ironically, "the writer's paradise."

The travelers though were not so cynical, and as the ship moved toward the United States, the group felt its excitement build. They had got what every traveler hungers for: proximity to heroes and events. The heroes were not precisely *their* heroes. Still, the meetings had had their effect. The travelers were now transformed from obscure analysts of the Soviet Union into bearers of news. Their victory was certified when Stalin published his long version of the interview; all of the Soviet Union, and, more important, all the progressive world, could now observe the star quality of the September 9 meeting.

Aboard the ship, the labor advocates and the academics raced one another to complete their contribution to one of the two volumes. Maurer, already a mayor, thought about his political ambitions. Within six months of the trip, he would be busy building a small-scale monument to his own vision of socialist reform: a new town hall for Reading. Chase wrote a big article in the *New York Times*. Tugwell worked on his agricultural contribution; he also thought about an article—it would eventually appear in *Political Science Quarterly*—arguing that while Russia needed more freedom, its concern for every man was worth serious study.

Any ocean liner arriving in New York was news, and reporters routinely met the vessels. Earlier that same September, the *Leviathan* had brought back Treasury Secretary Mellon and his daughter Ailsa from Europe. The headline had been: "Mellon Returns, Has Nothing to Say." Disembarking, the travelers from the Soviet Union gave the paper their own summary: the Soviet experiment was "meeting with success." If the United States was not friendlier, the returners also warned, Britain might succeed in driving an isolated Soviet Union to war.

In one way, upon their return, the travelers would have the influence they and Stalin wished for. As they and others talked about Stalin, more American policymakers began to take the possibility of recognition of the Soviet Union seriously. The travelers' positive reports validated the admiring view presented nearly daily in the *New York Times* by that paper's Walter Duranty. Among Duranty's readers was Roosevelt, the would-be governor. Most Americans in that period were divided into Germany people or Russia people; their like or dislike for one determined their attitude toward the other. Roosevelt had disliked Germany from his childhood days, and World War I had not altered that prejudice. He was therefore ready to take an interest in Russia.

None of the gently pro-Soviet messages, however, was taken seriously by the Republicans in power. As for the collectivist ideal, Americans were still generally not ready to share that either. The spring after his return, an enthusiastic Maurer was nominated as the vice presidential candidate beside Norman Thomas on the Socialist ticket in New York. Douglas, thoughtful as always, endorsed the Socialists; he felt the two mainstream candidates, Hoover and Al Smith, represented "sterile and corrupt groups."

Overall, the Socialists did not do well, pulling 300,000 votes, or less than 1 percent, and winning no electoral votes. Tugwell for his part concentrated on the main parties, focusing his energy on finding a way into the campaign of the Democrat Smith. He put out feelers to Belle Moskowitz, Smith's adviser and the mother of Carlos Israels. He advanced the concept that the government might pay farmers off to curtail supply of excess food. Moskowitz, though a reformer, "Smith's angel," still rejected Tugwell's ideas—as he would recall in his memoir—as "pretty drastic." How could even a mainstream Democrat hold up against campaign promises made or ascribed to Herbert Hoover, such as "a car in every garage and a chicken in every pot"?

The election did go to Hoover, who polled more than 20 million of the 36 million odd votes cast, taking 444 electoral votes to the Democrats' 87, even more electoral votes than Coolidge had in 1924. The Republicans also gained strongly in Congress, so that they now held a ten-seat lead in the Senate and nearly a hundred more seats than Democrats in the House. The progressives whom Stalin had watched so carefully did abysmally. Communists and Socialists together could claim less than 1 percent of the vote. For the moment, at least, the similarities between Hoover and the Left progressives were submerged, even though someone like Stuart Chase would work with the budding technocratic movement, which deified engineers. The travelers and intellectuals were leftists; Hoover was a businessman. They were ephemeral professors; he was the vigorous president. They were on the edge; he was at the epicenter. The only party as alienated as they were was Calvin Coolidge, who at first bridled at Hoover's request that a battleship be placed at his disposal so that he might cruise the coast of Latin America in the long interregnum. Take a cruiser, Coolidge said, "it would not cost so much."

They watched as President Hoover made his changes and established his rituals. Hoover inaugurated a morning regime, heaving about an eight-pound medicine ball with fellow members of his cabinet on a tennis court. He had a phone installed right at his desk, ending Coolidge's "no phone" rule in the Oval Office, and instituted a special switchboard that connected him directly to his prominent staffers. But he also added staff and erected barriers to reaching him; as he would later note, "The president is not open to everyone's call." Thus Hoover managed to achieve two habitual goals—maximum efficiency from within, and maximum defenses against unwanted intrusions or criticism from without.

Now the economy was still doing so well that it was hard for anyone to criticize Hoover without sounding morose or shrill. The election outcome brought home to the travelers two things. The first was that their ideas would not resonate in so prosperous a period as the late 1920s. The second was that if they were going to get their ideas through—even in bad times—then Tugwell had it right; they really would have to go with one of the two big parties. Nineteen twenty-eight made clear that from now on, a Tugwell or a Douglas could not attack the parties from outside. Reformers had to work from within—and do so more effectively than Tugwell had succeeded thus far.

In the meantime, therefore, the intellectuals retreated—back to the universities, the union halls, and the magazine offices, to talk and to write. If Russia had a five-year plan, perhaps the United States ought to as well. Tugwell finished an essay he published in a group of essays by the travelers, *Soviet Russia in the Second Decade.* While criticizing some aspects of the Soviet Union, the volume on balance was excited and favorable, containing an essay by Tugwell's friend John Bartlet Brebner titled "In the Ante-Room of Time." Counts trumpeted news of a "great social experiment"; Dewey, after his own trip, published a series in the *New Republic.*

Roger Baldwin for his part grappled with the ultimate question about Soviet Russia: Could there be freedom there? Baldwin thought there could in the future, even under a dictatorship, be something good like freedom. After his stay in Soviet Russia, he had not returned with the others but had stopped in Paris to write up his Russian experience, and hired Alexander Berkman, Emma Goldman's old partner, to translate some of the documents he'd brought out into English. ("He was gentleman enough to hide his view of me as naïve," Baldwin later told a biographer.) "I am confident that far greater liberties than are tolerated are consistent with the maintenance of the Soviet regime, and even with the Party dictatorship," he concluded. His thesis was the direct opposite of Coolidge's "All liberty is individual." What Baldwin was saying was that a higher liberty could be collective. He optimistically titled his book, published the next year, *Liberty Under the Soviets.*

Douglas discussed the Soviet Union in a symposium with the journalist Dorothy Thompson, who was shortly to marry Sinclair Lewis, the author of *Babbitt.* Both Thompson and Douglas were friendlier to the Soviet authorities that the third speaker, a former envoy of the interim Kerensky government named Boris Bakhmeteff: "I came here to sound a note of anxiety," Bakhmeteff told the audience on January 19, 1929. Nineteen twenty-nine turned out to be the year in which Stalin would begin a new stage of terror, systematizing exportation to concentration camps in what would later be known as the Gulag. Still, most Americans could not know this, and most New Yorkers certainly didn't: Walter Duranty, the *New York Times* correspondent, failed to convey the extent of the violence. Generally, the salons of Washington, Chicago, and above all Manhattan welcomed the travelers as celebrities.

Stuart Chase, too, would work on articles and a book about Russia, aimed at capturing the attention of the leaders in the established political parties. Interested in the future of cities, he was also imagining a new style of federal government far more ambitious than what had been before. The volume, published a few years later, would open with a reference to Keynes, the English economist who had approved of Hoover: "John Maynard Keynes tells us that in 100 years there will be no economic problem." To get to that point, though, Chase reiterated, the United States would indeed have to depart from free-market models. Once again, he sketched limits. "*Laissez faire* rides well on covered wagons; not so well on conveyer belts and cement roads," Chase wrote. Whatever the change that was happening, "it is going in the direction of more collectivism." Chase argued the key to the change was Russia. It might be a dictatorship, but it was, just as Steffens said, the future.

"Russia, I am convinced," Chase said, "will solve for all practical purposes the economic problem." Someday, the United States might begin its own experiment in central planning. The conservatives were having their day, and the planners would get theirs. After all, as Chase would ask in his final sentence, "Why should Russians have all the fun remaking a world?"

3
The Accident

October 1929
Unemployment: Heading toward 5 percent
Dow Jones Industrial Average (October 1): 343

"Closing Rally Vigorous," remarked the *New York Times* headline when the stock market crashed the last Tuesday in October 1929. Bankers noted that throughout the day periodic lifting spells had pulled the market up. Back on August 20, the Dow had closed at a high of 368. Then the market had gone yet higher, to 381. It made sense therefore that there had been spectacular drops all month, right up to this closing, on the twenty-ninth, at 230. As Treasury officials had told the paper the preceding Thursday, the losses did not matter so much anyhow—they were "paper losses." Things were moving into balance. Much of the nation shared this view. Hoover, six months in office, planned to spend the weekend at the White House with the author Samuel Crowther, who had published an adulatory biography of him during his campaign. In those days the treasury secretary was a member of the Federal Reserve Board. Mellon sat out the week, concluding, along with Fed colleagues, that no action was necessary. Paintings, as usual, were crowding out markets in his mind. That week he lent out one of his paintings, a Flemish primitive, so that it could appear in an exhibit that would benefit the distribution of free milk to babies.

In Chicago, Paul Douglas was preoccupied with a fierce municipal battle. He was trying to constrain Insull. Two reform-minded attorneys, Donald Richberg and Harold Ickes, were with him on the campaign. Richberg hired David Lilienthal, Frankfurter's former student, to work with him, and he could see that the young man would make his own name in public utility law. The admiration was mutual, and Lilienthal would later write of Richberg: "He is not simply a brilliant lawyer, but a social philosophy runs through everything he does." Douglas, Ickes, and Richberg had formed the People's Traction League to defeat state and city legislation that would consolidate Insull's control of the city's streetcars and elevated lines. Insull for his part was fighting back with vigor—and, as it would emerge later, with the support of both voters and legislators. The Windy City's uncrowned monarch was also preparing for the gala November opening of his opera house. Guests wearing ermine, sable, velvet, and brocade—among them Mellon's brother, Richard B. Mellon—would come to hear *Aïda.*

Now settled back in Massachusetts, Coolidge was busy earning up a storm writing freelance articles for *Cosmopolitan, Ladies' Home Journal,* and other periodicals, and netted a $65,000 advance for his autobiography. The first year out of office counted most in earnings for ex-presidents, and he was determined to make his own worth it. He also headed down to New York from time to time—the next year, for example, he would join Irita van Doren, a literary editor, along with Governor Roosevelt, Thomas W. Lamont of Wall Street, William Woodin (a financier), Ida Tarbell (the journalist), Henry Morgenthau Sr., and others in serving on the welcoming committees for the Indian poet Sir Rabindranath Tagore.

As for Tugwell, he certainly noted the crash; his thoughts about the economy were already dark. He was worried about the prospects for his father, who was expanding his business, working with a few small banks. But Tugwell was still, mostly, thinking in terms of agriculture, and his world was still a world of the classroom. "Of all the kinds of men, the farmer is the greatest speculator," he and coauthor Harry Carman would write several years later in their introduction to an eighteenth-century monograph, Jared Eliot's *Essays upon Field Husbandry in New England.* "He does not think of himself as a gambler, but he lives every day subject to such risks as would give a professional Wall Street operator nervous chills." Early that autumn Columbia had introduced a new series of adult education courses in midtown: Tugwell would sally down the West Side from Morningside Heights to teach.

Even the executives from the firms affected by the downturn were not so surprised. One was Willkie. Just that year he and Edith had come to New York so that he could join the legal team that represented one of the new utilities giants, the holding company Commonwealth and Southern. The Willkies' sudden trip from mid-America to Manhattan shocked them a bit—Willkie told an acquaintance that in Akron he hadn't been able to walk down the street without meeting a friend, but here, "there isn't a soul I know." Alfred Loomis and Landon Thorne, the pair who had created Commonwealth and Southern, had also created United Corp., another giant holding company. But Loomis, Thorne, and Willkie all understood that utilities' prices, which were astronomical, could certainly be expected to go down. Earlier in the year Loomis and Thorne had sold off shares after discovering to their shock that their new utility company, United Corp., was priced above even their own most ambitious estimates. Hoover's political opposition were beginning to consider whether the crash might be used to their advantage.

Franklin Roosevelt had proven himself a formidable campaigner the preceding year by winning the governor's office despite the Hoover sweep. The victory had been noted nationally for three reasons. The first was that Roosevelt was already a presidential name. The second was that New York was by far the nation's most important state, politically, with forty-five electoral votes—California, for example, a place that was mostly future, had only thirteen. The third was that Roosevelt had come back despite the crippling case of polio he had contracted earlier in the decade. Roosevelt was in perpetual discomfort—pain, often—and that normally would have made his fellow party leaders rule him out as a prospect for a post like New York. Governors of that state were elected every two years, requiring energetic figures willing to campaign without ceasing. But his fellow party leaders admired his spirit—he did not speak of his infirmity—and he had rewarded them. Now Roosevelt used the opportunity of the crash to jab at Republicans—Democrats certainly would have been blamed had they presided over similar stock drops. But Roosevelt also devoted time in his speech to another project: campaigning for greater public-sector involvement in the area of utilities.

On Wall Street, many investors were losing their livelihood. Most famous was the pair of men who committed suicide by leaping out of the window while holding hands: they had maintained a joint account. This was the kind of anecdotal tragedy that would come to be symbolic of the crash. But the despair was not uniform: indeed, on November 13, 1929, the city's chief medical officer reported that there had been forty-four suicides in the preceding four weeks in Manhattan, nine fewer than the fifty-three for the same period in 1928. As for banks, some were failing, but the rate was not outside the norm for the 1920s. Total commercial bank failures for 1929 would be lower than the same statistic for 1924, 1926, or 1927.

The nation's first impulse was correct. Washington might not have needed to do much. The miracle of the 1920s had followed a rough downturn at the start of the decade, and then a comeback. Such crashes—or panics, as they were known—did not make a lengthy slump an inevitability. Perhaps all that was needed now was for owners to sell their holdings, so that the market could find its own bottom. This was what Mellon would mean when he recommended that stockholders, banks, and farmers liquidate their holdings. The phrase "to liquidate" sounded harsh, but it also represented an old and important argument. Uncertainty was one of the market's problems. Only when stocks or wages were "marked to market" and found their bottom could they rise again.

While the market had indeed been high, it might rebound in the next few years. The increase in stock prices lately had not been the pure luck of roulette. It had reflected something genuine: productivity gains and the hope for future ones. New consumer products made everyone aware of the potential in the economy for profits. The utilities world of Insull, Loomis, Bonbright, and Willkie was a good example: electrification had dramatically brightened the 1920s, but 50 percent of American households were still waiting for power.

In this argument, the bull market of the 1920s was not an empty speculative bubble. Rather, it was the market's best effort to quantify the value to America of the potential of the recent and future innovation. At a time of breakthrough innovation like the 1920s, it was entirely rational for stock prices to break through old barriers.

Many economists at the time recognized this hidden value, though they often described it in different ways. One of them was Stuart Chase, who a few weeks following Black Tuesday was writing that "the stock markets will not affect general prosperity." Another was Irving Fisher, who in that Black October argued that stock prices were too low, and could move up shortly. At the time and later, this provoked ridicule: for decades people would laugh at Fisher, the last hold-out, and would note, maliciously, that he had also lost a fortune on his own forecast. Still, Fisher's surmise was based on facts: companies' profitable earnings reports, ratios of such reports to share price, and the large scale of companies' investments in research and development. The stock market reflected more growth, and less speculation, than the panickers said. And in fact the pattern of prices across the economy tended to support this view. While they had risen on Wall Street, they had fallen on Main Street, which meant that this was not traditional inflation. Traditional inflation, his area of study, shows up across the price landscape.

All this, of course, did not mean that highly priced stocks were "safe," or that they would not gyrate or fall. The same theory—that stocks suggested that new innovations promised further growth—would also have explained the Dow's volatility that autumn. Everyone knew there was value in new technology. But what that value was and when it might be realized—in 1929, or the next year, or a decade hence—no one knew.

As more bad news arrived, Fed officials, Mellon, Hoover, and many others began to reconsider. After all, they reflected, an adult who had begun watching the market in 1907 had had to wait two full decades for the Dow to double, until 1927, the year Tugwell and the others had sailed on the *President Roosevelt*. In 1927, the Dow Jones Industrial Average crossed the 200 mark. Yet from the 200 of the spring of 1927 it moved to 381—nearly doubling again—by the summer of 1929. This increase seemed too steep not to be somehow economically and morally suspect. Perhaps the country was indeed in the throes of a dangerous inflation.

As the days passed, more bad news came. London was concerned; at Shorters Court, where U.S. securities were commonly traded, chaos reigned until the closing bell. "Fear for our prosperity," wrote a London correspondent wonderingly, reflecting Britain's attitude. He quoted the *London Daily News* on its concerns that U.S. incomes would fall so much that Americans would no longer be able to buy European exports. Only prosperity had made Americans willing to buy expensive European goods, despite the extra tariffs. "The tariff wall will become temporarily insurmountable."

On the Continent, too, the potential for trouble was mounting: recently Gustav Stresemann, the fragile

Weimar Republic's best statesman and diplomat, had died. Who would hold Europe together now? "Successor a Problem," concluded the *New York Times.* Weimar Germany might fail now after all. The result was that gold, seeking a haven, flowed into the United States—$175 million in 1929 and $280 million the next year. If the Federal Reserve had followed the old gold standard rules, this would have increased money available to banks and citizens. But George Harrison at the New York Fed and others believed in the inflation theory, and thought that more cash would exacerbate inflation. The Fed therefore veered from the old gold standard tradition. It sterilized the effect of all that new money by selling bonds—in effect, soaking up money from the economy to offset any monetary expansion. Looking at a broader period, one could see the trend was even more marked: the actual money supply available dropped by nearly 4 percent between the end of 1928 and the end of 1930. The sterilization program more than offset some small interest-rate cuts taken shortly after the crash. Instead of loosening in a time of trouble, the Fed was tightening.

The more Hoover thought about it, the more he too liked the idea of a war against inflation. Market fever was in any case the sort of thing Hoover deplored; he had been complaining about abuse of credit for four years. Meanwhile, the market was beginning to look worse. Not merely the margin sellers but also the regular players were suffering on Wall Street. In the fall of 1929 the Dow's new utility index plunged as well. One night in November, Robert Searle, president of the Rochester Gas and Electric Corporation, gassed himself to death after confronting more than a million in losses in a month. The death seemed a metaphor for the failing of the utilities industry, all the more so because Searle had started out as Thomas Edison's office boy. Among the stocks that had disappointed him, the papers suggested, was Commonwealth and Southern. Willkie now had some 75,000 stockholders, and he was worried about their fate. The share price had been in the middle $20s in June. Now it was closer to $15.

At the beginning of that same month, November, Hoover received a confidential report from Fed officials that the market readjustment was not completed but would instead last months more. The mood was shifting to crisis, and Hoover felt energized—another rescue opportunity in the offing. He mulled over foreign policy, preparing an Armistice Day speech. Even as he moved to act, he savored the situation, commissioning the famed portraitist Leonebel Jacobs to produce pictures of himself and Lou. "The primary question," he later wrote, "at once arose as to whether the President and the federal government should undertake to mitigate and to remedy the evils." His conclusion was that yes, this was a job to be taken on: "we had to pioneer a new field."

Right away—in November 1929—Hoover pushed to expand an existing public buildings program by the healthy sum of $423 million on the theory that the spending would boost the economy. In Washington, builders put up great structures—a new agriculture department, for example. He asked his secretary of commerce, the man who held his old job, to establish a national system of cooperation among the states in public works projects. When Congress convened in December, the president called for "the expansion of the merchant marine, the regulation of inter-State distribution of electric power, the consolidation of railroads, the development of public health services, and departmental reorganization for greater economy."

But this was only the beginning. This time, he thought, perhaps the president could broker the recovery. "Words are not of any great importance in times of economic disturbance," he announced. "It is action that counts." The problem with the economy, at least as it was evolving, was mostly a monetary or an international one—Germany was already in depression. Yet at first Hoover focused on fixing it with domestic fiscal tools. And before a year would pass, Hoover had done damage that did matter on three fronts: by intervening in business, by signing into law a destructive tariff, and by assailing the stock market.

First came business. Hoover believed that business spending might make a difference. He thought he might cajole or bully Main Street, the industrial world, and labor leaders into pulling the economy back to recovery. Less than a month after Black Tuesday, on November 19, 1929, he therefore called a conference of railroad presidents in the cabinet room of the White House. Railroads mattered: they were at the time still the principal means of transport for both people and goods across the nation. The president asked the executives to sustain construction. Mellon came to the meeting. Later that week, industry leaders announced they planned a full billion dollars in outlays—an amount equal to more than a third of what the federal government had spent on all its budgeted projects in 1929.

Two days later, November 21, the cabinet room was the site of another meeting, this time of leaders from big industries. The guests included Treasury secretary Mellon, again, Henry Ford, Julius Rosenwald of Sears, Pierre Du Pont, Alfred Sloan Jr. of General Motors, and Julius Barnes, who was chairman of the board of the U.S. Chamber of Commerce. After hearing their views, Hoover did something radical. He noted that "liquidation" (layoffs) had accompanied all previous American recessions and that the federal government had allowed those liquidations to take place. This time "his every instinct" told him things must be different; wages must stay in place. Otherwise values would be "stepped down"; industry must help to "cushion down" the situation. At the worst, businesses in trouble might reduce hours to share jobs. But the general push must be to keep high wages and keep up employment.

That same day Hoover met with labor leaders including William Green of the AFL and his colleague Matthew Woll, whom Robert Dunn had maligned in the meeting with Stalin. From them Hoover collected an assurance that they would not push for an increase in wages above what was already being negotiated. The next day, the president had the construction industry in; the next morning, November 23, found him telegraphing governors and mayors not to cease public works, but to continue their activities so as to take up the slack in unemployment. Within a week Hoover had held two other conferences, one with national agricultural organizations and one with utility executives—Sam Insull of Chicago attended, though not, as far as we know, Willkie. Historians of the Hoover administration later recorded that everyone agreed with his ideas on the necessity for continued expansion except Insull, "who deprecated all such activities" and wondered aloud whether his industry, at least, was really in trouble. For good measure, Hoover created a temporary bureau to coordinate the expansion of public works among states.

Hoover's wage ideas sounded good to some. And they were indeed the opposite of federal policies in the last downturn. But they did not really make sense: to force business to go on spending when it did not want to was to hurt business. And in some areas—wages, especially—the president's policy was dramatically counterproductive. As the crash continued, profits began to drop. Yet businesses could not adjust: if they wanted to be good citizens, they had to keep their pledge to Hoover and sustain employment and wages. The president was, essentially, requiring that companies take the hit in profits instead of employment.

Later scholars would note the effect of the new precedent: wages of those who had jobs stayed the same. But many did not keep their jobs, or lost cash by being assigned part-time work—Hoover's job sharing. This was different from 1921, when companies had been able to cover their losses by cutting wages. But there was also, of course, an effect on employers. Their wage costs forced down the value of company shares, aggravating the downturn that Hoover had vowed to fight. Hoover's humanitarian policy sent a signal nationwide: do not lower wages. In the end, businesses had to choose between lowering wages and shutting down. Often, they shut down.

Some observers would, then or later, note the perversity. Coolidge, who had retreated to his home in Northampton, Massachusetts, the Beeches, would later rail against "these socialistic notions of government." Tugwell was preparing to pen an acid tract titled "Mr. Hoover's Economic Policy," noting that Hoover liked competition, but only in some cases; in others he backed the concept of companies working together. This, Tugwell said, amounted to "a desire to have his cake and eat it too." Albert Wiggin of the Chase bank argued that Hoover had his logic about wages backward. "It is not true that high wages make for prosperity," Wiggin would protest at one point. "Instead, prosperity makes high wages."

But Hoover proceeded, undaunted. He ordered governors to increase their public spending when possible. He also pushed for, and got, Congress to endorse large public spending projects: hospitals, bridges. The president documented meticulously all the positive responses he received from governors and senators when he asked them to increase spending. Among the telegrams came one from Franklin Roosevelt of New York, who wrote that he for his part expected to expand "much-needed construction work" in his state and that construction would be "limited only by estimated receipts from revenues without increasing taxes." By April 1930 the secretary of commerce would be able to announce that public works spending was at its highest level in five years. At the same time, Hoover went to work on another front: farm prices. These were at painful lows, in part because of production incentive programs advanced by Hoover himself earlier in the decade. The government had lured farmers into overproduction.

There was a monetary element to the problem as well. Looser money or credit policies could have limited the farmers' problems. So in fact could have more orthodox adherence to the gold standard—giving up sterilization. But Hoover chose to stick to the narrow challenge of price without regard to monetary factors. If farm prices were too low, he would raise them. Strengthening protection might bring them up. Protectionism had in any case been part of the Republican Party platform in 1928, in which the party had reaffirmed the tariff as a "fundamental and essential principle of the economic life of this nation." And on April 15, 1929, well before the autumn siege, the president had as good as promised a new agricultural tariff:

"Such a tariff not only protects the farmer in our domestic market, but also stimulates him to diversify his crop."

Now, with farmers in need, the tariff idea gained momentum. Lawmakers pushed for it. In the House, the leader was Willis Hawley of Oregon; in the Senate, Reed Smoot of Utah. In the end the legislation called for one of the highest tariffs in U.S. history. The new law made sense on an emotional level: America was in trouble, so America's domestic producers must be protected with fresh advantage. In the autumn of 1929 it became clear that a large new tariff would indeed pass the Congress—and that it would be up to Hoover whether to sign it.

Still, for the general economy the tariff was bad news. As Benjamin Anderson of Chase Bank would point out in an address the next March, the preceding fifteen years, going back to 1914, had seen an excess of exports over imports of $25 billion. America sold more than it bought in the international arena. Others agreed. A new tariff would shut U.S. sellers off from the world at a time when they badly needed customers. It would deprive foreign governments of trade. It would drive the prices of imports up for consumers at home. It would hurt other nations, nations that the United States hoped would become its markets. It would certainly hurt the worker. It would also, in the long run, hurt the farmer, by offering yet more—and greater—incentives to continue doing something that was uneconomical.

Meanwhile, the horizon darkened further. At the time unemployment data were not collected as they are today, but later analysis suggests that unemployment went from something like 3 percent in the fall of 1929 to 9 percent by the new year. One of the state officers who did the best job was Frances Perkins, the industrial commissioner in New York. Perkins announced early in the year that joblessness in her state was "very serious"—worse than in fifteen years. Hoover's labor department blithely told the country that things would normalize, but as *Time* would note in February 1930, "Communists stirred hungry, cold, jobless men and women to demonstrations which required no statistician to interpret."

Of the problems confronting the economy, the tariff threat seemed most urgent, and easiest to stop. There were disputes, but some said it was the heaviest tariff in American history. Yet Hoover egged the lawmakers on to complete the legislation. In May 1930, one thousand and twenty-eight economists signed an open letter urging the president to veto the tariff legislation—and published the letter in the *New York Times.* Irving

Fisher was among them. So was James Bonbright, the expert in utilities finance. And so was Rex Tugwell, whose name came after Bonbright's on the list. The language of their protest was strong:

> *We are convinced that increased restrictive duties would be a mistake. They would operate, in general, to increase the prices which domestic consumers would have to pay. By raising prices they would encourage concerns with higher costs to undertake production, thus compelling the consumer to subsidize waste and inefficiency in industry. At the same time they would force him to pay higher rates of profit to established firms. . . . Few people could hope to gain from such a change.*

The economists went on to predict that "many countries would pay us back in kind." As for unemployment, they reminded the Republicans, "we cannot increase employment by restricting trade." They pointed out that farmers, the bloc whom the lawmakers were aiming to please, would also suffer from Smoot-Hawley. "The vast majority of farmers would also lose. Their cotton, pork, lard, and wheat are export crops and are sold in the world market." Among the other signators were seven professors from Hoover's alma mater, Stanford, including the dean of a business school that Hoover had helped to found. Without free trade, the golden passportless world would fade.

Herbert Hoover of Tientsin, China, and Hyde Park, London, should know this. Thomas Lamont of J. P. Morgan was watching his bank shrink; it would lose half its net worth in the Hoover years. "I almost went down on my knees to beg Herbert Hoover to veto the asinine Hawley-Smoot tariff," Lamont recalled.

Washington received 106 wires from forty-nine General Motors overseas officers in fifteen countries. GM's European director, Graeme K. Howard, sent a telegram whose message was as terse as it was clear: PASSAGE BILL WOULD SPELL ECONOMIC ISOLATION UNITED STATES AND MOST SEVERE DEPRESSION EVER EXPERIENCED.

Le Quotidien in Paris published an editorial entitled "Can Mr. Hoover Limit the Catastrophe Which the American Protectionists Are Preparing?" The daily went on to write that if "the Yankees" brought in a tariff, "there will be nothing for us to do but to resort to reprisals, and that would mean war." Writing from Paris as well on May 19, the *New York Times*'s Carlisle MacDonald predicted trouble. French prime minister Aristide Briand had proposed a novel idea: a United Europe, including a common market, also then a new thought. The author speculated that such an entity might not like the idea of Smoot-Hawley and would become a "medium for counteraction" at some point.

In the United States there was also another kind of concern—concern that civilized Europe would be lost if the United States did not trade with her. Stalin as a momentary revolutionary was one thing, but now that the Soviet leader was hardening his grip on Russia, many Americans, even some economists, felt alarm. The *Los Angeles Times* announced to its readers that leaders of all religious faiths would hold a protest meeting at the Trinity Auditorium, for, as the paper noted in shock, it was now clear that 70,000 churches, mosques, and synagogues had been closed in the Soviet Union. "For the first time in history, a nation has undertaken a general crusade against religion," the *New York Times* wrote, noting that a 1929 Kremlin decree had included a step no one had imagined: forbidding churches to "hold special meetings for children." There were to be no more church reading rooms. The Bolsheviks had recently demolished the storied Simonov Monastery and replaced it with a so-called Palace of Culture—what was that? On March 7, some 3,500 representatives of all Christian denominations filled the Cathedral of St. John the Divine in Morningside Heights to pray for Russian souls; 3,000 more filled St. Patrick's in

New York's midtown. With everything else, "few had realized what was going on in Russia," Bishop Manning told New Yorkers. A number of city leaders were hosting another event, at the Metropolitan Opera, also in the name of pointing out religious persecution in Russia. Among them were Matthew Woll, the AFL leader, and Nicholas Murray Butler, Tugwell's boss and the president of Columbia.

The medicine ball was in Hoover's hands, and this time, he dropped it. He announced that it was nationally important to have a tariff, and also important for the executive to play a key strategic role in formulating it. He wanted to stop "congressional logrolling"—the congressional game of setting tariffs on specific industries to please specific constituencies.

The position he therefore ended up adopting was Hooveresque. He would not oppose the new tariff but would battle to make it fairer. He therefore advocated the engineering of a "flexible tariff" that would be controlled by a bipartisan commission made up of a precise fifty-fifty breakdown of commissioners from each party. The commission would achieve an important goal of Hoover's: they would take tariffs as far away from politicians as he could get them. It would be a "definite rate-making body acting through semi-judicial methods." The commission would then set tariffs based on a rational review of costs and prices at home and abroad. There was one other thing: the executive would then have the authority "to promulgate or veto the conclusions of the commission." The progressive and engineer in him triumphed over the international merchant.

Congress gave Hoover what he wanted. Hoover not only signed the legislation; he signed it ceremoniously, in June 1930, using six gold pens, one each for the Republican lawmakers Smoot, Watson, Shortridge, Hawley, Treadway, and Bacharach. But by focusing on winning the battle of flexible tariffs, Hoover lost the more important struggle: to right the ship in a storm.

For the economists proved right: Smoot-Hawley provoked retaliatory protectionist actions by nations all over the globe, depriving the United States of markets and sending the country into a deeper slump. Dozens of nations acted, as it became clear the tariff would become law, or after the formal signature. France imposed an auto tariff; so did Italy. Australia and India legislated new duties. Canada raised tariffs three times. The first tariff, an emergency retaliation, hit 125 classes of U.S. products. The Swiss, furious at a duty on watches, boycotted U.S. imports to their country.

There were indirect international consequences as well. Foreign governments still owed considerable debts to the United States. Some of those debts were denominated in gold. To get the gold to pay those debts, the governments and their people had to be able to sell in the United States. The tariffs made this necessary task more difficult. At a time when the country could have pulled itself out of a slump through trade, Washington was buttressing the walls preventing that trade.

But this was not all. Hoover was also intervening on a third front: markets. What the stock market at that moment needed was clear rules and pricing, Mellon's "liquidation." This was what everyone expected in any case, for at that time Washington did not regulate the stock market; the exchange was a New York corporation.

Still, Hoover could scold, and he did. In his first annual message to Congress, delivered in December 1929, Hoover railed against the "wave of uncontrolled speculation" that he saw as a cause of the crash. Over the course of the winter and the next year he would speak out, too, against short selling. In a short sale, a trader borrows a stock and sells it at a certain price, in the hopes that by the time he must deliver the stock, he can buy it himself even more cheaply. Hoover believed that this was not logic but roulette at its worst. The game was dangerous because it moved away from the value of the underlying asset—shares in a company—and into the racy world of betting. Without short contracts, he reckoned, the stock market would not experience such violence ructions. The shorts, to his mind, put downward pressure on a market that might in some instances otherwise do fine. Now he wanted new rules to limit shorting.

But the argument against shorting had a flaw. For every short seller—the man who was exerting the downward pressure—there was always a long buyer—the man who bet he could get the stock for cheap under the arrangement, and then sell it himself, for more.

One reason this logic did not penetrate was that many of the messengers who carried it were flawed. Wall Street in the 1920s had felt like a gamble, and some of the players had been irresponsible or worse. One was Richard Whitney, the new president of the New York Stock Exchange.

Whitney, a patrician, could make the free-market argument as well as any. At a meeting in October 1930 at the Stevens Hotel in Chicago, Whitney criticized the idea of blanket legislation to restrict short sales and other forms of speculation: "The Exchange is convinced that normal short selling is an essential part of a free market in securities." How could a market exist if it was not allowed to place such bearish contracts? "Such a contract to deliver something in the future which a person does not own is common to many types of business. When a builder contracts to build a skyscraper he is literally short of every bit of material." Yet no one, Whitney pointed out, considered that builder a criminal for signing the contract. Whitney was making precisely the same point that Tugwell had made in his introduction to the old book on animal husbandry: everyone engaging in any kind of commerce was placing a bet of some kind.

The trouble was—as many Wall Streeters knew even at the time—Whitney himself was more than flawed, a compulsive gambler and a liar. It would later become clear that even as others were losing their homes, Whitney's Wall Street allies were sustaining him with friendly loans. He would eventually do prison time for covering up illegal loans with the aid of his loyal brother George.

To have a Wiggin or a Whitney as their spokesman hurt defenders of the market at a time when their argument was crucial. For it was not wrong that a restriction on short selling would scare the market by depriving it of a vehicle for hedging its risks. That fear alone might even trigger big drops in stock prices. And there would no longer be the countervailing pressure of the short buyer. Mellon's "liquidate" phrase sounded harsh but was far less constraining than the president's restrictions on short selling. When a man marked your stocks to the market price and sold, everyone knew what everything was worth. The dread uncertainty of a further decline would diminish, and stocks might begin to move up again. Whitney's colleagues outdid one another in their efforts to demonstrate to Hoover that they could handle matters without Washington. A week after Whitney spoke in his city, Insull announced that employees and managers at his group of public utilities would each contribute one day's pay a month to workers idled by the crash.

Hoover had attacked a practice—speculative short selling—not a person. Congress was less conceptual. Legislators took the president's signal to mean they were free to turn on Wall Streeters. From the winter of 1929, they made short sellers and speculators generally targets for investigations, prosecutions, ridicule, and shame. Hoover believed that the regulation of this problem still remained with the stock exchanges and the states where they were located, but pressured the exchanges into suppressing "illegitimate speculation." Though no new law on this issue passed, the sense that the market would not be left alone to right itself disturbed investors. After rising early in 1930, the market was drifting downward, passing below 200 in October.

Franklin Roosevelt, who was now running for re-election, also took the ad hominem approach. On the day the Dow hit 193, Roosevelt motored on snowy roads from Elmira to Buffalo to give a speech assailing Hoover directly, charging that since the market crash twelve months prior, "nothing happened but words." Roosevelt went on to charge that Hoover had failed to expand public spending sufficiently. On the causes of the crash itself, however, Roosevelt out-Hoovered Hoover: inflation, he insisted, was the problem. Indeed, the governor charged Hoover with presiding over a false prosperity that had actually been an "inflation orgy." In his own state, New York, Roosevelt had provided millions extra in public works spending. A president should do the same.

Hoover could not stand to think about the troubles mounting across the country. From the beginning of 1930, he had withdrawn repeatedly to the presidential retreat, Rapidan, to fly-fish and contemplate the economy. He organized nursing for his son, who had contracted tuberculosis, at the camp. Ambivalent about his inability to control the national economy, he created his own small world. He built a school at Rapidan, with desks of modern steel, finding that he was, at least in a small corner of Appalachia, the hero. In front of the fire at the "town hall" he had created, he debated the locals. His secretary Theodore Joslin later recalled that Hoover "would reason patiently with an opponent or a recalcitrant. The very force of his arguments would invariably influence them."

In November 1930 Governor Roosevelt gave one last campaign speech at Brooklyn's Academy of Music. He reminded voters that he supported Prohibition and touched on the topic of power, noting that the taxpayer suffered "when you pay six dollars a month for electricity instead of two." But the area he truly scored on was joblessness: "Not only is the dinner pail empty, but millions are eating out of it at home because there is no place of employment to carry it to." Roosevelt annihilated his opponent, district attorney Charles Tuttle, winning 1.7 million votes to Tuttle's 1 million. The fact that the victory was so resounding, and that it came in all-important New York, meant that Hoover now knew his likely opponent in 1932—Roosevelt. Across the nation, voters gave more seats to Democrats. In the end, Republicans hung on to their majority in the Senate by a thread, and lost the leadership of the House. Hoover soured. He kept his distance from the Hill when it was led by his own party; it was even harder to work with the opposition.

That December, the Depression took on a new seriousness. Heretofore, most of the banks to fail had been rural banks. Now an important bank in a big city ran into trouble, one that was a member of the Federal

Reserve System. It was the young Bank of United States, the one that served so many immigrants—half a million depositors. The trouble at first did not appear so bad: unlike many others, the bank could count more than $200 million in deposits. Something like half of depositors were small fry—low earners. And the Bank of United States was an important symbol in the city. It was still not in the state club of established banks, the New York Clearing House, but it was making its name. Earlier that year, the Bank of United States basketball team had beaten bankers from the establishment Bank of Manhattan before a crowd of 3,500 to capture the championship of the Bankers Athletic League.

The New York state superintendent of banks, Joseph Broderick, organized various potential rescue mergers with Manufacturers' Trust and with Public National. But the Clearing House banks killed the mergers. At the time, many observers saw the bank's problems as a consequence of class differences between the working class and immigrants on the one hand and Anglos on the other. The Establishment believed that the Bank of United States was marginal, and that this was a moment when only the strongest banks, as in Darwin, deserved to survive. But the Bank of United States had relatively strong books—at least as strong as many that were propped up by fellow banks. The problem was not so much individual weakness as bad monetary policy, inconsistent credit policy, and sheer bigotry. The bankers who turned against the Bank of United States were acting like Victorians.

Broderick begged for the bank's future:

> *I said it had thousands of borrowers, that it financed small merchants, especially Jewish merchants and that its closing might and probably would result in widespread bankruptcy among those it served. I warned that its closing would result in the closing of at least ten other banks and that it might even affect the savings banks. The influence of the closing might even extend outside the city.*
>
> *I reminded them that only two or three weeks before they had rescued two of the largest private bankers of the city and had willingly put up the money needed. . . . I warned that they were making the most colossal mistake in the banking history of New York.*

The bank did suspend payments to depositors, the largest bank in America ever to do so. A leading banker described the attitude that motivated other banks' decision to abandon Bank of United States: "Let it fail, draw a ring around it, so the infection will not spread."

On December 11, the sixty offices—sixty emblems of hope—closed their doors. Later in the month, a crowd of 5,000 depositors and protestors gathered at the Freeman Street branch in the Bronx to protest the change: "They Robbed the Poor," a sign said. Broderick arranged a deal with Clearing House banks whereby depositors at the Bank of United States might borrow cash from the other banks. But that was not the same as the assurance that they would get back their money. The story reminded Jews that Peter Stuyvesant's old contract was still in force. On December 22 at 4:00 a.m., a line of 400 began to form at a Second Avenue branch. Some 2,000 Bank of United States depositors gathered at its branch on Forty-second Street so that they might begin to collect loans from the other banks. Only a fraction were served. Another Jewish-owned bank, Manufacturers' Trust, was absorbed by non-Jewish banks.

It shortly became clear that sacrificing immigrants' banks would not confine American depositors' demand for currency. The infection that the banker had described was too large to draw a ring around. By 1931, panics at the larger banks began in earnest. The lucky engineer was running out of luck.

4
The Hour of the Vallar

September 1931
Unemployment: 17.4 percent
Dow Jones Industrial Average: 140

One late summer day in 1931 in Salt Lake City, the money ran out. Not just the money in the banks, and not just the money in town coffers—the money that citizens had to spend. Locals reached into their pockets and, finding nothing, began to trade work and objects. Barbers traded shaves and haircuts for onions and Idaho potatoes. From there, the trading spread to other products. Life in Utah had always been a desert when it came to water. Now it was a desert when it came to money, as well. People in Utah knew how to survive in a desert. Maybe they could find a way to manage in the money desert as well.

A short drive north in Ogden, a banker instructed the employees of several family banks in the art of bluffing. The Ogden State Bank had closed its doors. His bank would not go down the way it or Bank of United States had if he could help it. "If you want to keep this bank open, you must do your part," the banker told his staff. "Go about your business as though nothing unusual was happening. Smile, be pleasant, talk about the weather, show no signs of panic. . . . Pay out in fives and singles, and count slowly." The man's name was Marriner Stoddard Eccles, and the next week he would turn forty-one. He was a leader in his community, the firstborn from the second marriage of a wealthy Mormon patriarch. He had pushed hard to ensure that a company connected with his bank—Utah Construction—got a part in the Colorado River dam project.

Others in Eccles's community were also thinking and improvising. A real estate man in Salt Lake City named Benjamin Stringham began to organize the barter trade more formally. He pulled together workers without jobs, then shipped them out to farms to work for the day. They returned with their pay: peaches, eggs, pork.

The improvisation was not confined to Utah. Communities across the country were beginning to find new ways to get through the trouble. Out in California, city people were beginning to think about moving to abandoned farms, taking up plows, and trying to make a life independent of money. Back east, Ralph Borsodi, an author and social thinker, was readying a book titled *Flight from the City*, about his own family's effort to live on the land an hour and three quarters outside New York. Borsodi concluded that self-sufficiency of the family was the new ideal, that with his poultry yard of fat roasting capons, his self-built swimming pool, and his apiary, he had found the solution to downturns like that of 1921 or 1929. The family ought to be the next factory. He wrote that "domestic production, if only enough people turned to it, would not only annihilate the undesirable and non-essential factory by depriving it of its markets" but also "release men and women from their present thralldom to the factory." Within a few years the director King Vidor would make a film that offered a similar vision, calling it *Our Daily Bread*. The message in *Our Daily Bread* was not only that the land could provide, but also that moving to the country improved the character of corrupt urbanites.

In Chicago, Paul Douglas would shortly draw on his Russian experience with food cooperatives to create a system for saving money in his own community: a food co-op. Groups in Hyde Park, the neighborhood around the university, began to purchase food in bulk in order to cut back on prices. But they were so short on cash that even this efficiency did not help them. Douglas advised them to start a co-op retail store for members, or shut down. They opened the store.

In New York, there was a sense of solidarity among the old Wall Streeters. Bill Wilson, drunk, had watched as the price of his favorite stock, Penick & Ford, slid toward nothing. He had taken to sleeping on Livingston and Schermerhorn streets in Brooklyn Heights. Still, he felt a curious sense of excitement; now the whole country was like him, down like an alcoholic. Wilson found a job with a Canadian firm, Greenshields, and taking his wife, Lois, headed north. He joined the country club, rented an apartment on the Côtes des Neiges, and enjoyed the view of the St. Lawrence River from its windows.

Even the very poorest communities, including the blacks, found their own response to joblessness and hunger. In Washington, Solomon Elder Lightfoot Michaux, a radio preacher, reached millions with his "Happy Am I" aphorisms. Michaux fed the hungry and maintained apartment houses for those evicted. Another figure in the black community to respond was Father Divine on Long Island. He began to expand the Sunday banquets served at his Sayville residence. What stood out about Father Divine's meals was that they were the opposite of apples on the corner or soup kitchen food. Father Divine's meals were luxurious. The coffee percolated; the roasts—chickens, ducks—were plentiful; the vegetables were splendid. "We charge nothing," Father Divine ordained. "Anyone, man, woman or child, regardless of race, color or creed can come here naked and we will clothe them, hungry and we will feed them."

The playwright Owen Dodson later remembered a wonder he and his brother had seen at Sayville—an unending supply of milk, like a fountain, from a spigot. Studying the setup, the boys eventually discerned that "the source of infinite supply" was two boys pumping at a small machine beneath the table. What was especially striking about Father Divine's "heaven" were the images of plenty and the clear message that there was to be no shame about hunger.

Still, Eccles worried. Faith and improvisation alone could not help. Charity work was not enough to feed all those without jobs. And bluffing, Eccles observed, was not saving enough American banks. Eccles was a man with a great sense of responsibility. At night in bed he ran through the assets of the national economy as if they were those of his own household. Even though the Colorado River dam was proceeding, the rest of the country seemed in need of shoring up. In this downturn he had had a sort of revelation: "I saw for the first time that though I'd been active in the world of finance and production for seventeen years and knew its techniques," he would later remember, "I knew less than nothing about its economic and social effects." The same year, 1931, would be Eccles's turn to read the work of William Trufant Foster, one of the two authors who had developed a new theory of the economy. "When business begins to look rotten, more public spending," Foster and Catchings had prescribed. Maybe government spending—including the new Federal Reserve's providing cash liquidity for the banks—was the way out. Now he wondered when the nation's leaders would be able to face the "fundamental facts" of the currency problem.

The money drought that America was suffering from had a technical name: deflation. Deflation meant that the currency was becoming more valuable every day, rarer and scarcer. Deflations can be good for lenders; the money they are owed in the future is more valuable than it was when they wrote the original contract to lend. But deflation is terrible for borrowers, whether they be countries, banks, businesses, or families. It means they must pay back more than they originally contracted to borrow. Inflation taxes savers. Deflation taxes risk takers and punishes leveragers. It makes paying mortgages, as well as property taxes, especially difficult. It goes against the American sense of promise, punishing those who dare to hope they might move ahead.

Today we know that the Treasury and the Federal Reserve might have done much to alleviate the deflation problem of the early 1930s. They could have allowed the gold-standard mechanism to function—money would have been created automatically with the gold inflows. Or the Fed could have taken what we call countercyclical action. If the economy is strong, monetary authorities nowadays put on the brakes. If it is weak, they help out by greasing the wheels, pumping money into the economy one way or the other. That was what Fisher believed—he was now writing Hoover about money.

But in the early 1930s the Fed and its member banks lacked tools and knowledge. They did the opposite of countercyclical action. They acted pro-cyclically—tightening and tightening in the face of a downturn. One of the reasons for the mistake was a rule known as the real bills doctrine. Under the doctrine, the young Fed system favored banks that carried substantial commercial paper—business loans of short term (one year or less)—on their books. Commercial paper was regarded as the best form of hedge against the risk involved in demand deposits. Banks that carried such paper therefore were deemed prudent and worth saving. The more business they had, the more the Fed was ready to lend to them. This was called serving "the needs of the trade." Mortgages—which tended to have maturities of somewhat longer periods—won less approval from the banking system. This was mainly because their worth was harder to gauge; they were individual contracts, and not traded as they are today.

The effect of all this was that banks tended to make loans to businesses in periods of expansion. In periods of contraction, the banks made fewer loans. Yet those same bad periods were the very times when the banks most needed an infusion of cash from the Fed. Now, when they could have used help so much, the Fed denied them on the theory that they did not need the money.

The newness of the Fed—it had only been created in 1913—was a big part of the problem, especially for small banks. Most of these were state-chartered banks that were not part of the first Federal Reserve System. These banks did not have much commercial paper in their portfolios. They served farms. Other banks did not regard them as especially worthy of rescue. And because they were not part of the Fed system, they were not the Fed's responsibility. So whereas other banks might have rescued them before, now everyone hesitated, and no one did. And when a bank died, money died with it, worsening the deflation.

One casualty in the banking disaster was turning out to be Rex Tugwell's father. As Tugwell would write, "His business was paralyzed along with the rest; his chain of small banks discovered that investments of depositors' funds in railway and public utility bonds, in Peruvian or other foreign issues, could not be recovered. After twice replenishing capital out of their own pockets, the directors themselves were bankrupt." To Tugwell, it all seemed the confirmation of his suspicions and also, especially, of the perniciousness of the middleman taking his cut. Big business was wrong, too: Of his father and colleagues he concluded, "They had done the honorable thing as small businessmen, but the big businessmen they had trusted had let them down."

The banks' money problem played out everywhere. The market crash itself had not at first hurt Insull, whose brother Martin had completed some crucial financing of various Insull projects *after* the crash. But the deflation did, for the Insull empire was heavily leveraged. Now Insull's long-standing philosophy of "Take on debt to grow as fast as you can," of leaning one's sail into the wind as far as it would go, was threatening to capsize his business. Insull was confident in his market: he believed that consumption of electricity would continue to grow, even through a downturn. In the first half of 1931, cash was still pouring into his operating companies. But you could not keep buying up your own shares forever. New York banking houses finally had their chance for revenge against their old competitor. They vengefully drove down his shares. By the end of March 1931 Insull's creditors owned his bank portfolios.

Banks had loans on their books, but their books did not reflect the families and businesses they turned down as money and credit became scarce. The worst hurt were the most hopeful: "farmers, small firms."

For the wage earner, the lengthy deflation also had a peculiarly depressing effect. Since he was more likely to be a borrower than a lender, the deflation made it seem as though life were stacked against him. In inflation, wages rise—though often of course followed by price rises. In deflation, that figure more important to the worker than any other—his wage—does not move for years, or even drops. In Harlan County, Kentucky, wages were below those of a decade prior. A wage cut of 10 percent by a desperate coal company in early 1931 sent workers into a fury—one killed a deputy sheriff, another a worker who chose not to strike. "I've orders to shoot to kill," the sheriff, Johnson Henry Blair, told

Time. One of the miners' wives then wrote a bitter song. "They say in Harlan County, there are no neutrals there," wrote Florence Reece of the battle between the unions and the sheriff. "You'll either be a union man, or a thug for J. H. Blair." The title of the song, "Which Side Are You On," captured not only the worker-employer division but also the division that deflation was causing.

But by far the most dramatic place that the deflation played out was in American homes. In those days home loans were not traded in bundles; it was hard for a bank to use them as collateral. Mortgages represented smaller shares of home values and carried shorter maturities, five or ten years. Still, most mortgages had a contract with a bank or savings and loan that said if one couldn't pay, the bank got the house—it was as simple as that. And with the economy declining, house prices were also moving down, so some owners found themselves under water—their loan cost more than what the house was now worth. This experience, the experience of deflation, caused a chain reaction. A grim Senate witness would tell a subcommittee:

> *There will be this situation. There will be three mortgages in a block on all equally valued property. One mortgage may be for $3,000 on a house,*

> *another for $4,000 and another for $5,000, on houses that sold originally for $7,500, which are cut down in value now to $4,500. The holders of the mortgages buy the properties in. The man who holds the $3,000 mortgage on the first property wants to get his money. Someone comes along and says, "I will give you $2,500 for it." He replies "Make it $2,750" and the deal is closed on that basis. That fixes the value for the whole row.*

The money drought also meant that it was harder to borrow to cover one's losses. Worst of all, however, was that it slowed growth. In good times, the housing industry and the average family buoyed one another up. Now people were seeing the nightmare of the cycle in reverse. Home construction was down, hurting families; families were down, hurting home construction. Home equity, for example, was one of the most important ways that members of the construction trade financed the opening of new businesses. Unable to borrow against their homes, they could not work. "Very often," a witness testified of a would-be small businessman, "he is absolutely prohibited from going into debt because the banks, speaking generally, will not give him any mortgage accommodation under any circumstances—not even a 20 percent mortgage on his proposed home. That means, to my mind, a throttling of the building trades, the building industry, labor, and everything else that goes into the building game. They are strangling today." And what good could barter money do the home owner if the bank would not accept it for a mortgage?

Other components of the downturn worsened the deflation. Each day proved the Cassandra economists right anew: in the two years following Hoover's Smoot-Hawley legislation, U.S. imports dropped more than 40 percent. Though people were unable to quantify the change at the time, economists later estimated that a share of that decline was due to the tariff. Retaliation by other countries was taking its toll. Unemployment had risen in 1930. Now, by 1931, the guess was that the national figure was something like 16 percent and rising again. In New York alone, there were 14,000 or 15,000 homeless men. Others were not officially homeless but still walked the streets evenings. Wilson's Canada venture did not play out. He soon found himself back in Brooklyn, wandering about. His wife's mother died, but he missed the funeral, going on another bender.

Writing in the *New Republic*, Bruce Bliven had already noted that every man was different. Some men could bear to spend a night in a shelter without the event bringing them down. "But there are others,"

Bliven wrote, "who pay dearly. This winter differs from previous ones in the exceptionally high number of men who have never before had this sort of experience, for whom it is a personal tragedy too deep for words."

By far the hardest hit of any urban group were the blacks of the North. Now the unemployment rates between the races diverged. In the cities, something like one in two blacks was unemployed. Women, whose work as domestics often provided the most important family income, were similarly unemployed. In Harlem, a street became known as "the lung block" because its tuberculosis rate was high. Many churches did not know what to do for their parishioners: "God is mad with the people," one minister in New York summed up. Blacks, the historic stalwarts of the Republican Party, began to feel not only economic but also political desperation. "My friends, go turn Lincoln's picture to the wall. That debt has been paid in full," ordered one black newspaper's editor.

Hoover still did not entirely understand; the failure he confronted was too great. Coolidge had retreated into his dark moods, and now Hoover retreated into his. He even suggested, improbably, that Americans were profiting by gouging one another, selling apples at high prices: "Many persons," he would write, "left their jobs for the more profitable one of selling apples."

The attitude came out of his sense of futility, but to citizens it seemed too much. Abroad, all that foreigners knew was that Americans were hungry or worse. That year donors in the Cameroons of Africa shipped over $3.77 to help "the starving."

One group was meanwhile becoming emblematic of Americans' wronged virtue: World War I veterans. In the 1920s, Congress had promised them a bonus pension, to be handed out in 1945. The idea was to provide the equivalent of a federal pension. Now, the veterans were losing their houses, their businesses, and their farms. It seemed reasonable therefore to ask Washington for their money—or at least an advance on the bonus. In Congress, representative Wright Patman of Texas was leading a fight for a new bond issue to raise the bonus money for the veterans. Mellon and his undersecretary at the Treasury, Ogden Mills, opposed it in the name of balancing the budget. It was one of the Hoover administration's weakest moments: "We loan millions to wealthy shipbuilders at 2 percent but we charge veterans 6 percent," a congressman pointed out.

There was something amiss with any country that did that. Was it the money, or the leadership? At a conference of disabled veterans in Wilkes-Barre, Pennsylvania, Patman called for the ouster of the man who was becoming the symbol of all that was wrong: Mellon.

But what might fix it? Mellon still believed that, however hard things were now, they could rebound. Like Hoover, or Coolidge, he tended to turn inward in the face of onslaughts like 1931. He reverted to his old business self, which knew what to do in a time of downturn: buy. As treasury secretary, he was constrained when it came to investing in companies or bonds for his own profit—indeed, Patman was after him for holding stock in a shipping company. But he could, he believed, buy art for himself. The best moment to buy art, of course, as he had learned at the bitter dissolution of Frick's estate, was in a distress sale.

And as it happened the greatest distress sale in the art world was on: the Soviet Union was selling. Stalin needed hard currency badly. He was therefore offering a transaction obvious to both him and Mellon: Mellon would purchase some of the art from the Hermitage. The buying had already begun. Mellon did not talk about it, but in September 1930, the *New York Times* carried reports that a Mellon representative—Knoedler—was in Paris arranging the purchase of Jan van Eyck's *Annunciation* from a Bolshevik regime broker, A. V. Lunacharsky, the former Soviet commissioner of art. The price was rumored to have been $800,000. Both Mellon and the Soviets promptly denied the transaction, the Soviets because, as the

Times put it, such a purchase would be "an admission of moral as well as financial weakness from which the Soviet Union is not now suffering." But the rumors had not died, and the next month, October, saw reports that Frans Hals's *Admiral* and Rembrandt's *Portrait of Sobieski* had exchanged hands, along with the van Eyck. Now, in 1931, Mellon moved again, making a series of purchases that formed the nucleus of a new collection. From "time to time," David Finley, then Mellon's staffer at Treasury, recalled later, "cables would arrive, saying that Botticelli's *Adoration of the Magi,* Jan van Eyck's *Annunciation,* Perugino's *Crucifixion* . . . could be bought if Mr. Mellon gave his approval." Mellon gave his approval. He still saw Russia as a wreck, and himself as retrieving some good from what was otherwise a general disaster.

Now Hoover began to stir. He could see the consequences of the international debt and the tariff, for both the United States and Europe. In June 1931 he announced a plan that would bring relief to Europe: a moratorium of interest payments on the German debt. Mellon negotiated on his behalf. Keynes of course liked the idea but announced his own, more comprehensive five-year plan. The stock market in Berlin exploded in joy, moving so fast that officials could not track and record the price rises; perhaps the shaky Weimar Republic might have a future after all. In Chicago, wheat climbed five and a half cents per bushel. At home, Hoover set to work creating the Reconstruction Finance Corporation, to help banks that sustained the homeowners and create the very sort of network that Eccles had seen as lacking.

On other matters, Hoover stayed firm in his old positions. The dam on the Colorado ought to be hurried along, but only because it was constitutional. Not however the government operation of Muscle Shoals, the vast power and nitrate plant built on the Tennessee River at the end of World War I. Washington's operation of Muscle Shoals was to Hoover's mind still wrong; government in the power business was still wrong, at least in peacetime. A few months earlier, even as he was mulling over legislation put forward by Senator Robert Wagner to create employment agencies across the states, Hoover spent a day reviewing Muscle Shoals legislation sent to him by Congress. Senator Norris—the Muscle Shoals champion and a fellow Republican—sent emissaries to the White House to try to talk Hoover into going along even as he publicly mocked Hoover as "the great engineer." But Hoover vetoed the laws on principle, writing that such projects "break down the initiative of the American people." He summed up: "I am firmly opposed to the Government entering into any business the major purpose of which is competition with our citizens. There are national emergencies which require that the Government should temporarily enter the field of business, but they must be emergency actions and in matters where the cost of the project is secondary to much higher considerations." Government shouldn't get into the power business, or any business—"that is not liberalism, it is degeneration." Traditional federalists marked it as his finest hour.

Even as Hoover made his point, however, broad interest in the concept of government control was expanding among Americans. American magazines had carried 112 articles on economic planning in 1928. By 1930, that figure had been 210, and as economic trouble deepened, it would reach 365 in 1931. Cole Porter had not yet written the song with the line "You're the top, you're Mussolini." Mrs. Hearst had not launched the sort of barrage of publicity that the Soviet junketeers had. But attention to Mussolini's model was indeed also intensifying. The U.S. press reported Mussolini's every action, including the fact that the premier had now reached a trade agreement with Germany's chancellor, Heinrich Bruening, and would sell the Germans surplus Italian oranges and lemons.

The Soviet model was also under a spotlight. Suddenly, at least in the intellectual world, it looked as if Mellon's contempt was the minority view and frank interest in the Soviet effort was mainstream. It was an exhilarating turn: the intellectuals were hot after all. Between 1920 and 1931, some eighty books by American authors on the new Soviet experiment had been published in the United States. Lately the pace had accelerated—Tugwell, Chase, Douglas, all were now publishing books or articles. (Chase also had another non-Russian book out, about Mexico, illustrated by a new artist named Diego Rivera.) Russian authors were being translated and published as well. In one week in July 1931, around the time Eccles was worrying about his banks, the papers reported news of the publication of all the following new books: *The Success of the Five Year Plan*, by V. M. Molotov; *Red Villages*, by Y. A. Yakovlev; *The Volga Falls to the Caspian Sea*, by Boris Pilnyak. As for George Counts, he had himself a national best seller in a Russian children's book that he had translated and edited. And this book, unlike some of the earlier ones by the junketeers, was published by a mainstream press: Houghton Mifflin. The book, Mikhail Ilin's *New Russia's Primer: The Story of the Five Year Plan*, offered lessons from Russia, right down to the construction of hen coops. It suggested that each peasant in Russia could be sure of owning "two good laying hens," both a doubling and a mockery of the offer Hoover had made in his campaign just three years before.

Many Americans wanted to see Stalin's experiment for themselves. Some 2,500 had visited the Soviet Union in 1929. The next year that figure doubled, and it more than doubled the year after that. In 1931 Amtorg announced that it had 6,000 skilled jobs to fill, with 100,000 applications flowing into the office. Sometime at the end of September or in early October 1931, the editors at *Business Week* sorted through a morning's sample of 280 applications and found that there were two barbers, one funeral director, two plumbers, five painters, fifty-eight engineers, and a dentist among the group. Two of the top three reasons the applicants cited for their interest in the move were unemployment and "disgust with conditions here," the magazine reported. The third was "interest in the Soviet experiment." Immigrants were especially disillusioned with the United States; nearly all stated that they planned to stay once they got to the Soviet Union.

The next month—November 1931—the left-leaning *Nation* inaugurated an ebullient series titled "If I Were Dictator." The author of the first article in the series was the prolific Stuart Chase. Chase seized the opportunity to present an agenda that was simultaneously modest and far-reaching. It included abolition of the protective tariff, except for new industries, and an end to war debts. Wine and beer were to be made legal.

He also wanted to recognize Russia "at once" and then sell her a billion dollars in American goods. He wanted federal relief for the unemployed, as well as a "complete system of old age pensions." He advocated spending freely to promote growth, notwithstanding what that did to deficits—"the effects on the federal budget will not disturb me in the least." More important, however, Chase wanted to set up a "long swing program" to plan the U.S. economy, dividing industry into state trusts, regardless of what old antitrust law might say about that ("the Sherman anti-trust law is of course declared a piece of antiquated lumber").

To run his fantasy economy, Chase appointed a fantasy board of planners: Robert Lynd, the author of *Middletown*; Walter Lippmann, the journalist; Bernard Baruch, who had run the War Industries Board in World War I; John Dewey, the philosopher and educator; and two friends from the Russia trip, Paul Douglas and Rex Tugwell. The article made the cover of the issue of November 18, 1931; underneath ran the headline for another story: "Mr. Hoover Gets Notice to Quit."

There were a number of others who thought that now was the moment for revolution, or at the very least communism. The author John Dos Passos wrote that given such conditions, "becoming a socialist would have just the same effect on anybody as drinking a bottle of near beer." Many of the progressives, likewise, continued to believe that to pin all hopes on Norman Thomas, James Maurer, and their Socialist Party would be to fail. The country had not changed that much. So they began to hunt for political leaders within the big parties: Newton Baker, a New York lawyer who had served as mayor of Cleveland before World War I, attracted a number of them, including Adolf Berle and Wendell Willkie. Baker, too, thought a lot about Russia. In the *New York Times* of the following year, Baker would outline what seemed now a permanent problem: child homelessness. Thinking of the children on the streets in revolutionary Russia, Baker would write: "America's vagabonds, however, share this quality in common with Russia's wild children; having tasted the poison of a wandering life they find it difficult to give up."

The reason that both Willkie and Berle preferred Baker to New York's Franklin Roosevelt was that the governor had a split reputation, as a supporter of Tammany Hall and as a progressive heir to Theodore. Many political observers disdained him. Some found him arrogant. Walter Lippmann said Roosevelt was "a pleasant man who, without any important qualifications for the office, would very much like to be president." Still, the fact of Roosevelt's victory in the 1928 gubernatorial contest was something his opponents could not get past. With the 1930 census, New York would add two electoral votes to its already stupendous forty-five. Maybe specific policy positions didn't matter if one was talking about someone who had won by a landslide in New York, a state with forty-seven electoral votes.

And Roosevelt for his part was leaning more toward the intellectual progressives. In 1928, a year after the Sacco and Vanzetti case, Felix Frankfurter had begun writing regularly to Roosevelt, at first to congratulate him on his nomination as a candidate for governor for the Democratic Party. Roosevelt wrote back, and from then on the two corresponded regularly, Frankfurter's letters filled with lavish flattery. "By holding out on your water power policy for New York, you have vindicated courage in government," Frankfurter had, for example, written to Roosevelt in 1930 in regard to a battle he was conducting with the chairman of the Niagara-Hudson Power Corporation.

Recently, Frankfurter had also sought to arrange a meeting between the New York governor and Justice Brandeis. Frankfurter thought that Brandeis might advise Roosevelt on an area of interest to them both: public utilities. Such encounters would strengthen Roosevelt's commitment to fighting for the public power issue, Frankfurter hoped.

When a group backed by Wall Street put together an effort to develop the power resources of New York State, Roosevelt fought back: he was now convinced, as a governor, that public power was the answer. ("It is our power," he had said in his first inaugural.) Martin Insull, the brother of Samuel, had only increased that conviction when he penned a series of nasty articles about Roosevelt. Sam, aware of this, counseled his brother that he was making a mistake. "In our business we can and do attack politicians bitterly in the abstract without making enemies, but we cannot attack individuals," Sam would say later, regretfully.

Governor Roosevelt had also come to lean on Frances Perkins, who as a social worker had spent time at Jane Addams's Hull House. Perkins had her duels over unemployment data with Hoover. But she also held conferences with state labor commissioners to talk about the importance of creating new forms of unemployment insurance, if only as "safeguards against the dole." She brought in Paul Douglas, still at the University of Chicago, to organize the roster of speakers at one conference; the governor attended and took time to speak with each expert in turn. Later Perkins herself traveled on the *Rotterdam* with her daughter to Britain to study that nation's system for unemployment insurance.

The British plan seemed to her quaint: senior bureaucrats in spectacles climbed up high on painter's ladders to retrieve individual documents for each worker. Still, Perkins liked Britain's model: "You saw their government's extraordinary skill in handling a human situation." The British bureaucracy's magnanimity shone through in a way that Washington's did not.

By the winter in which 1931 became 1932, state and county governments were beginning to give up on handling the hunger and homelessness. The people were beginning to give up as well. If their leadership could not understand the details of the monetary problem, then how could they? Many Americans did not even think that a new institution like the Fed could do either much good or much bad. "Federal Reserve Board decisions and pronouncements were read by very few," wrote the journalist Mark Sullivan of the period. "Bank officers who would be obliged to conform to them. Among businessmen, a small proportion . . . A few scholars in monetary theory and economics. Of reading by the general public, there was almost literally none."

All people saw was that things were not working. In Utah, notwithstanding Eccles and his allies, thirty-two of one hundred and five banks had failed. Wages dropped to nearly half their old level—for those who still had a job. The unemployment rate for the state was rising into the 30 percent range. Whatever the initial missteps, the deflation compounded them. What's more, across the Midwest and West, there was now a genuine drought—and a bad one: in all of 1930, 1931, and 1932, the rainfall was less than average.

Tugwell, looking west out the windows at Columbia's Morningside Heights, could now see trouble as bad as what he had viewed abroad: "a sprawling Hooverville soon spread along the riverbank across the railroad tracks. Before long there were thousands of shacks put together out of orange crates, beaten cans, old pieces of rubber, leather, or cloth, their denizens gone back to caveman status, scrounging in neighboring garbage heaps for food or fuel." This downturn, he was beginning to conclude, was worse than any other. "It felt," he would write, "as though a sense of jeopardy was about to open."

Part of his conclusion came from personal evidence—the abiding business troubles of his father upstate. Though some might recover, "it was different for my father," Tugwell would later write. His father's canning business, Tugwell and Wiseman, was failing. Tugwell suspected his father's decline might be permanent. Later, he would decide that the Depression had done his father in: "My father was almost as poor as when he started, and now he was old. . . . He would live on, broken and helpless, for another fifteen years."

Father Divine's popularity grew. So did the conspicuousness of his Sunday revival meetings in otherwise all-white Sayville. After neighbors complained to police in November 1931, the intimidated town summoned sheriffs and the district attorney. When the assistant DA pushed his way into the house on Macon Street, he was punched unconscious by a Divine follower, "St. Peter." Divine was arrested for disturbing the peace. "Yes, my success and my prosperity disturbs you," Father Divine retorted to the community. The next month, as Long Island awaited his trial, Father Divine spoke at a number of rallies, one of the larger at Rockland Palace in Harlem, where a hall that held 5,000 overflowed. Father Divine seated fifty people up front, about half of whom were white. His point about integration could not be overlooked.

Later, when the case of *Sayville v. Divine* was heard, he would be convicted in his municipal case—and the judge would, by strange coincidence, die a few days later. Father Divine was exonerated upon appeal. Father Divine allowed observers to speculate that his persecution by the judge and plaintiffs had somehow led to the judge's death. In the course of the proceedings another former DA, James C. Thomas, had taken up Father

Divine as a civil rights cause. "To allow an incident of this nature to go unchallenged is to weaken the foundations of democracy in the United States," Thomas's statement read. Earlier that very year, authorities had arrested nine black teenagers for the alleged rape of white girls on a Southern railroad freight train, and suddenly there was a new consciousness in the country about even small incidents such as Sayville's. The whole story served to increase Father Divine's status in the black community as an independent and race-blind leader. Still, even he and figures like him were not creating jobs. And the new gap between black and white unemployment persisted. In Pittsburgh, blacks had been 38 percent of the unemployed in the first half of 1931, even though they were only 8 percent of the population. In Chicago blacks made up 16 percent of the unemployed, though they were only 4 percent of the population. For all groups, the problem became not merely unemployment but the duration of joblessness—one year, two years now.

On Christmas Eve in 1931, the *New York Times* carried what seemed a new kind of story. A couple from New York City had retreated to a stranger's empty cottage on the edge of the Catskills to die. The young pair had made their way to the area in search of employment, investing all but twenty-five cents of their cash on the journey, but had failed to find work. "Finding none," the *Times* reported, "they went into the cottage, preferring to starve rather than beg."

A constable discovered the man, Wilfred Wild, and his wife in the lakeside cabin after three days, "at which point the wife, age 23, was too weak to walk." As the *Times* wrote, the officials reacted in the manner typical of the period: "An effort is being made to obtain employment for them, but if this fails they will be sent back to New York."

Just a month later, in January of 1932, author Florence Converse, a Wellesley grad, published a poem in the *Atlantic Monthly* asking:

"What's the meaning of this queue,
Tailing down the avenue,
Full of eyes that will not meet,
The other eyes that throng the street . . .
To see a living line of men
As long as round the block, and then
As long again? . . ."
"All lines end, eventually—
Except, of course, in theory."

The poem was about unemployment, but it had an additional point—a deadly one. American common sense was failing. The downturn was proving the falsehood of pragmatism's old thesis: that "all lines end, eventually." Perhaps the country was now entering new territory, the realm of theory, where lines did not end and must always be addressed.

And now, as the country began to feel panic, the world of theory—the world of the pilgrims—began for the first time to have political potential. In Congress, lawmakers began to search for scapegoats. Wright Patman, the Texan who lobbied for the veterans, now put forward the resolution to impeach Mellon that he had warned of the year before. "If we get rid of Mellon we'll have a chance to restore prosperity," Patman said. The basis of Patman's attack was an old codicil, ancient legislation, Section 243 of a law from 1789, that had been created by the first Congress of the United States and limited the amount of commercial involvement permitted for cabinet officers. Fellow congressmen were shocked at the boldness of the move: in a flurry "page boys moved like shadows about the chamber, looking for law and reference books," the *New York Times* reported. Patman charged that the fact Mellon owned shares in shipbuilders and companies violated the law. But previous Treasury secretaries had owned stock in various companies and had not been accused. What's more, Mellon had renounced his corporate directorships upon first becoming secretary. Mellon had been in office ten years; his holdings had not changed, but the times had.

The same day that Patman called for impeachment, 15,000 unemployed descended on Washington, many carrying pup tents to camp in. This was not the Bonus Army but rather, quite simply, a group of desperate workers. They were led by Father James R. Cox of Mellon's own Pittsburgh. There was a new national feeling that somehow the United States could not go on, that the time had come to target some of the wealthy. In the same issue of the *New York Times* that carried the Mellon impeachment report, Governor Roosevelt spoke out against the "increasing concentration of wealth and power."

That same month the populist from Louisiana, former governor Huey Long, was sworn in as a U.S. senator: Long wore purple pajamas when he entertained the press before the ceremony. Once on the Senate floor he cast his first vote in support of legislation barring government loans to companies whose presidents were paid more than $15,000 a year.

By February 1932 Mellon had resigned; by April he was on the ocean liner *Majestic*, headed to Southampton to serve as ambassador to the Court of St. James.

Though Patman did not know it, this banishment may have been welcome, for it led to a new and rewarding phase in Mellon's life: a phase in which he could think about art. David Finley, his Treasury colleague and art adviser, sailed with him. Mellon took his own collection—at least the ones whose transfer would not be damaged by the change in climate—to adorn his new home at Prince's Gate. His son-in-law David Bruce suggested he have a look at a painting of the Indian princess Pocahontas owned by Francis Burton Harrison, a former congressman. Mellon bought it. In 1931 he had written his son Paul, who was studying at Cambridge, that he hoped Paul was "having some time to spend at the National Gallery as it will be useful for you to have some knowledge of the important pictures in view of the contact you will have with works of a similar character in the future here." The senior Mellon was hinting at something. The older he grew, the more Mellon valued pictures; Washington had nothing like that gallery. He had already created a trust into which to put art that would later be some sort of gift to the government. In his retirement from the Treasury, Mellon's mind unfolded and ranged.

And Roosevelt's mind focused. He had decided, by now, that he was running for president. He wanted a new kind of adviser. Of the industrialists, the financiers, the political leaders, his adviser Sam Rosenman said to him, "I think we ought to steer clear of those. They all seem to have failed to produce anything constructive to solve the mess we're in today." Instead Rosenman had a novel idea: "Why not go to the universities of the country?" Roosevelt wanted to stick with the bright thinkers he already knew; and he wanted a small group of scholars, including some new ones, to write him up memoranda on agriculture, tariffs, the important issues of the crisis. Rosenman suggested Ray Moley of Barnard, whom the president had known and worked with since his friend Louis Howe had introduced the pair in 1928. Moley in turn brought his own list of advisers: it included Adolf Berle, newly married and cutting quite a figure in the world of corporate law—Berle could do credit. "Agriculture" was also down on the president's list as a topic, Moley later remembered, and "'Agriculture' suggested Rex Tugwell." But whom should the governor meet first? "I suggested Tugwell first," Rosenman later recalled. "Roosevelt wanted to lay great emphasis on the sad plight of those who lived by agriculture."

The governor liked Tugwell—an upstate man, after all. And Tugwell now gave up his interest in Belle Moskowitz and Al Smith and switched to the Roosevelt team. His ebullient personality pleased them. "Tugwell was like a cocktail," Moley would later recall. "His conversation picked you up and made your brain race along." The fact that there was what Moley called "a rich vein of melancholy in his temperament" made Tugwell seem intriguing. Roosevelt probably also liked Tugwell's love of discussion, his joy in writing policy. The following year Tugwell would choose as an epigraph of a book on industry a quote from the president of the Massachusetts Institute of Technology, one of the great believers in the science of economics, Francis Amasa Walker: "the long debate of reason resulting in the glad consent of all." The phrase corresponded to Roosevelt's attitude about his new experiment. If they discussed and analyzed enough, he and his new friends might solve the big problems.

What a change of fortune this seemed to Tugwell and the other misfits, who had believed themselves doomed to abide at the edge of the world. Before they knew it Moley, Tugwell, Berle, and Rosenman were working morning, afternoon, and night composing briefs for the governor. Roosevelt, now an official candidate, was away in Warm Springs; Rosenman brought the briefs down. Roosevelt read others when he came back. Tugwell believed that "Roosevelt was the first governor who had really understood the hill farmers' troubles," and thought of his own father. At one point Roosevelt, in a conversation with a reporter, referred to his advisers as the "brain trust" and coined a phrase. When it came time for instruction, there would be a meeting, often at the governor's baroque mansion: "The governor was at once student, a cross-examiner, and a judge," Moley later wrote.

The energized professors wanted to fashion for Roosevelt a dramatic message, certainly something that went farther than the old antitrust arguments, such as Louis Brandeis's thesis of the "curse of bigness." "We are no longer afraid of bigness," wrote Tugwell. "Unrestricted individual competition is the death, not the life of trade." Marketing too was important. Franklin Roosevelt had to offer something more than his cousin TR's "Square Deal." Stuart Chase, aware that Roosevelt now represented the best possible bearer of the progressive message, was writing articles in the *New Republic* on the importance of reform and the new sense of experiment. He was also publishing his book, the one that he had been thinking about since the Russia trip. He still had in mind the question that had come to him while traveling with Tugwell: Why should Russians have all the fun? Hunting for a title, he decided on a phrase that had been in the air, both in the context of Roosevelt and elsewhere. The title played off the Square Deal of Teddy Roosevelt: *The New Deal.*

Roosevelt was not an ideologue or a radical. He tended to think in political, legal, moral, or military terms; his formative experience, outside New York government, had been as assistant secretary of the navy. He thought a lot about agriculture, too, and conservation, and water, just as his cousin Teddy had. Economics had to be in the mix, of course, but it nearly always came last. This time, though, he thought the pilgrims and progressives might be right, and he could see that their ideas resonated. What's more, he enjoyed some of them, like Frances Perkins, a bluestocking like his wife Eleanor, a good deal. He might not agree with the bright bulbs on everything, but their spirit and willingness to experiment cheered him out of his physical pain and discomfort, at least for an hour or two.

Meanwhile the economy continued to flounder, and both Roosevelt and the other obvious candidate for the Democratic nomination, Smith, assailed Hoover on the tariff. Their assaults turned out to be so similar that it proved embarrassing. The *New York Evening Post* one night carried under the headline "A Deadly Parallel" the following quote from Smith: "The consequences of the Hawley-Smoot Bill have been tremendous, both directly and indirectly." Beside it was a quote from Roosevelt: "The consequences of the Hawley-Smoot Bill have been tremendous, both directly and indirectly." Europeans must be stopped when they moved to "raise their own tariff walls." It turned out that one brain truster, Lindsay Rogers at Columbia, had read aloud to Belle Moskowitz of the Smith campaign the same material on tariffs that he had provided to the Roosevelt team.

And what policy—beyond opposing tariffs—would come out of that team? In April 1932 the new candidate Roosevelt gave his first national radio speech, a coast-to-coast hookup organized by the National Broadcasting Company and sponsored by the cigarette company Lucky Strike. FDR and Raymond Moley, who worked on the speech, knew that they wanted to talk about the underdog. The next week Moley recalled his effort at formulation of the text in a letter to his sister Nell: "When I was working on it with him I was trying to suggest the ideas, words, and phrases that would make that picture of him over the radio and would fix the image in the public consciousness." In the end, Moley wrote, "I scraped from my memory an old phrase, 'The Forgotten Man,' which had haunted me for years."

Moley probably did not realize it, but the phrase also could be found in William Graham Sumner, Fisher's mentor at Yale. Sumner's "forgotten man" was the hidden taxpayer, the average citizen—not someone who received, rather someone who paid in. But that did not matter; who, at this moment, would be pedantic enough to point out the provenance of a quotation? Rosenman, however, remembered Sumner and realized that this use of the phrase was a shift: "the philosophy of the 'forgotten man' speech was entirely contrary to the philosophy that had prevailed in Washington since 1921, that the object of government was to provide prosperity for those who lived and worked at the top of the economic pyramid, in the belief that prosperity would trickle down to the bottom of the heap and benefit all. Roosevelt believed that prosperity did not 'trickle' that way . . ."

On the air, Roosevelt laid out his program. The common man was "the infantry of our economic army." America needed "the forgotten, the unorganized but indispensable units of economic power." It must offer "a real economic remedy" to that man. Public works were merely a "stopgap"—that was what Hoover had done. The 1929 tariff was disastrous, for it had killed trade. He would stand up for "the forgotten man at the bottom of the economic pyramid." As Sam Rosenman recalled later, the novelty was not merely the message but the messenger. Most rich people believed that wealth came from the top, that it trickled down. But Roosevelt was showing himself to be different. The candidate also did not mind assailing the wealthy, a feature rare in the loyal culture of his class.

All this horrified some Democratic colleagues. Al Smith gave a speech at the Democrats' Jefferson Day dinner, taking issue with the way Roosevelt and others seemed willing to assail others: "We seem to seek negative victory rather than affirmative victory," he told his party. "I will take off my coat and fight to the end against any candidate who persists in any demagogic appeal to the masses of the working people of this country to destroy themselves by setting class against class and rich against poor." The statement was striking, even when one took into account that Roosevelt was supplanting Smith, a man who had been born poor and had dropped out to work at age fourteen. Smith was rejecting a class fight while a man given the advantages—Roosevelt of Harvard and Groton—chose to take one up.

The power of Roosevelt's rhetoric and the medium through which it came trumped Smith. Hoover had had his luck. Now Roo-sevelt had his: the luck to be a great radio speaker born into the era of radio. Where Hoover had been brusque, Roosevelt inspired. His advisers were astonished at how convincing his broadcasts sounded. Roosevelt was inventing a new kind of public speaking. Shouting and superlatives were not so necessary now that there was a microphone. He showed that one could chat into listeners' ears and convince them that way. "The great strength of President Roosevelt as a propagandist, for example, lies in his reasonable voice," a professor of public speaking talking about radio generally would point out, several elections later.

Not only the medium but the message suited the national mood. Before the country's very eyes, there was a visual of the Forgotten Man to accompany the audio that Roosevelt had provided—the Bonus Army marcher. The idea that the government was holding back the dollars for some retirement day seemed increasingly absurd to the crowd. And now, veterans began to head for the District of Columbia. At hearings the year before, one of the men to testify had been Joseph T. Angelo, the man who had rescued George Patton from death in 1918, the veteran who had walked down to Washington from Camden, New Jersey. "I came to show you that we need our bonus," he told the lawmakers. He was not even asking for the full bonus, just part of it. Angelo let the committee see the pocket watch that Patton's wife had given him. In February, Hoover vetoed a bill; his veto was overridden. Under the new law, the bonus stayed on its old schedule, but veterans might now borrow against it. Still, neither Wright Patman nor the veterans found this compromise satisfactory. In late spring 1932, the marchers started to head for Washington. By summer, somewhere between 10,000 and 20,000 marchers would be camped out in the capital city.

Hoover, without calling attention to his action, was supplying this army food, clothing, and tents. He even offered to see the marchers. But as usual, he wanted communication only on his terms—and the Bonus Army veterans were not as passive as southern flood refugees or malnourished Belgians. What's more, they had an ally in Washington to egg them on: lawmakers like Patton on Capitol Hill. The Bonus Army squatted in Anacostia Flats, unwilling to give ground. He eventually sent troops to corral the protesters out of the city; as a believer in the rule of law, Hoover did not like to see vagrants within the District of Columbia. The veterans threatened the lawmakers, occupying the Capitol steps; members fled through subterranean tunnels.

The troops were led by a trio associated with another chapter of U.S. history: Dwight Eisenhower, Douglas MacArthur, and Patton again. MacArthur, stubborn and independent as always, felt that there was "revolution in the air." He charged on the men, who in turn set their camps in Anacostia Flats afire. As Dwight Eisenhower would remember later, it was "a pitiful scene, those ragged discouraged people burning their own little things." Hoover's PR success of flood rescue was now matched by an equally spectacular public relations failure. It did not matter that FDR too had opposed advancing the bonus money to veterans.

Here Roosevelt won his first victory over Hoover. "Hoover's failures," Tugwell would write, "were more of an asset than Roosevelt's promises." As a candidate, he could say he stood for these forgotten men in a way that Hoover had not. The Forgotten Man theme began to carry the campaign. Hoover could not keep up; his mind was still on longer-term programs. In mid-May architects announced completion of a new home for his old department, Commerce. Less than ten years ago, when he had taken the Commerce job, observers had joked about the irrelevance of the bureau. This structure, on Fourteenth Street and Constitution Avenue, was Hoover's rebuttal, with thirty-five acres of floor space. "Hamilton, gazing at it," wrote a critic, "would realize that his theories of government had triumphed."

As May turned to June, even as the Bonus Army waited restlessly in Anacostia Flats, Hoover misstepped again. Instead of visiting with the soldiers, he decided to walk over to Congress and plead with the lawmakers to raise taxes. Hoover was worried about foreign countries and banks, and the fear that they would cause a run on the dollar unless the budget were balanced. He read in what the papers described as "a low, barely audible voice." For once, Congress cooperated, and a week later, Hoover signed into law a large tax increase: the Revenue Act of 1932.

In the tax increase Hoover had Mellon's support, at least in name. Mellon had already been on his way out when he had devoted a few words to the support of the new bill: "We are convinced that in the long run lower rates are more productive than higher ones. But these are not normal times." To Paul, Mellon acknowledged the contradiction of his action: the government was going to "impose additional taxes upon industry and commerce that are in no condition to bear additional burdens. . . ." It was Mellon's worst moment. By going along with Hoover, Mellon was betraying many of his own principles. The tax that most Americans were having the most trouble with at the time was not the income tax, it was the property tax. The deflation made mortgage obligations unbearable, but it also made it hard to come up with the cash for property tax bills. Delinquency in both instances led to the same bitter conclusion: foreclosure.

The Revenue Act did not increase everyone's taxes. At that time the income tax was young, and the average citizen did not pay it: it was said the tax was a "class tax" and not a "mass tax." Still, tax increases generally were like interest-rate increases: bad news at a time when, just as Mellon said, neither citizens nor the economy could handle it. Starting a business and hiring both became harder. The deflation made the whole tax story even worse: at a time when every dollar counted more, the law took dollars away by raising rates. Added to all this was the fact that Hoover's was not just any tax increase, but a giant one: an increase in the top rate from the mid-20 range to 63 percent. Such increases were the sort the country had heretofore thought possible only in wartime. The maximum top rate when the income tax was first introduced, less than two decades back, had been 7 percent, and that was only on incomes over half a million dollars. The House had already gone Democratic in the midterms, but Republicans still held the Senate and the White House. Yet this tax act was such an antiwealth gesture, it seemed a sort of symbol of Republican capitulation to a coming Democratic moment.

That moment had its true beginning later that June, at the Democratic National Convention in Chicago. Roosevelt offered yet more ideas—again, contradictory, and more political and moral than economic. On the one hand, he stuck to old and conservative policies. He talked like Hoover about how "government, of all kinds, big and little, be made solvent." He complained about high taxes: government "costs too much." On the other hand, he made expansive statements whose import was hard to gauge. The country, he believed, had grown too fast: beyond "our natural and normal growth." The problem was that there had been "an era of selfishness." There existed "throughout the nation men and women, forgotten in the political philosophy" of the last years. These people "look to us for guidance and for more equitable opportunity to share in the distribution of national wealth." This language sounded new—it was that of the pilgrims and the progressives. Roosevelt also assigned blame to Hoover and Coolidge for the inflation that they both wrongly believed was doing the damage. And now Roosevelt made fun of the Republicans for assigning the blame for the Depression to international causes, even though he himself had acknowledged this might be part of the story through his earlier attack on tariffs.

At the same convention speech, the future president signaled something else: that he would ignore Al Smith in regard to the scapegoats. The Depression, FDR said, was the result of "lack of honor of men in high places" and "crooks." More generally, he assigned blame to a moral fault: national greed. "Let us be frank in acknowledgment of the truth that many amongst us have made obeisance to Mammon, that the profits of speculation, the easy road without toil, have lured us from the old barricades." But it was Roosevelt's finale that struck everyone most. "I pledge you, I pledge myself to a new deal for the American people." Roosevelt stealing the mantle of reform for the Democrats from the Republicans. All the reformers and Brain Trusters were impressed, including Willkie, who was probably in the audience; he had come to the convention as a floor manager for Newton Baker. This routine of targeting class enemies in the name of reform would become Roosevelt's hallmark. While on his way to accept the nomination, Roosevelt had a long conversation with Frankfurter, which Frankfurter wrote down afterward. Frankfurter told Roosevelt, "As soon as I get your free mind, I have some things of importance to tell you." Roosevelt responded that Frankfurter ought to get hold of a document on economics produced for him by Tugwell, Berle, and Moley.

After the convention, the intellectuals redoubled their efforts, sensing that they now had their first real chance since Wilson of taking the White House. Roosevelt was curious about Russia—perhaps it was time for recognition. He hosted Walter Duranty, the *New York Times* correspondent, at the governor's executive mansion in Albany in July. Duranty had just won a Pulitzer Prize for his coverage of Stalin's first Five-Year Plan, and the meeting was one between expert and politician. "I turned the tables," Roosevelt told a reporter. "I asked all the questions this time. It was fascinating." Roosevelt's primary interest was trade and political leverage; having the Soviet Union as an ally would help the United States. But he was also interested in the Soviet spirit of experiment. "Consults Walter Duranty in Regard to Suggestions That Our Attitude Should Change," read the subheadline on the news report. Roosevelt also hosted the Yale economist Irving Fisher, also in Albany: "Now is the time to educate him," Fisher wrote to his wife.

Frankfurter, still a professor, then wrote Molly Dewson, an old acquaintance who had been at the convention, asking "to acquire detailed knowledge about Roosevelt's work" as governor, including work on financial policy, social legislation, water power, other public utilities, unemployment relief, and agriculture. There were some differences among advisers. Tugwell, like Insull, liked the idea of big companies and thought that, as he would later write, "modern concentrations" could be taken advantage of, that government could become a "senior partner in industry-wide councils." Frankfurter, like Brandeis and indeed Woodrow Wilson, believed in trust-busting as the progressives' rule. In their discussions Roosevelt's advisers cited a book that Berle would shortly publish with coauthor Gardiner Means, *The Modern Corporation and Private Property.* The thesis of the book was that U.S. industry would be controlled by ever fewer hands. Tugwell found himself annoyed by the outrage of some of the Roosevelt advisers over this idea; he thought assaulting big business and trust-busting a "diversion" from more important economic work.

But the battles among the progressives did not have to be fought yet—the campaign was still on. Reporter Anne O'Hare McCormick—the same reporter who had attended the Stalin interview—now landed an interview with Roosevelt at the governor's mansion, a Victorian, in Albany. Roosevelt did not know exactly what he was, but, he assured her, he was no engineer. He told the writer that the presidency was "more than an engineering job, efficient or inefficient. It is preeminently a place of moral leadership." Her article conveyed the mood of excitement at the house. Callers came and went; books lay strewn about in the hall and in the offices. The day she was there, McCormick noted, "a reporter picked up a copy of Stuart Chase's book, *A New Deal*, and discovered fifty new one-dollar bills between the pages sent as a campaign donation" by a citizen.

Now the Brain Trust team set to their work with new energy. They also vied for FDR's attention, and, inevitably, jobs. At points in the campaign Roosevelt declared himself for "sound money," by which most people assumed he meant a gold-backed currency. But he was ambivalent. Beatrice Berle wrote about it in her diary. "The first debate, I believe, was on inflation, Roosevelt at that time being in favor of it. Bernard Baruch then stepped in and wanted FDR to pronounce himself against inflation under any circumstances. A. [her husband, Adolf] did not want this and apparently pacified Baruch.

"From that time on, A. has been in great and constant demand at Albany and Hyde Park. Moley is the constant companion but A. seems to be the inspirer! . . . The triumvirate of Moley, Johnson and A. remains. A., certainly had a frabjous time!" As for Beatrice herself she noted, with irony, that Sara Roosevelt, the governor's mother, liked to have her at her table in Hyde Park—perhaps because Beatrice too was descended from the old Dutch aristocracy along the Hudson.

In September 1932 Berle and Beatrice sat down to write an address that Roosevelt would give at the Commonwealth Club of San Francisco. The pair worked hard to find the right phrases for Roosevelt to use when he attacked the wealthy. "A. dictated a first draft which I thought very sloppy so I chewed the pencil, and did a powerful lot of pruning and rewriting. By that time he had clarified his thinking and wrote off the end in fine shape thus at last getting off 'the princes of property.' A. has the draft in our joint handwriting." They sent it to Moley in Portland, where the candidate was giving another speech laying out his ideas on power. A few days earlier, the U.S. attorney's office had moved to investigate Sam Insull's network of utilities companies; the aim was to uncover any criminal violations of federal law, and to that end, authorities sifted through postal documents and tax returns. Now, in the Portland speech, Roosevelt convicted Insull before a trial, referring to the "Insull monstrosity."

Within days the candidate Roosevelt gave the second, more general speech in Hoover's own home territory, downtown San Francisco. Laissez-faire had functioned, he argued, as long as the American frontier was open. "Depressions could, and did, come and go; but they could not alter the fundamental fact that most of the people lived partly by selling their labor and partly by extracting their livelihood from the soil, so that starvation and dislocation were practically impossible." But now, building on the ideas of his old professor Frederick Jackson Turner, Roosevelt reminded his listeners of the importance of the fact that "our last frontier" had long since been reached. It was time for the "princes of property," the wealthy, to share their resources. Growth would not provide for the poor; only redistribution could.

The new situation had made the trusts all the more dangerous. The speech cited Berle's academic work:

"Recently a careful study was made of the concentration of business in the United States. It showed that our economic life was dominated by some six hundred-odd corporations who controlled two-thirds of American industry. Ten million small business men divided the other third." Many of those big magnates were not acting in the citizens' interest. The Berles had written of an emblematic "Ishmael," the malevolent figure in the economy who turned against the small man. Roosevelt changed the text so that it also, again, attacked Insull. He spoke of the "Ishmael or Insull, whose hand is against every man's."

This vision was a darker one than had prevailed in the 1920s. Where Americans—even the very poorest of Americans, such as Father Divine's constituent souls—had believed in a future of plenty, Roosevelt believed in a future of scarcity. The paradox was that he presented the message in a framework of optimism, to the music of the tune "Happy Days," and with, simultaneously, an unspoken offer of an end to Prohibition. More and more voters began turning to Roosevelt.

Willkie gave up on Baker and lined up behind Roosevelt, donating $150 to the governor's campaign. Willkie's industry suffered too with the Depression—C & S dividends had dropped from seventy cents to one cent—and Willkie thought he might team up with a new administration to change things. In his own state, Indiana, families were losing their farms; later Willkie would buy up distressed farms and have family members, friends, or employees turn them around. If Roosevelt was prosecuting someone else in his industry—Insull—that was perhaps just as well.

That same September, Treasury Secretary Ogden Mills traveled to a key state that held its congressional and gubernatorial elections two months early—Maine. Mills had been at Harvard like FDR, and he would try to woo back the voter by reminding him of the other definition of the Forgotten Man. Reading aloud from Sumner's original essay, Mills recalled Sumner's "forgotten man" and pointed out that he was different from Roosevelt's. The true forgotten man was not "inert and helpless" but rather "the backbone of the nation" and certainly not "industrial cannon fodder." Then Mills asked the voter not to change horses while the country rode through a storm. The remarks fell flat. Democrats did so well in Maine that their main debate point, among themselves, was not whether Roosevelt would win the country but by how many millions.

The prospect of a Roosevelt victory had a mixed meaning for the national economic outlook. Inauguration in those years did not come until March. Before,

Hoover had assumed that did not matter, because he assumed he would be reelected in November. But if Roosevelt won, for four long months no one would be in charge—a devastating situation for already teetering banks.

Perplexed, yet courageous, American towns and neighborhoods rallied one more time. They made a last effort to solve the money problem on a local level. Salt Lake City had now gone further than barter. The townspeople had banded together and created a group, the Natural Development Association (NDA), that made its own money. They had given their unit the reverberating name of the vallar.

Citizens could work to earn vallars. They came in different denominations: V5, V10, V15, V20, and V25. They then in turn could use those vallars to buy and sell oil, soap, coal, food, furniture, meals at a restaurant, and even medical treatment. A music company had sold forty pianos for vallars.

The vallar traders created their own newspaper, the *Progressive Independent,* whose masthead blared its purpose: "A New Economic System: For Human Welfare, Man Above Money." Salt Lake City banks cleared the new money. By the end of 1932, some 10,000 people would be, somehow or other, in the vallar system.

Salt Lake City had a "daylight restaurant," because the local power company would not accept the scrip.

Ventura, California, Minneapolis, and Yellow Springs, Ohio, were all also making some form of scrip. In Arizona, the state's governor enforced a three-day bank holiday—banks were closed. The legislature, by a special act, ordained a state scrip, to be issued in denominations up to $20. Three million in scrip was to be lithographed by a private firm. As it happened, this scrip was not used. But that did not stop other private firms from generating their own. The *Nogales Herald*, which had the advantage of possessing its own printing press, issued its own bills. In areas near the border, Mexican pesos began to trade at a premium; the peso, at least for a short moment, had become another form of American money.

Obscure nonprofits and citizens' groups, towns and businesses, were all creating money: the Business Men's Club of Oak Hill, West Virginia, issued coins. The Lane Bryant Store issued money in Indianapolis, not fifty minutes from Willkie's hometown. City and state governments also got into the money business. The state of Washington issued money, as did the Port Authority in New York and the Village of Chatham, New York (its money was orange). Sometimes towns paid their workers in scrip; Hawarden, Iowa, was one. A clothing store owner reported that he "even paid a life insurance premium to a Hawarden agent in scrip." In Inwood, a "Mutual Exchange" on upper Broadway created credit tokens for trading among other exchange members. Thus, for example, a farmer near New Hope, Pennsylvania, asked the exchange to send him workmen. They came and were paid for their construction job with 134 bushels of winesap apples. Some of the apples were then sold for credits, in order to increase the credits' circulation; some were even converted to jelly, promptly labeled "Barter Brand." "Barter apples, the best you ever ate."

In Manhattan itself, there was the Inwood office's affiliate, the "Emergency Exchange of New York City." Its board was supervised by an individual no less respectable than the assistant chairman of the New York State Power Authority, Leland Olds—who had also had a hand in FDR's speech about power, in Portland. The exchange opened its first temporary office at 52 Vanderbilt Avenue, just around the corner from Grand Central Station. People who owned homes were losing them because the barter did not help with the bank, but at least the emergency exchange might get them something to eat. One of the economists to join those working on the barter money project in New York was Stuart Chase.

The barter systems kept growing; by the spring there would be some 150 barter and/or scrip systems in operation in thirty states. Tens of thousands of people—perhaps hundreds of thousands—used barter money. Barter enthusiasts claimed the number of those engaging in some form of barter had hit a million.

The notion of scrip seemed enormously satisfying. Holding one of the slips of paper, one could feel the pleasure of people who, lacking a basic thing once supplied by a faraway bank, had now crafted it for themselves on a small scale. There was also the pleasure of establishing value where there had been only paper before. Citizens liked the idea of reverting to pioneer mode, of confronting economic problems and working them out themselves.

Still, trading in kind, especially when one did not live on a farm, did not feel like progress. Even vallars could not keep mortgage holders from losing their homes. People were beginning to realize that the problem was simply not something they could solve in the neighborhood, or even in the state. The hour of the vallar was merely that—an hour.

From Northampton, on November 7, Coolidge did his duty and issued a final plea for reelection. "All the teachings of common sense require us to reelect Hoover," he told radio listeners.

As for Roosevelt's advisers, they were now in an intense negotiation for jobs in his administration. Henry A. Wallace, the son of the old agriculture secretary, published a puff piece in the *New York Times* about the merits of Roosevelt's outreach to farmers; Hoover was staged, while Roosevelt had "genuineness," he wrote. There was another set of Henrys: Morgenthau, a Democratic leader, badly wanted his son, Henry Jr., to have a role in the new government. Henry Jr. was trying out a life at farming, in Roosevelt's Dutchess County. Both men hoped that Roosevelt would name Henry Jr. to the post of agriculture secretary as the second Jewish cabinet member, the first having been Oscar Straus under Teddy Roosevelt. The chances looked pretty good—the Roosevelts and Morgenthaus were neighbors in Dutchess County.

At Albany, and in New York, other job applicants were gathering. Lilienthal, perhaps unaware that Frankfurter himself had been playing catch-up as recently as the convention, nursed his contact with Frankfurter. He also tried to build connections with any other New Dealers he could throughout the 1932 campaign. October found him in Washington interviewing with Berle, the speechwriter. Lilienthal also saw Frankfurter and Justice Brandeis. And that same month Lilienthal delivered a speech—a sort of public job application—arguing that holding companies represented "a social loss to the community." Frankfurter reported back that his words had earned "high praise." Whatever came after the election, Lilienthal wanted to be part of it.

Early in November, Roosevelt defeated Hoover by seven million votes. The victory was so thorough that Hoover was shocked. Even Boulder City, the town created by the project that he had brokered, wanted him out: 1,620 out of 2,074 voters went for Roosevelt. Hoover nonetheless headed out west to look at the one project he still had the most hopes for. Around the world, the papers trumpeted the new U.S. president. Among the most pleased were the Soviet leaders. "Russians Hopeful of 'a New Deal,'" ran the headline of a dispatch from *New York Times* correspondent Walter Duranty on November 10. The Soviet Union's leaders hoped that people like the travelers would now convince FDR, finally, to recognize Russia.

But in America, at least at the outset, there was little sense that Roosevelt would create a leadership of the Left. Instead the view was that this crisis would need men from all political backgrounds. Now a number of Wall Streeters joined the brain trust and gathered around the new president, knowing that only Washington and Wall Street together could fix the monetary situation. Among those who visited with the president were James Warburg, the young banker who had walked away from Hoover's Commerce Department offer, and Alfred Lee Loomis, the venture capital man. On Thanksgiving, the victorious FDR struck an especially intimate tone with his radio audience when he referred to a section of the Episcopalian's Book of Common Prayer that was intended for worship within the family: "Remember in pity such as are this day destitute, homeless, or forgotten of their fellow-men." The effect of the phrase was subtle; Roosevelt was suggesting his motto had a provenance—not the Sumner essay, but the prayer book. Sumner, himself trained as an Episcopalian clergyman, would have approved heartily of the use of the phrase in a religious context, meaning the man deserving of charity, for example. But Roosevelt was not so much thinking of religion as allowing that the religious impulse of charity should find expression in the political sphere.

Hoover, meanwhile, was still trying to absorb what had happened to him. He paid a surprise visit to the Hoover Dam—"I never in my life saw a man look so worn out," recalled an observer at Boulder City. Recognizing, better perhaps than any man, that something had to be done over the course of the winter to slow an alarming spread of the bank crisis, he made a point of showing himself to be a good soldier. He wrote memos. He telegraphed Roosevelt to ask for a meeting on the international debt problem—but also, really, in the hopes the two could be partners over the coming months. His moratorium was coming to an end and Britain and France were begging for reconsideration. Hoover, his

Treasury Secretary, Ogden Mills, Roosevelt, and Ray Moley met at the White House, all so nervous that they smoked—Hoover had a fat cigar. Fixing his eyes on the presidential seal, or occasionally Moley, Hoover spent an hour reviewing the problem, and while they agreed on some things, he did not convince Roosevelt to join him in a plan to work together throughout the interregnum.

But there was always hope Roosevelt would shift. While Hoover waited, he laid the cornerstone for yet another large building—the new Labor Department; the masons who helped him used the same trowel and gavel used by Washington himself at the construction of the Capitol in 1793. He spoke of the wonders of American labor, which functioned well in a country with no "class distinctions."

Early in January Roosevelt joined Senator Norris at Muscle Shoals, a clear signal from him that the Tennessee River was replacing Niagara Falls in his thoughts.

Days later, Coolidge died suddenly at his home in Northampton, Massachusetts, the Beeches. Mellon, aboard the SS *Majestic* on one of his transatlantic crossings, sent a radio note of condolence: "Coming in the vigor of his activity, it is indeed a loss for all our people." The effect of the death was to seal the conviction that the country could not go back to the 1920s, even if it wanted to. Willkie at this time had some news of his own. He became president of Commonwealth and Southern. He was also doubtless concerned about what was going on at home: farmers from Indiana, like those from six other states, were striking at the state capitals to win moratoria on their crushing debt and tax burdens.

The uncertainty of the interregnum took its toll. When the banking crisis grew yet worse, Hoover tried contacting Roosevelt, even sending at one point a lengthy personal letter. Historians would later note that Hoover's sense of urgency even showed up in the way he addressed the letter, misspelling Roosevelt's name "Roosvelt." Roosevelt was cruising the coast of Florida; Hoover expected to meet on his return. But Roosevelt was not interested in cooperation. We will never know all his motives, but it was clear that a crisis now could only strengthen his mandate for action come inauguration in March. Hoover became incensed at the silence, and took to documenting his own goodwill in the name of an accurate history. In late February, one of the lowest moments—as the Dow stood at just above 50—a manufacturer would leave a phone message for Hoover. He had had a meeting with Tugwell, who had confirmed that the new administration had no interest in cooperating. Hoover wrote a formal letter to the manufacturer to put on record what the man had reported to his secretary: "I beg to acknowledge your telephone message received through Mr. Joslin, as follows: 'Professor Tugwell, adviser to Franklin D. Roosevelt, had lunch with me. He said they were fully aware of the bank situation and that it would undoubtedly collapse in a few days, which would place the responsibility in the lap of President Hoover.'" Hoover added his analysis: "When I consider this statement of Professor Tugwell's in connection with the recommendations we have made to the incoming administration, I can say emphatically that he breathes with infamous politics devoid of every atom of patriotism. Mr. Tugwell would project millions of people into hideous losses for a Roman Holiday."

The brain trusters were ferociously busy and in an expansive mood, a fact that their opponents were quick to take advantage of. Hoover was not the only one to quote remarks that Tugwell imagined he was making in private. In addition to his lunch with the manufacturer, Tugwell had granted an interview to a journalist who seemed to him "a kind of minor Lincoln Steffens"—a muckraker. He had not yet had too many experiences with newspapermen, and told the man "we were talking off the record." The journalist interpreted this to mean he could write what Tugwell said and attribute it to him, but not quote it. On January 26, the *New York World Telegram* printed Tugwell's thoughts, calling them "an authoritative outline of what the new administration plans": "Drastically higher income and inheritance taxes," large changes in agriculture, and other dramatic proposals.

"I was appalled," Tugwell wrote—not about the content of the article so much as about the high position he appeared to give himself, and about betraying Roosevelt's plans too early. But Moley reassured him, he later wrote, that these things happened. Besides, Moley added, what Tugwell had laid out "wasn't a bad program." But in his diary, Tugwell made a vow: "I shall never trust another reporter."

Tugwell suggested that as agriculture secretary Roosevelt appoint Henry A. Wallace. "Rex, I really ought to be working for you," Wallace said when he formally offered Tugwell the job of assistant secretary. Stuart Chase was not in the Roosevelt coterie but was finding a role as the brain trusters' scribe and herald. In February 1933 Chase was traveling the country, arguing that purchasing power would help to restore the country. In the restaurant of the Hotel Utah at Salt Lake City—one of the scrip towns—he met Marriner Eccles.

Over lunch, Eccles asked Chase about the brain trusters, and Chase asked Eccles for his views on the economy. Eccles lectured Chase on the need for spending to forestall inflation. Deficit spending could conceivably slow or reverse a downturn. This argument was a limited one, especially for its time. Even in 1932, government spending from Washington was such a small share of the economy that increasing it would not matter much. Still, Chase believed the idea that the economy needed more money was essentially accurate. "Why not get yourself a larger audience?" Chase, impressed, asked.

Eccles already had a date to testify before the Senate Finance Committee. He would tell the committee that Washington needed a $2.5 billion spending plan, as well as higher estate and inheritance taxes. But Chase offered something that might lead to a longer stay in Washington than the overnight visit of a hearing witness: a letter of introduction to Tugwell, still in New York.

Tugwell agreed to receive the banker from Utah at Columbia; he was, as Eccles later reported in his memoirs, late for the meeting because of a dentist appointment. The pair headed off for lunch at a drugstore booth. Eccles was amused by the venue choice; "the setting of food and pills was appropriate to the talk we had about the nation's ills." Tugwell was gloomy, and his gloom seemed justified: by inauguration day, a few days hence, most banks in the land would be closed or under some form of restriction. The early months of 1933 were seeing some of the worst joblessness of the slump—or in memory. At that point Washington did not quantify unemployment as it does today, but conservative estimates suggest that then, in the fourth year of the slump, up to three in ten workers were unemployed.

In January, Germany held elections; this time, it looked as if Hitler would indeed be able to form a government. At the end of the month, president-elect Roosevelt received a letter from 800 professors and university presidents, a missive from the academy on a scale with the letter Hoover had received in regard to Smoot-Hawley. This time too the issue was the "critical world situation"—a phrase that referred to Germany but also, likely, to trade. The signatories included George Counts and John Dewey. The proposed solution was the academic club's boldest gesture on behalf of the Soviet Union taken to date: Roosevelt should do what preceding presidents had not, and recognize Russia. Recognition became the central news story about Russia, obscuring other events there, including the news that Stalin was moving forward in the North Caucasus with the collectivization of agriculture.

William Green of the American Federation of Labor instantly repeated his opposition to the prospect. But the world was changing. The same day the *New York Times* carried the report of the professors' petition and Green's objections, January 30, the paper also told of the meeting between Germany's president Hindenburg and Hitler that would lead to Hitler's ascent as chancellor. Several months later Mrs. Corliss Lamont, the daughter-in-law of Thomas Lamont, one of the top executives at J. P. Morgan, would announce the creation of an additional committee to recognize Russia. Another group of women also signed a recognition petition that went to the president—signers included Mrs. Lorado Taft, Paul Douglas's new mother-in-law, and Jane Addams of Chicago, as well as Amelia Earhart, Ida Tarbell, and Irita van Doren in New York. It was not yet clear whether Roosevelt would actually act, but it was clear that he would take the idea of recognition far more seriously than had Hoover, Coolidge, Harding, or even Wilson's secretary of state, Bainbridge Colby.

Meanwhile, as a cultural accompaniment to the political theme of the Forgotten Man, Warner Brothers was readying a film for the inaugural year, titled *Gold Diggers of 1933.* At the end Joan Blondell sang a song about what the Depression had done to World War I doughboys:

Remember my forgotten man,
You put a rifle in his hand;
You sent him far away,
You shouted, "Hip, hooray!"
But look at him today!

Over the lengthy months before the March inauguration, and certainly through the weeks after, the country indeed saw Roosevelt as a savior—as his voice on the radio, and the money problem, had convinced them to do.

"What matter," Prince Edward would ask of Britain in these years, "if some trifling blunder is committed here, or some project fails there? The very attempt of the community to achieve some social betterment for the sake of the workers in their midst will lift the general level of hope and make easier every national solution by statesmen and economists." The Britons placed their faith in royalty, and the Americans were placing their faith in Roosevelt. The scrip makers began trying out new designs. In at least two places, Albion, Michigan, and Evanston, Illinois, the new faith would shortly become explicit. The scrip craftsmen there placed a likeness right on the front of their paper bills: the visage of FDR. In February, assassin Giuseppe Zangara shot at the president-elect in Florida. The bullet hit Chicago mayor Anton Cermak instead, and Cermak would die of the wound. Yet FDR bounced back quickly. "I have never in my life seen anything more magnificent than Roosevelt's calm," Moley wrote later. Moley, a criminologist by training, interviewed Zangara and determined he had acted alone. Roosevelt refused to let the Secret Service introduce a protective glass barrier between him and the crowds at the inaugural parade. Mrs. Roosevelt for her part opened herself to the country too, inviting citizens to write her once she arrived in Washington. On an auspiciously fair inauguration day, FDR pronounced his now famous phrase, "The only thing we have to fear is fear itself." Here was a president who would not barricade himself at Rapidan. He was right there with them in this time of crisis. The pilgrims could not believe their luck. Tugwell, Chase, Moley, Berle, Frankfurter, and the others had waited so long for a chance to try out their ideas. Now they had one.

5
The Experimenter

October 1933
Unemployment: 22.9 percent
Dow Jones Industrial Average: 93

They met in his bedroom at breakfast. Roosevelt sat up in his mahogany bed. He was usually finishing his soft-boiled egg. There was a plate of fruit at the bedside. There were cigarettes. Henry Morgenthau from the Farm Board entered the room. Professor George Warren of Cornell came; he had lately been advising Roosevelt. So did Jesse Jones of the Reconstruction Finance Corporation. Together the men would talk about wheat prices, about what was going on in London, about, perhaps, what the farmers were doing.

Then, still from his bed, FDR would set the target price for gold for the United States—or even for the world. It didn't matter what Montagu Norman at the Bank of England might say. FDR and Morgenthau had nicknamed him "Old Pink Whiskers." It did not matter what the Federal Reserve said. Over the course of the autumn, at the breakfast meetings, Roosevelt and his new advisers experimented alone. One day he would move the price up several cents; another, a few more.

One morning, FDR told his group he was thinking of raising the gold price by twenty-one cents. Why that figure? his entourage asked. "It's a lucky number," Roosevelt said. "because it's three times seven." As Morgenthau later wrote, "If anybody knew how we really set the gold price through a combination of lucky numbers, etc., I think they would be frightened."

By the time of his inauguration back on March 4, everyone knew that Roosevelt would experiment with the economy. But no one knew to what extent. Now, in his first year in office, Roosevelt was showing them. He would present it all in what came to be known as the Hundred Days, that first frenzied period of legislative activity.

Some of the projects were mere extensions of Hoover's efforts, no matter what Hoover said. Roosevelt asked for war powers to handle the emergency, just as Hoover had suggested in a note during the interregnum. Hoover had called for a bank holiday to end the banking crisis; Roosevelt's first act was to declare a bank holiday to sort out the banks and build confidence. Now Roosevelt's team worked with Republicans to write the first emergency legislation to stop the bank runs. Hoover had had Ogden Mills; Roosevelt had another respectable man as treasury secretary, Will Woodin. Ray Moley would later write of that period, "Mills, Woodin, Ballantine, Awalt, and I had forgotten to be Republicans or Democrats. We were just a bunch of men trying to save the banking system." In this period Washington asked the two banks in Detroit, Father Coughlin's hometown, to merge, and Woodin placed or took a call with Coughlin to win his support. It succeeded, with Coughlin supporting the administration over the air. "Secretary Woodin asks me . . . ," listeners heard.

There were further commonalities. Hoover had spent on public hospitals and bridges; Roosevelt created the post of relief administrator for the old Republican progressive Harry Hopkins. Hoover had loved public works; Roosevelt created a Public Works Administration and assigned Insull's old enemy, Harold Ickes, to run it from the Department of the Interior. Hoover had known that debt was a problem and created the Reconstruction Finance Corporation; Roosevelt put Jones at the head of the RFC so that he might address the debt. Indeed, it was through the RFC that Roosevelt was making these gold purchases. Hoover had wanted to pass legislation to help farmers. So did Roosevelt. "When it was all over," Tugwell would later write, "I once made a list of New Deal ventures begun during Hoover's years as secretary of commerce and then as president. . . . The New Deal owed much to what he had begun."

Yet other projects were mere gentle departures from Hoover. Hoover had encouraged families to tend "subsistence gardens" so that they might feed themselves with their own vegetables. Roosevelt instructed Ickes to develop a subsistence homestead project where families might feed themselves on new farms. Hoover had signed a Glass-Steagall Banking Act in 1932, to expand credit; Roosevelt now prepared his own Glass-Steagall Act.

Hoover had deplored the shorting of Wall Street's rogues; Roosevelt set his brain trusters to writing a law that would create a regulator for Wall Street. The new Securities and Exchange Commission would turn the stock market from a free-for-all with hazy rules into a more comprehensible game, one in which the small player had a fairer shot. Hoover had expanded public works to create jobs; Roosevelt too would create job and relief programs. Hoover had not cared much about Prohibition, and neither did Roosevelt; he now sought an end to it.

Perhaps, some thought, Roosevelt understood about the uncertainty problem. Mellon, the departing ambassador to London, paid a courtesy call in which he discussed with Roosevelt a plan to insure bank deposits, of which Mellon disapproved. Roosevelt assured Mellon that he agreed with the sentiment 100 percent, and Mellon left confident. But Roosevelt was feeling bold, ready to create a new country, or give the impression of doing so. Shortly afterward—to Mellon's shock—he reversed himself, signing a bill that included deposit insurance. Shortly after, other differences emerged.

The main tasks Roosevelt assigned himself were simple. The first was that there be a broad sweep of activity; Americans must know Washington was doing something. If there were contradictions between experiments and within them, well, that did not matter. Partly this came out of the restlessness of the invalid; Roosevelt had risen politically but he still could not stand unaided. But partly it came out of a grandeur of spirit. "Do I contradict myself?" Roosevelt seemed to be asking, as Walt Whitman had. "Very well then, I contradict myself. I am large, I contain multitudes." The second goal was to get prices up, without much regard to whether the methods applied to achieve that goal made sense. The country was in no mood in any case to put Roosevelt's concepts under a microscope.

What mattered was change: like an invalid, the country took pleasure in the very thought of motion. Roosevelt invoked wartime powers, and to the people it seemed that he, like Abraham Lincoln in the Civil War, needed those powers.

"Dominant note of courageous confidence," summed up the *Chicago Tribune* after the president's inaugural speech. "Country will back him," said the *News* of Dallas. The *St. Louis Globe*'s headline called Roosevelt a man "who has will to do." As the comedian Will Rogers said, "The whole country is with him. . . . If he burned down the Capitol, we would cheer and say, 'well we at least got a fire started anyhow.' "

The first great project was the National Industrial Recovery Act. Its purpose was clear: to drive prices up and "put people back to work," as Roosevelt announced happily. The act established Ickes's Public Works Administration on the thesis that spending would fix the economy. Second, the act created the new labor rights that Perkins, Douglas, and Bob Wagner had been seeking, on the thesis that a worker with more pay would spend more and strengthen the economy, a theory that owed something to both Henry Ford and Waddill and Catchings. Roosevelt named Perkins his labor secretary, the first woman in an American cabinet. There was a third part that made the other two look modest:

the National Recovery Administration. General Hugh Johnson, who had spent evenings at Hyde Park, would lead this like a military campaign: a campaign in which industries and trades would join hands together with workers so inefficiencies—including labor unrest—might be diminished. The National Recovery Administration even had its own emblem: a blue eagle. The eagle was a clear invocation of war; it would inspire the economy to march.

The NRA was the consummation of a thousand articles and a thousand trends. It was the ideas of Moley, the trade unions, Stuart Chase, Tugwell, Stalin, Insull, Teddy Roosevelt, Henry Ford, and Mussolini's Italian model all rolled into one. The law worked on the assumption that bigger was better and that industry, labor, and government must work together, as in Italy, or risk staying in depression. It advocated both greater productivity and greater efficiency while forbidding price cutting, in order to nudge prices up. There was little escaping the NRA. Some 22 million workers came under its 557 basic codes. There would be a consumer advisory board, just to be sure the consumers' thoughts were included—Paul Douglas, who "rejoiced" at the emergence of the more activist Roosevelt, would sit on that board. More than a hundred industries would establish codes of business, codes that included minimum wage rules, child labor rules, maximum hour rules. There was a long roster of other requirements, including health requirements or standards regarding customer choice.

The authority of the NRA ranged widely, and the New Dealers were hoping to use it in original ways. In the oil industry, for example, prices were in collapse—a barrel of oil went for four cents in east Texas by May 1933. Under the NRA Ickes had authority to set production quotas, an authority he used to curtail supply in the name of driving up price. In other industries, the NRA rules were equally specific. NRA code determined the precise components of macaroni; it determined what tailors could and could not sew. In the poultry industry the relevant line of code had barred consumers from picking their own chickens. Customers had to take the run of the coop, a rule known as "straight killing." The idea was to increase efficiency. If smaller businesses died out, that might be for the best anyhow.

"Must we," Roosevelt himself asked the people as he announced the NRA on June 16, "go on in many groping, disorganized, separate units to defeat or shall we move as one great team to victory?" General Johnson and FDR picked an old progressive to be the NRA's general counsel: Donald Richberg, who had hired young Lilienthal in Chicago and who with Douglas had battled Insull over those tram lines.

Roosevelt and the brain trusters also created a twin for the NRA: the Agricultural Adjustment Administration, to sort out farming. Henry Morgenthau Jr. had been Roosevelt's conservation commissioner in New York State and had been "crestfallen" about not getting a cabinet post, his son later recalled—he had watched his father get the news while visiting him at boarding school. But he still had plenty to do helping out farms. At the Agriculture Department, Henry A. Wallace advocated a weaker dollar and price supports for farmers. Tugwell would serve as well, though Columbia's president, Nicholas Murray Butler, had been flatteringly reluctant to let him go, telling Tugwell he might come back after a year's leave or so. He would be Wallace's assistant secretary and work on developing his own notions of planned capitalism. Americans had already retreated to the land in some respects with the onset of the Depression: Roosevelt was now blessing that move and assuring the country that he would seek to make it financially sustainable.

Wall Streeters thought nearly any Roosevelt policy would be better than the uncertainty of the winter, and flocked around. One new adviser was James Warburg of the banking dynasty, who rode to the rescue, believing that Roosevelt needed help since "no president ever took over a nastier, stickier, more complicated mess." Of their first meeting Warburg wrote: "I vividly remember my first impression of FDR's massive shoulders surmounted by his remarkably fine head, his gay smile with which he greeted his guests, and the somewhat incongruous, old-fashioned *pince nez* eyeglasses that seemed to sit a little uncertainly on his nose." Irving Fisher, the Yale professor, had a similarly transporting experience. He had made a career of getting close to presidents; over the course of his life he had met with Teddy Roosevelt, Taft, Wilson, Harding, and Hoover. It was not hard for a charmer like Roosevelt to win Fisher over, and he did. Leaving a meeting with the president that year, Fisher would write, "It was the most satisfactory talk I ever had with a President and the most important. . . . They were all very nice to me but I never felt I got as good a reception of my message before." In April, thinking Roosevelt would go along with his plans for inflating, Fisher euphorically wrote his wife: "I am now one of the happiest men in the world. . . . I feel that this week marks the culmination of my life work." The intoxication of being with Roosevelt changed yet another life.

Another adviser was Alfred Lee Loomis of Bonbright, who had hired Willkie and sat on his board. Lew

Douglas, likewise a conservative, became Roosevelt's budget director. A final conservative was William Woodin, who filled the spot of treasury secretary: in the first scary week of the bank holiday, Woodin cajoled not only Coughlin, but many banks into staying open when he believed they should. As for the undergraduates of Vassar, Mary McCarthy would note, they were thrilled: their trustee, their own neighbor by the Hudson River, was now in the White House.

The next stage, governing, came quickly, with the largest item on the agenda being farm prices. In the spring of 1933 agricultural prices stood at 40 percent of their 1926 level; farmers threatened a general strike. Henry Wallace and others told Roosevelt that this must mean there was too much supply and too little demand. They suggested correcting supply. If farmers sold less, their prices would go up. The AAA's official purpose was to be to "relieve the national emergency by increasing purchasing power"—especially the power of farmers. Unlike many offices in the New Deal, the AAA did hire a number of true leftists, including people who later turned out to be Communists, or spies, such as Alger Hiss. The politics of the individual staff members were not especially evident—Tugwell would later write that if Hiss was a Communist, he was "so far underground that even his close associates did not know it." The AAA was itself radical, both in the scope of the legislation and the authority it gave officials. The man named to lead the AAA was George Peek of Moline, Illinois. Peek declared that the AAA was not about helping farmers; rather, it would enable them "to do something for themselves that they have been prevented from doing."

The act established that farmers deserved the level of price they had received prior to World War I, labeled the fair price. The AAA began paying farmers to produce less. The government also encouraged farmers to sell less by offering them favorable loans in exchange for restraint. To fund all these changes, the AAA taxed middlemen—distributors—an idea that made particular sense to Tugwell. To him middlemen were the profiteers in a system; if a retail price for food was simply too high, then it was the middlemen who, through their cut, were pushing up retail prices. An old folk song line seemed truer than ever: "the middleman's the one who gets it all."

Other big changes were also coming for farmers. Agriculture business required a license, and anyone who operated without one could be fined $1,000 a day. In the new system, commodity buyers might work together in ways that had heretofore been illegal. If in the first year the recalibration did not work, then the AAA would recalibrate until farming and farm prices found their balance.

At the Agriculture Department, Wallace had his concept for a mechanism to align supply and demand properly: a perpetual mechanism known as the "ever normal granary plan." When there were surpluses, farmers would receive commodity loans to keep those surpluses off the market and store them. In times of dramatic surpluses, the government would impose production controls. When shortage came, the routine would reverse. The whole idea had an appeal, like the story of Joseph in the Bible teaching prudence to Pharaoh.

Tugwell, now Wallace's assistant secretary, was pursing another policy project—limiting overfarming by retiring excess acres. The plows of the 1920s had loosened the fields of the plains and caused floods, including the great flood of the Mississippi. He would replant the country. He focused on conservation and argued that in any case "our trend is from cultivated crops to meadows, lawns and pastures." He wanted to end the slums in the cities, make parks of them, and relocate entire communities to new suburban centers.

Some doubted whether Tugwell would get his plans through. Only months into his job, he was already quarreling over the scope of the AAA with Peek. Tugwell irritated Peek, who told people that he thought some of his colleagues' plans and dreams too ambitious: the name of his department was the Department of Agriculture, not the "Department of Everything."

Still, Peek saw that Tugwell was close to the Roosevelts. He drove out of Washington when he could to visit Louis Howe, Roosevelt's oldest adviser, at his small country house. He traveled out to Quantico to join Roosevelt on his yacht, the *Sequoia.* He sat at Eleanor's dinner table. Eleanor's support was especially crucial, as he would later write in a memoir of the period: "No one who ever saw Eleanor Roosevelt sit down facing her husband, holding his eyes firmly and saying to him 'Franklin, I think you should . . . Franklin, surely you will not . . .' will ever forget the experience. . . . It would be impossible to say how often and to what extent government processes have been turned in a new direction because of her determination." Eleanor liked many of Tugwell's ideas—and that fact, he knew, would help him enormously.

Agriculture was only the first part of Roosevelt's experiment, and the most controlled part. There was also the gold standard, which literally made dollars expensive—and, at that point, scarce. Wall Streeters deemed the gold standard in need of adjustment. Loomis wrote a memorandum advising a more flexible version of the gold standard. If the United States devalued the dollar, raising the price it would pay for gold from its old $20.33 level to a new fixed level, that too would raise prices at home. Devaluation would also raise those farm prices, so crucial politically. Observers guessed that Roosevelt would do one of two things. He would stand by the old gold rules, or—more dramatic but still possible—devalue decisively and establish a new monetary order.

At first, Roosevelt gave signals that this guess was correct. He sent to Congress legislation titled "A Bill to Maintain the Credit of the United States Government," one which Loomis supported. Living up to its conservative name, the bill cut government salaries by $100 million, then a significant amount. It also cut both government pensions and veterans' compensation payments—a bold position, less than a year since Hoover's stinginess toward the Bonus Army had hurt him in the campaign. Congress and the vice president saw salaries cut 15 percent. What people noted most however was the extra power the new law gave the executive branch when it came to payments. "President to Prescribe Degrees of Disability," read a newspaper summary of Section 3 of the law, which laid out new rules that curtailed payment. The line seemed telling given Roosevelt's own health challenges. "Claims Once Disallowed Not to Be Reopened," read another header.

Yet a third headline seemed especially striking: "Administrator's Ruling Declared Final." It all seemed responsible—the Associated Press noted that the average age in FDR's cabinet was older than that in Hoover's by one and a half years. Representative John McDuffie of Alabama pushed the legislation through the House within two hours.

Wall Street's confidence strengthened. Roosevelt might be a less expensive president than Herbert Hoover. Whatever else the new executive did in 1933, the market reckoned now in its collective way, he really was reducing uncertainty of the lengthy interregnum as hoped. The fact that his senior advisers included Wall Street wise men was good news. Especially important was the inclusion of James Warburg. The elder Warburg, after all, had been important at the time of the founding of the Federal Reserve. Over the course of the spring, investors applauded by buying stocks, and the Dow began the first steps in what would later be known as the Roosevelt Rally. Industrial production climbed as well.

But even the conservatives understood that Roosevelt might take more dramatic action; the pressure from the West was great. Silver was more plentiful than gold. Many, especially from the West, were arguing for a currency backed by silver as well as gold. Bryan had died, but never had the Cross of Gold seemed more punishing or his arguments against it more compelling. After the inauguration it became clear that inflation of some sort enjoyed majority support in Congress, no matter what Roosevelt did. The old financial hands in the administration despaired. The news from Europe threw them off balance. Hitler was only growing stronger in Germany, a country whose central bankers they knew and whose economy they had once expected would recover. Treasury Secretary William Woodin was an accomplished songwriter. When the financial team grew tired, he would play the violin. Late in the nights of 1933 Woodin composed a tune he titled "Lullaby in Silver." The point, as he told a colleague, was "to get this silver talk off my mind before I go to bed."

Caught in a jam, Roosevelt decided to waffle. His advisers pondered the scrip in circulation and whether to put it on a par with true dollars. There were ways to loosen the gold standard without officially going off it, and he tried them first. He asked the treasury secretary to call in all the gold in the country, forcing citizens to sell their gold to the Treasury for dollars. There must be no hoarding and no exporting. Citizens must hand in their gold and take dollars in exchange. This was a weakening of the gold standard, which guaranteed one could always buy and sell gold from the government at the set price. And many gold holders had more to lose than those who had placed the image of Roosevelt on their scrip. But Woodin and other spokesmen made it clear to the press that this was not "going off the gold standard." Roosevelt called the change "temporary," and loyal Will Woodin told the press outright that it was "ridiculous and misleading to say that we have gone off the gold standard."

Next Roosevelt set to work invalidating gold clauses in contracts. Since the previous century, gold clauses had been written into both government bond and private contracts between individual businessmen. The clauses committed signatories to paying not merely in dollars but in gold dollars. The boilerplate phrase was that the obligation would be "payable in principal and interest in United States gold coin of the present standard of value." The phrase "the present standard" referred, or so many believed, to the moment at which the contract had been signed. The line also referred to gold, not paper, just as it said. This was a way of ensuring that, even if a government did inflate, an individual must still honor his original contract. Gold clause bonds had historically sold at a premium, which functioned as a kind of meter of people's expectation of inflation. In order to fund World War I, for instance, Washington had resorted to gold clause bonds, backing Liberty Bonds sold to the public with gold.

Now, in the spring of 1933, upon the orders of Roosevelt, the Treasury was making clear that it would cease to honor its own gold clauses. This also threw into jeopardy gold clauses in private contracts between individuals. The notion would be tested in the Supreme Court later; meanwhile, bond and contract holders had to accept the de facto devaluation of their assets. The deflation had hurt borrowers, and now this inflationary act was a primitive revenge. To end the gold clause was an act of social redistribution, a $200 billion transfer of wealth from creditor to debtor, a victory for the populists. Announcing the legislation before it passed, Senator Elmer Thomas of Oklahoma said, "No issue in 6,000 years save the World War begins to compare with the possibilities embraced in the power conferred by this amendment." The rich had the money, and "because they have it the masses of the people of this Republic are on the verge of starvation—17,000 on charity, in the bread line." Now the debtor would, through devaluation, see his debt reduced.

Even those in Roosevelt's entourage who disapproved went along, in the hope that all these moves were genuinely temporary, one-time events, and that if and when Roosevelt officially went off gold, he would quickly replace the old rule with a new one—exchanging one promise to deliver gold for dollars at a certain rate for another promise to deliver gold or silver for dollars at another, different rate. The operation would be like resetting a dislocated shoulder—painful but quick, and followed by the euphoria of relief.

Yet the cries from the West did not abate. And Roosevelt felt he must do more. One April evening that spring—shortly before guests from Britain were expected—he met with members of his cabinet. Secretary of State Cordell Hull was there, along with Treasury Secretary Woodin, Herbert Feis (a holdover from the Hoover administration), budget director Lew Douglas, Ray Moley, now at the State Department under Hull, Jimmy Warburg, and others.

"Congratulate me," Roosevelt suddenly said to them. He had gone off the gold standard. There was a political plan—by supporting the Thomas Amendment in Congress, which allowed him to set the price of gold, he would signal to farmers his strong backing for higher prices. But there was no economic plan for what to do now. The men in the room reacted so strongly as to make themselves seem ludicrous: "They began," as Moley recalled in amusement, "to scold Mr. Roosevelt as though he were a perverse and particularly backward schoolboy." First of all, they reacted simply at the shock of what he had done—the country had not been off the gold standard in peacetime since before the turn of the century. But far more unusual was the way the president had done it. Why was there no new price of gold? This move was different from simple devaluation. Instead of reducing uncertainty, the president was increasing it.

Roosevelt was merry, which gave him the advantage. He even teased his guests about their devotion to the gold standard. Even the gold standard had worked, after all, only when men said it must work. Cordell Hull, Moley later recalled, looked as if he had been stabbed in the back when FDR removed a ten-dollar bill from his pocket, examined it, and said "Ha!"—the bill was from a Tennessee bank—"in your state, Cordell. How do *I* know it's any good? Only the fact that I think it is makes it so."

The country seemed to like Roosevelt's attitude. Someone had to deal with the money problem, and he had been brave enough to do it. The papers were reporting that Roosevelt had already, in one way or another, put more than a million people back to work. The Dow was now heading toward the 90s, up from its tiny base. Legislators and southern agricultural commissioners were already busy quantifying the amount of acreage to retire—10 million acres of cotton fields, for example. Farmers began receiving their payments. Peek would be able to announce that checks to a million farmers to pay $110 million on their contracts to take more than 4 million bales of hay out of production had already been sent. To many, this seemed odd, outrageous even. The $110 million that went to farmers more than offset the $100 million in savings the government had gained by cutting its employees' salaries. In a year of hunger—the year that the pair had starved in the cabin on the New York lake—food production was cutting back, and additional food was being withheld.

Still, Peek was calling this "the most amazing period in the history of American agriculture." Of course, the political solution of buying off the farmers in this way was a short-term one. Come spring, the farmers would still be there, and they would want payments again. But this did not seem like an unattractive change, at least politically: it gave the politicians an annual chance to rescue their farmer constituents.

Roosevelt, for his part, was still busy with the money question, and with an international economic conference to take place in June. In a style that his advisers were already beginning to recognize as typical, he sent emissaries with differing missions to the conference. One was Hull, a great free trader, who imagined that the conference would be an opportunity for the United States and other nations to talk about mutual tariff reduction, undoing some of the damage of Smoot-Hawley.

That in turn could open the way to recovery. Roosevelt was also sending his old professor Oliver Sprague, now an adviser, to the Bank of England. Another emissary was James Warburg, who was to work on a new monetary agreement, the sort that would unite countries so that they would be less likely to put up tariff barriers in the future. The best possibility, or so traditionalists among his advisers believed, was that together with Britain, which had earlier gone off gold, Roosevelt's representatives might put together a new international gold standard after resetting the dollar—at, of course, a new, fixed level.

Hull and Warburg each traveled to London in the hope of realizing their projects. Both men believed the president backed them. Besides—and thank heaven—freer trade and a stable dollar were two goals that need not contradict each other. Peace and prosperity would be negotiated by sixty-six nations in June at the Geological Museum in Kensington. The journalist and historian Ernest K. Lindley would later write that the Roosevelt administration now appeared "in the eyes of the public at home and abroad as the chief sponsor of the World Economic Conference and the international approach to recovery." Hull prepared a strong speech about how the era of economic nationalism had passed.

In London, George Harrison of the New York Fed, Hull, Warburg, and others promptly began work on a plan to maintain exchange rates among the United States, the UK, and France, which was still on the gold standard. The move would be a return not to the U.S. gold standard but something like one, with France's golden franc as anchor.

Still, Roosevelt had not made up his mind about what outcome he sought. Shortly, he sent a message that crossed the Atlantic like a torpedo: he would not accept a France-UK-U.S. agreement as his own team had negotiated. The participants reeled. Building consistency and trust was what such summits were all about, yet their president was now undermining his own emissaries. After a two-hour conference with Moley aboard the schooner *Amberjack II* the president next made a show of dispatching Moley to London. Moley took out of the meeting that the president did want to stabilize. The press speculated that Moley was to replace his boss, the free trader Hull. Moley's job was to suggest a third plan: prices must move up globally. The president had told Moley that when he arrived in London, "the essential thing is that you impress on the delegation and others that my primary international objective is to raise the world price level." The U.S. delegates in London were confused; one spoke out for low tariffs, another for higher tariffs. After Moley arrived, Roosevelt telegraphed yet a third and a fourth policy position, all variants on the question of monetary arrangements, budgets, and international relations. Finally, the representatives in London cobbled together a document that they believed addressed Roosevelt's concerns for the struggling farmer. In fact, as Moley later recalled, "the most fanatical inflationist could not have objected to this statement." And, after all, Roosevelt had suggested to him on the *Amberjack II* that he would stabilize. Now Moley and the Europeans expected Roosevelt would sign off on the document—indeed, the Europeans were ready to agree to almost anything they thought Roosevelt would sign. Roosevelt rejected it.

Still, neither Moley, nor anyone else, reckoned on the last and most dramatic Roosevelt missile, the one later known as the "bombshell." It was sent from aboard the steamship *Indianapolis.* The president now attacked the effort to return to the gold standard—the very effort that, the public and his advisers believed, was the purpose of the London conference—as a "purely artificial and temporary experiment." The effort, Roosevelt said, was a "specious fallacy." International cooperation did not matter. The only thing that mattered was "the sound internal economic system of a nation." It was wrong to be misled by "fetishes of international bankers"—a clear reference to Montagu Norman of Britain. Reduced government costs at home and other domestic concerns were what was important to the United States.

Roosevelt's vacillation had partly to do with trouble at home. So many families were defaulting on their mortgages that he was now contemplating a home version of Hoover's debt moratorium, a national plan to prevent any more Americans from losing their homes. What he felt he needed was to buy time, and if stalling on the international monetary question helped that, he would do it. And he was a comfortable punter—as he would confide later to Morgenthau in regard to another debate, "Strictly between the two of us, I do not know. I am on an hourly basis and the situation changes almost momentarily." But the president was also inconsistent because he saw no cost to being inconsistent. Lately, moreover, he had been listening less to Wall Street. Increasingly, a new man had his ear: George Warren, a professor of agriculture to whom Morgenthau had introduced him. (The threesome had apparently discussed trees in their first meetings. "How different life would have been had Franklin and Henry not met those arboreal experts!" Mrs. Morgenthau is said to have mused at one point.) Warren was telling Roosevelt that domestic prices were the most important thing. If he could get those up while paying lip service to other goals, the rest of the U.S. economy would follow, and so would the rest of the world.

In Britain, John Maynard Keynes declared that Roosevelt was "magnificently right." Perhaps if the United States and Britain "put men to work by all means at our disposal," then prices would rise to a level to match existing government debts. Also pleased were western lawmakers—finally, a president who was concerned about deflation, and not Wall Street.

But many Americans, and most of the British leaders, were furious. The United States might want to grow, but it could not grow without foreign markets and stable exchange rates. Oliver Sprague disapproved so intensely that only loyalty kept him from resigning. "What is it that occurs in the Holy Writ about keeping the hand to the plow?" he asked newsmen. Ramsay MacDonald, UK prime minister, threw up his hands, telling Moley, "I give up now. I can do nothing." The king told people that he would not have foreigners "worrying my prime minister this way."

The Europeans felt that Roosevelt had made fools of them by inviting them to negotiate with his envoys and then undermining both sides with a contradictory statement. The advisers were in agony—it was as though the doctor had jerked at the dislocated shoulder but stopped short of setting it right.

Warburg composed an icy note to Hull, who was staying at Claridge's, announcing his resignation to the secretary. He had flown for the navy in World War I, and he selected images that would mean something to Roosevelt: "We are entering upon waters for which I have no charts and in which I therefore feel myself an utterly incompetent pilot." Other economists were aghast. So were the statesmen. Hull begged his subordinate Moley, who was closer to the president, to ask the president one thing: "not to change his policies again, because his sudden turns had been exceedingly embarrassing." Lindley heard one correspondent call Hull "a stricken man."

The Dow crossed 100, then dropped back to around 90 at the end of July. The stock market knew that London, Paris, and Berlin mattered. Roosevelt had declared himself an internationalist, but he was preventing the sort of internationalism that the market believed would help recovery.

Over the summer, Roosevelt increasingly concentrated on farms and the land. In July, while the monetary conference was still dissolving, came the news that the first Civilian Conservation Corps camp—Camp No. 1 at Big Meadows, Virginia—was up and running. The Labor Department found young men to work there; the army ran the camp; and the Forest Service managed the woodsmen. In August, Tugwell, Wallace, and Ickes met up with Roosevelt for a day of inspection. They ate lunch, as Tugwell later recalled, at a rough board table; Ickes recorded in his diary that they ate "so-called apple pie"—it was not his responsibility, since it came from the army. Roosevelt and Tugwell were happy. Their own farm dreams were now being shared with hundreds of thousands of young men, coming off the street to live in woods and forests for the first time. The camp was one of more than a thousand that would rise up in the coming years. The CCC would serve youths and men who otherwise would be unemployed; CCC programs would also give hope and work to the sorry Bonus Army marchers and other veterans.

That same month Roosevelt was already claiming success on the farm front. Supply had been curtailed, and agriculture prices seemed to be duly moving up. Lately, he had spent more and more time with Warren of Cornell. Now he and Warren together presented to the press a chart demonstrating that the price of agricultural goods had already returned to the level of 1914. As for Henry Wallace, the name most associated with the AAA, he was popular. The *Des Moines Tribune* would shortly suggest that Wallace might shepherd the agricultural sector out of "the wilderness of mortgages and overproduction." He had proven that Washington could really do something for the farmer. Wallace was in his forties, and he had time yet "to inscribe his name among the greats of the nation's history."

Roosevelt believed he was also making progress on the money issue. By September, the Dow was back over 100 again. After taking advice from dozens of anxious bankers and lawmakers, Roosevelt finally found a monetary plan that he liked: and it came not from the Fed, or Wall Street, but from the professor, Warren.

The plan was beautiful for its simplicity. The government, the president, using his new war power, would buy gold on the market. That would drive the price of gold up and the value of the dollar down. This was a matter of supply and demand. And if the price of gold went up, then so would the prices of everything else, especially farm goods.

The theory had some validity. With the gold purchase program, the White House was introducing cash into a cash-poor economy. Finally, the government was making an attempt at correcting the money shortage. Reflation generally was what Irving Fisher had been begging the president to undertake, this was why Fisher was so enthusiastic. And there might indeed have been an increase in the overall price level subsequent to such action had the country still been on the gold standard, under which gold had set the worth of the dollar, and therefore everything that dollar bought.

But Roosevelt himself had snipped the link to gold. For the moment, gold was just another commodity whose price seemed too low. More important, the increments of the money that Roosevelt was spending on gold in the Warren case were so small that they could not affect the overall economy by themselves. What Roosevelt was doing, under Warren's tutelage, and to Morgenthau's applause, was like pouring glasses of water into the ocean in the hope of raising the sea level. The discretionary aspect of the project, however, was the worst thing about it. On money, the executive needed to send a clear signal—pick a certain price—if markets were to follow him. Fisher's goal was establishing stability, not undermining it. Here FDR's playfulness was at its most destructive.

None of the Wall Streeters who had originally rallied to Roosevelt's side—not Warburg, not Woodin, not Dean Acheson, undersecretary at the Treasury—liked the plan. The president had needed emergency powers over money—but again, to set the money right with one quick twist, not to prolong the agony all autumn. Some of the brain trusters were also puzzled. Moley was surprised. Tugwell, still an economist for all his dreaming, was frustrated at the illogic and at the rising power of the two Henrys, Wallace and Morgenthau: "The catch in the plan had seemed plain enough. It was that the currency, in fact, was no longer so closely related to gold that the value of dollars would be greatly affected by what happened to its price." The UK press was more direct. The *Times* of London wrote that "most businessmen in the United States and elsewhere regard his currency policy, in so far as they understand it, as impracticable and likely to result in a loss of confidence in American currency, and thereby to hinder rather than assist the process of recovery." Editorialists at the *Times* wrote apologetically, trying to cover for Roosevelt: "In a sense, all currencies are 'managed.' . . ."

But Roosevelt, like many bosses, was choosing to change advisers rather than hear criticism. By that autumn Sprague was resigning noisily, the news making page one in the papers. "Sprague Quits Treasury to Attack Gold Policy," read a headline. "It's a free country," shot back Morgenthau, now becoming the "go-to" man. "I think father wanted to be his own Secretary of the Treasury," Roosevelt's son James would later write. Roosevelt told Morgenthau he wanted to buy gold to force up the price. Morgenthau—who of course knew Warren as well—asked whose idea the project was. "Mine," the president blithely answered. Woodin was ill, and beginning to fade from the scene.

In a last-ditch effort, Warburg traveled to Hyde Park and debated Warren at a luncheon attended by the president and presided over by Sara Roosevelt, the president's mother. Each man talked about his ideas; the Roosevelts listened. Warren and Warburg left Hyde Park together, sharing a taxi. "Well, I guess you ruined my plan," Warren said to Warburg. "On the contrary," Warburg answered. "You have won. Wait and see." Warburg was correct. Shortly thereafter Fisher was at Hyde Park as well, and Roosevelt told him he would have none of Warburg's gold-standard talk. Fisher wrote a letter to his wife documenting the president's words about Warburg: "J. W. wants me to fix a definite price of gold, etc., as people now can't make future contracts," Roosevelt had reported. He went on: "I said 'that's poppy-cock. The bankers want to know everything beforehand and I've told them to go to h——.' " Exhilarated, Fisher transcribed the whole meeting at the Poughkeepsie train station. Roosevelt was leaning decisively to Warren, and Morgenthau would help him do that. The dollar was now like a sail flapping in the wind, and Roosevelt was sure he knew how to bring it in line.

Over the course of the autumn, the results of the first big farm experiments were coming in, and they were mixed. Morgenthau, by the end of the year, was lending $1 million a day, a total of $100 million, at low interest rates. That was helping some farms. But neither the gold price experiment nor the AAA was having the hoped-for effect on prices. The government, as even Fisher had to admit to himself, had tried to address a macroeconomic problem, money, through a micro-economic format, tinkering with supply and demand for agricultural goods. "It's all a strange mixture," Fisher wrote to his son in Europe. His joy at contact with the president was moderated by his perception of Roosevelt's illogic. "I'm against the restriction of acre-age and production, but much in favor of reflation. Apparently FDR thinks of them as similar—merely two ways of raising prices! But one changes the mon-etary unit to restore it to normal, while the other spells scarce food and clothing when many are starving and half naked!" The subsidy might help farmers, but it could not help the overall price problem.

And it was not even helping all the farmers. To larger farmers, the new AAA payments were welcome. Food and cash from another New Deal agency, the Federal Emergency Relief Administration, reached many of the poorest farmers. Small-farm owners, however, found the AAA regimen challenging. And tenant farm-ers were stunned. Landowners had historically hired sharecroppers because they themselves made profits from their share of the crop that the tenants planted and harvested. That relationship had become more tenuous as crop prices came down, and there was less for landlord and tenant to share. The tractor, a new arrival, was already obviating the need for the sharecropper—and now the AAA was paying the landowners not to farm that land. Removing the tenants began to make sense, especially when prices for crops were still not high. The next spring a professor from the University of Tennessee would write to the cotton section of the AAA:

> *We already have many cases before us in which evictions have been ordered, apparently in direct defiance of the contract. . . . In one case we find that the owner of a large plantation has recently acquired a new tract of land upon which lived eighteen sharecropper families. He has evicted nearly all these families. . . . The acres rented by the government are not being turned over to the tenants on the terms contemplated by the contract, if at all.*

Tenant farmers overall "fared badly," wrote one agricultural historian, "in part because the AAA programs were built around commodities instead of people."

The AAA got its first serious negative publicity after Americans learned that a total of six million young pigs were killed before reaching full size over the course of

September. "It just makes me sick all over," one citizen would write, "when I think of how the government has killed millions and millions of little pigs, and how that has raised pork prices until today we poor people cannot have a piece of bacon."

The move did drive pork prices up—a bit—but other agricultural products were not behaving despite such efforts. In October 1933 the commodity reports that Warren and Roosevelt watched so closely edged down or stayed flat. The Dow for its part repeatedly touched the 100 level it had seen earlier in the year, but refused to go much over it, retreating back into the 90s several times. Events overseas were also pressuring stocks. From Germany reports of Hitler's new detention camps continued to flow. That month a German who made it over the French border reported that Hitler had locked up the ex-head of the German Broadcasting Company, his executives, the former president of the court of justice in Berlin, and a number of former mayors at Oranienburg.

But Roosevelt was not concerned about markets. He relished the fact that he had three and a half years in office still ahead—it seemed a luxury after the two-year terms of a New York governor. Relaxed, he persevered in his dollar experiment. After all, no one, even he, could fix everything at once. And at least one element of his money experiment was working. The gold price was rising; the dollar was down to a low against gold unimagined just a few years ago. Gold now cost $31 an ounce, up by half from its level under the gold standard. Roosevelt proceeded with vigor, though he was at moments concerned for the health of the members of his administration after such a blistering first few months. When Tugwell seemed tired, Roosevelt dispatched him to look at conditions out west. He traveled to Glacier National Park by train, taking his two assistants, Grace Falke and Fred Bartlett. They saw the new CCC camps and agricultural experiments, "apple, citrus, date, wine, rubber, erosion control," Tugwell wrote in warm memory later. Falke kept notes on the trip.

On October 22, Roosevelt gave a Fireside Chat in which he told listeners that his aim was "to establish and maintain continuous control" of the dollar. The commodity market responded instantly, with wheat futures prices rising nearly 40 percent in the course of his broadcast. Historians report that in late October Roosevelt told George Harrison at the New York Fed that it was "imperative to get agricultural prices up before Congress meets and that if we did not, he was fearful of what Senator Thomas and the other inflationists might do." On October 25 Roosevelt began the gold purchase program from his White House bedroom. When his attempts to influence the U.S. gold market failed, he and Warren went for the worldwide market in their purchases.

But some observers were becoming restive. July's high of 109 for the Dow had not been repeated; a plateau pattern of around 100 seemed to be becoming a rule of life. Al Smith of New York, still a hard-money man, spoke out against Roosevelt's "baloney dollars." And toward the end of November, the city of New York staged an evening's version of the national debate over money. At Carnegie Hall the pro-gold forces rallied—professors from New York University, Wall Streeters, and once again, Matthew Woll of the AFL. Inflation, Mr. Woll lectured, was hardly in the interest of the worker—if only because of the instability it caused; the good was "to stabilize our currency as soon as possible." The same evening at the city's Hippodrome, Father Coughlin, the national radio star, hosted a pro-Roosevelt rally, arguing that Roosevelt must "be stopped from being stopped." Six thousand people were in the hall, and thousands more, who had tried to get in, listened in through loudspeakers on the street. Roosevelt, Coughlin said, had done the right thing and "brought the dollar down to speaking terms with the English pound." One of the guests at the Coughlin event was Henry Morgenthau, the father of the acting treasury secretary. Coughlin had spoken in the past publicly of "modern Shylocks." Morgenthau declared the country's policy "in excellent hands."

The news about homeowners continued to be bad. A January 1934 report would show that in all of twenty-two cities surveyed, at least two in ten households with mortgages were either late on their payments or in default; in two cities, Birmingham, Alabama, and Indianapolis, at least half of mortgage holders were in trouble. In Cleveland, the rate was six in ten households.

Keynes was nonplussed. He penned an open letter of criticism to Roosevelt, saying he understood that in the terrible state of the U.S. economy, the currency and exchange policy should be "entirely subservient to the aim of raising output and employment." Nonetheless, whatever he had written during the London conference, an uncertain and changing currency was also bad for both those goals. And the Roosevelt regime was causing the dollar to gyrate. Keynes, like Warburg with his images of the lost pilot, was so anxious to convince Roosevelt that he stretched for a metaphor he thought would impress the president. As the United States happened to be in the final stages of Prohibition's repeal that autumn, Keynes told Roosevelt that his gold purchase program looked like a "gold standard on the booze."

Roosevelt did not seem to react. To most of the country, Keynes was just another economist, and not an American one at that. By the time Keynes spoke, Roosevelt was in any case working on helping the economy by another means.

For years now, Roosevelt had been reading Duranty in the *New York Times* on Russia. The godlessness troubled him—the purge of the churches. He told Perkins about his meetings with Maxim Litvinov, the Soviet envoy. "Well now, Max, you know what I mean by religion. You know what religion gives a man. You know the difference between the religious and the irreligious person." He went on: "Look here, sometime you are going to die, Max, and when you die you are going to remember your old father and mother—good, pious Jewish people who believed in God and taught you to pray to God." Roosevelt told Litvinov that religious freedom was important if the United States was to recognize the Bolsheviks, and he told Perkins he thought he had made an impression on Litvinov.

It seemed time to give Russia a chance. In the autumn he ended a sixteen-year U.S. policy toward Russia, and recognized the Soviet Union. "Huge Trade Orders Wait Our Credits," blared the *Times* in a subheadline. The Soviet government owed the United States hundreds of millions of dollars; it would not pay all the money, but it could give the country credits with which to buy Soviet goods. At a moment when people were still hungry, the deal seemed pragmatic—as president, Americans understood, Roosevelt had to choose the lesser evil. In Danzig, a port city, Nazis that winter were tossing newspaper editors in jails; they were also taking over the courts of the Saar. Recognizing Russia was also a way of counterbalancing the Germans.

Finally, unexplainably, Roosevelt shifted on the money question. After numerous conversations on the phone with George Harrison of the Federal Reserve, he slowly edged away from his gold purchase program. Professor Warren faded. In January came proof that, without ever really conceding, Roosevelt was switching tactics: the president submitted a bill to return to the gold standard on the fifteenth. The new dollar would be $35 an ounce. Fisher grumbled at the time Roosevelt had wasted on his experiment, writing to his son in March of 1934, "Yes, the gold act was a fine birthday present [for Fisher]; but I can't help feeling that the president could have gone much faster if he hadn't mixed in so many things which were holding us back instead of getting us out of the depression." But now Roosevelt was inflating, and devaluing, and giving up on the day-to-day control. The dislocated shoulder was back in place. Those who wanted to do business with America—and in America—once again, knew the terms of the arrangement.

As it happened, the tragic news from Europe brought about an irony. Chancellor Hitler in Germany was moving past revolution to what he called a period of reconstruction, which Roosevelt could see was simple and effective retrenchment. Hitler was banning the Catholic Center Party, more evidence of descent into barbarism. The French were growing warier. As had been the case at the time of Gustav Stresemann's death, desperate foreigners sent their gold and their money to the United States. American investors who had sent their cash and gold overseas also began to bring it back. The country's gold stocks doubled in six months.

On a single heady day in February 1934, banks would report, $100 million in gold arrived at New York piers on two ocean liners, the *Paris* and *Europa.* The gold on board was, the papers noted, "expatriated capital which was turned homeward by the restoration of the gold standard." As the supply of gold reserves in the American banks rose, so would the amount of money in circulation. What Roosevelt had tried and failed to achieve with his gold purchase program would come to pass—in part because he had recognized his error, and in part because of foreign events entirely outside his control. The New Deal recovery seemed to have a chance again. Roosevelt was indeed now the one with the luck.

6
A River Utopia

November 1933
Unemployment: 23.2 percent and heading down
Dow Jones Industrial Average: 90

My name is William Edwards
I live down Cove Creek way,
I'm working on the project
They call the TVA
The government begun it
When I was but a child
And now they are in earnest
And Tennessee's gone wild.
—American folk song

In November 1933, the sound of hammers rang through a small town in Tennessee that had, until recently, been known as Cove Creek. Workmen were already completing the roof of a dorm for the workers on the New Deal's most dramatic project, the Tennessee Valley Authority.

Roosevelt might be a politician on monetary policy; he might waffle on the gold standard. But there was a question for which he already did have an answer. It was the question of control of natural resources, and especially the hydroelectric power that could be generated by dams. Here, his progressive side, his Roosevelt side, his navy side, came to the fore. As governor, he had seen the issue as one involving the state authority and public control of the St. Lawrence River. As president, he saw the matter as Washington's task. Power resources generally were too important to stay in the private sector. They belonged to the people. The president had an ally in George Norris in the Senate, who for so long had fought for public power. Felix Frankfurter, after all, had been teaching about this for years at Harvard. Most of the new president's advisers agreed. Tugwell described himself as a "gas and water socialist," writing: "I saw no reason to hesitate because the United States was larger than New York." The press around the advisers produced supportive articles: Chase was writing these days about the importance of regional planning for power. There was even a way around the old constitutional obstacles to control by Washington of natural resources. You could circumvent the whole problem by creating a new regional level of government, and expand from there.

Not everyone found the idea of arranging the American economy by river basin intuitive. At his house in Tuxedo Park, New York, Alfred Lee Loomis was put off. He and his partner Landon Thorne had already retreated from Wall Street. Projects like the TVA convinced Loomis that he had better stay out. "He thought it would destroy the business world," his son's first wife later recalled. Loomis was a gifted scientist as well as a financier. In World War I he had invented a new way to measure the velocity of flying shells, later known as the Aberdeen chronograph. Now he was converting his Tuxedo Park compound into a scientists' lab and think tank; with all the changes going on, this seemed the more productive way for him to work for the time being.

Still, the times made the argument for the TVA seem more logical than it might have in the 1920s. So many industries nowadays seemed unable to complete what they had started. Perhaps that might also prove true for utilities and the electrification of the country. The price of power was coming down, but the poor could not afford to be lavish with electricity, even when they did have it—there would be a radio, but not a washing machine. Only government, many argued, had both the capital and the goodwill to rectify this situation.

Roosevelt decided now that whatever government control of electric power there was would remain; the government would also begin to control power in new areas. He had four goals. The first was to provide electricity to homes and farms—many farms were still without. The second was to increase use of electricity in all homes, providing Americans with a better standard of living. The third was to reduce the cost of electricity to the average consumer. And there was a fourth, more ephemeral goal: that through the electricity industry the New Deal might create a new and more prosperous form of society.

The ideal demonstration project was already in construction, the Hoover Dam on the Colorado River. The Hoover Dam would be a dam like a pyramid, something that matched Stalin's Dnieprostroi. All it would take to make the Colorado dam seem like Washington's project would be to legally and formally convert the multistate format of the Colorado project to a national one.

But there was a problem: the project was Hoover's, hard for Roosevelt and the New Dealers to claim. At the Department of the Interior, Harold Ickes had even made a stab at taking credit for it. Not long after the election Ickes issued a special order changing the dam's name—henceforward it was to be the Boulder Dam, after the original site picked by engineers.

It had been a bold move. After all, there was the Theodore Roosevelt Dam in Arizona, completed in 1911 as the first multipurpose project built by the Bureau of Reclamation. There was the Wilson Dam at Muscle Shoals. There was even the Coolidge Dam, in Arizona. The renaming of the Colorado River dam turned out to be so controversial that both publicly and privately, Ickes repeatedly found himself forced to justify his action. In his diary, Ickes complained that "a number of insulting letters, some of them anonymous have been coming attacking me for changing the name of Hoover Dam to Boulder Dam." After noting some of the details, he went on: "I have always called this the Boulder Dam myself, as do many people and I have continued that usage since I came to Washington. I consider it very unfair to call it Hoover Dam. Hoover had very little to do with the dam."

FDR meanwhile was already turning his attention to the Tennessee Valley, which he knew from his many trips south. This was territory untouched—intentionally untouched, owing to those Hoover and Coolidge vetoes.

Roosevelt knew well the frustration of the local officials, the sort who had written the telegram to Washington in the fear that electrification would pass the South by. He would not only reverse his predecessors' decision on Muscle Shoals and build up the Wilson Dam there, he would create an intermediate-level entity on that regional level. A public authority, it would manage power, rivers, and economic development—all three—throughout the Tennessee Valley. Whereas Hoover had one dam, Roosevelt would have nine. And though Roosevelt had allowed Ickes to take Hoover's name off the dam in Colorado, Roosevelt also allowed his advisers to give a politician's name to one of his new dams: the Tennessee Valley Authority's dam at Cove Creek would be Norris Dam, after George Norris.

The more he thought about it, the more Roosevelt warmed to the project. It had that expansive feel of reinvention that he liked. The three directors at the TVA would report directly to Roosevelt, an unusual arrangement that circumvented Ickes and other cabinet members. The men therefore could hear out Roosevelt on his specific thoughts—and he could hear them out, too.

From the start, the TVA had a utopian feel to it. It would create new towns. The most important of these, at Cove Creek, was approved at a meeting of the TVA board in the summer of 1933. It would be a model community; someone attending the meeting suggested that an appropriate name would be New Deal, Tennessee. But the board decided that the town, like the dam, should be called Norris. That was not all: the TVA would flood over a vast area, 153,000 acres of land, and buy out 3,000 farm families to create space for a reservoir behind the dam. There would be disruption, but it would be worth it, for the Tennessee Valley Authority would provide power cheaply. The idea was to create a "yardstick" against which the rates charged by private companies could be measured. And this project might in its turn be a model for similar structures in other regions. After all, the switch that controlled electric power might also be the switch that controlled the economy. Roosevelt wasn't actually sure how it would all hold together. But that did not necessarily matter for the moment.

Neither Tugwell, nor Chase, nor even Paul Douglas got a big role at TVA, but all three had something to say about it. And the management of the project did go to other progressives. The senior of the three directors was to be Arthur Morgan of Antioch College in Yellow Springs, Ohio. Decades earlier, Morgan had read Edward Bellamy's *Looking Backward*, and he had attempted to create his own little educational utopia at Antioch, dividing students' days into blocs of study and of work on a farm or in a factory. Antioch students founded companies—the Morris Bean Company, the Antioch bookplate company, the Antioch Press. Morgan was also an engineer, and had overseen the construction of the dams along the Ohio Valley. He had voted for Hoover—another engineer, after all—but what Roosevelt cared about were his progressive ambitions: "I like your vision," he told Morgan in 1933 at their interview. The second director was another university president, and another Morgan, Harcourt Morgan. The former president of the University of Tennessee, Morgan was expert in agriculture and entomology. This Morgan also had hopeful theories: he spoke about the environment as if it were a seascape, and envisioned a "common mooring" of air, soil, and water.

The third director was David Lilienthal, at thirty-three the youngest by more than a generation. Lilienthal was junior, but he had already collected credentials with the leading utilities reformers of the country. After his time at Frankfurter's Harvard, and in Chicago with Richberg, Lilienthal had proposed and written utilities law for Governor Phil La Follette, the son of "Fighting Bob," in Wisconsin. Now Lilienthal himself had determined to expand the tradition. "You see," Lilienthal wrote in those years, "I have a very strong feeling that if we cannot control our basic industries, and certainly nothing is more basic than the utilities industry, then we have no government in fact, merely a pathetic fiction of government." Like the La Follettes, and Roosevelt for that matter, Lilienthal feared that once private power conquered the electricity market, it would abuse it. This was the sort of thing he and "FF"—Frankfurter—had discussed when Lilienthal saw him at the university, or at Frankfurter's house at 192 Brattle Street in Cambridge.

While as serious as the two Morgans, Lilienthal was fundamentally different. If the two were dreamers of the nineteenth century, Lilienthal was a twentieth-century dreamer—the sort who believed that having adequate statutory language, the sufficient bureaucratic authority, and the sufficient budget were prerequisites to realizing a dream, not something one went out and retrieved as an afterthought. This difference between the Morgans on the one hand and Lilienthal on the other was emphasized by their appearance: the two older men wore round horn-rims and seemed a pair of owls; the younger man had glasses but did not wear them in all the early photos. From the start, Lilienthal worked hard to set himself off as less doctrinaire and more practical than the Morgans—"A river has no politics," he would say again and again.

In throwing these three men together, Roosevelt was replicating what he had done by sending advisers with conflicting philosophies to the London conference. Again, he wasn't sure whose model would predominate, or how the three would get along, or how the project would work. That did not matter for now. He would postpone his judgment, and so would the South. The project itself might be "neither fish nor fowl." What mattered, he told Tugwell, was that "whatever it is, it will taste awfully good to the people of the Tennessee Valley."

And Roosevelt was correct on the last. In the autumn of 1933, the New Deal project seemed worthy. The Tennessee River was what the experts called a "flashy" river. In a minute or two the waters of a calm creek could rise into a giant wall of water, almost as the Mississippi had in 1927. There were highs, there were lows, and there were surprises. Around Muscle Shoals, for example, the water dropped 140 feet in 30 miles, as much as the Niagara Falls of Roosevelt's own New York.

As for the people of the Tennessee Valley, they were among the poorest in the country, truly forgotten men. Their land was poor and overfarmed. Exactly how overfarmed, Tugwell, now assistant secretary of agriculture, was demonstrating in a report on Grainger County, part of the area to be purchased in the TVA plan. The report warned that the conditions of the land were now so bad that it might be impossible to restore it, and that the farming methods of the citizens were "little better, if any, than that of the early colonial farmers in the tidewater sections of Virginia and Maryland." Tugwell concluded that the lands in the river basin of Wilson Dam "were approaching the limits of arability."

The TVA, everyone hoped, would change all this. Even as the news of the TVA circulated, homeless families or families led by unemployed men started appearing around Muscle Shoals to watch as the first bits of construction started. Many did find work. Within a month after the start on Norris Dam, construction began on Wheeler Dam. The TVA management hired thousands of people. By the middle of 1934, there would be more than 9,000 employees at TVA. As locals watched, workmen erected seven more bunkhouses, a cafeteria that would serve 3,000 meals daily, a theater facility, a library, even Norris's own post office. In the area around the valley, men from the CCC planted 3 million trees and laid 2.6 million square yards of brush to keep the soil in place. TVA workers in Norris received a wage above those in the area, and they could eat all they liked for twenty-five cents in the Norris cafeteria.

Arthur Morgan, who at first ran the whole project—especially the details at Norris—recognized that it was the greatest opportunity to build a community he would have in his lifetime. He and his wife, Lucy, poured energy into each detail of Norris. Houses had electricity then. The walls of the houses in Norris—or at least some—were made of attractive local stone. Pedestrians would walk not on hard concrete sidewalks but rather on softer, curving footpaths constructed of natural earth.

The soul of the TVA worker also concerned Morgan. While riding on a train one day in 1933 from Washington back to Yellow Springs and Antioch, Morgan roughed out a moral code for the staff of the Tennessee Valley Authority. It included rules such as absolute openness. "A man who will lie for me will lie to me," taught Morgan, accurately enough. TVA employees could not take gifts from contractors; employees should live modestly as well.

Morgan served workers in the Norris area with a mobile library. But he worried that those farther out cutting timber must also need books. Morgan's team came up with an answer: "Why not place a box of books beside the tool box and make the saw-filer its custodian?" The saw filers then played the role of librarian to their fellow woodsmen. Observers noted that the demand for children's books was the greatest—the average woodsman had only made it to the seventh grade. Children's books proved the most popular.

The principle was that good and industrious living would also be better living was even applied when it came to building a school. The planners created a large brick building—to be heated entirely with electricity—to house a school that would be open twelve months a year. Children would be instructed not only in traditional subjects but also in enterprise, so they might be more self-sufficient than their Appalachian parents.

Morgan had critics. (One was a Chicago minister who noticed that the town, so minutely organized, was built "without provision for religious worship"—it would be "Godless Norris.") But he also had support very high up, for the president was constantly on the hunt for ways to defend the TVA and other New Deal projects. In Britain for the year, Frankfurter wrote Roosevelt that he ought to take heart and put all criticism in context by looking up what attacks had been leveled against Teddy Roosevelt or Wilson. Roosevelt gave an assignment to Tugwell in a memorandum: "Do you think you could get someone to dig up the expressions and utterances made against T.R., Chas. E. Hughes, and Woodrow Wilson, when they were trying to get through their exceedingly mild type of legislation?"

Still, most people were impressed with the results. The Norris Dam proceeded rapidly. On September 14, 1933, only a few months after the TVA directors had held their first meetings in the Farragut Hotel in Knoxville, Lilienthal was ready to announce the TVA's first rate schedule. The TVA would provide power more cheaply than any other company. Citizens of Muscle Shoals requiring small amounts of electricity—enough for a few lights, a percolator, a toaster—would be charged $1.50 per month for fifty kilowatts, compared with $2.94 in Chicago, the city of Insull, or $4.50 in New York, Thomas Edison's old home territory. For families who used more electricity, the new low prices were even more dramatic: $4.50 a month for 200 kilowatt-hours, whereas the price was $7.44 in Chicago.

Behind the spectacular headlines, there was trouble among the TVA heads. Of the directors it was Lilienthal who was rapidly becoming the public face of the TVA—he gave public address after public address. His very youth, his frank speaking, captured both the crowds in the South and his most important audience, the TVA's immediate boss, Roosevelt. To the Americans of the era, well versed in the Bible, it was not hard to think of this David as something like the young David of Scriptures, the precocious hero who went up against more powerful figures.

The first of such figures was his own partner, Arthur Morgan. For as Lilienthal began to succeed, he began to find the fact that Morgan outranked him irritating. Early that first summer of 1933 the three TVA heads divided up their work: Arthur Morgan in charge of the dams, Harcourt Morgan in charge of agriculture, and Lilienthal in charge of power and rates. Still, even with such a clear division, Arthur Morgan and Lilienthal began to quarrel—Morgan playing his usual part, the utopian, and Lilienthal the constructive intellectual, heir to Frankfurter and Oliver Wendell Holmes.

Nominally, their struggle was over the project. Lilienthal did not like the way Morgan did things. But it was also a struggle for political control of the TVA. Harcourt Morgan, the oldest of the three directors, interested Lilienthal less. He was over sixty-five, and focused on farms—less of a potential source of revenue for the project. It was A. E. Morgan, the dam builder, who dominated. If Lilienthal could prevail over A. E., he could rule the TVA. The very idea seemed as improbable as the TVA itself had ten years ago. But of the two, Lilienthal was the cleverer player and by far the more tenacious.

Another who would now encounter that tenacity was Wendell Willkie of Commonwealth and Southern. Before the TVA became law, Willkie had briefed the

House Military Affairs Committee on the concern that TVA might supplant Commonwealth and Southern, and that that could threaten $400 million in shareholder equity. "I have not much at stake in this except my position, and that does not amount to anything, but I do feel a great urge as a trustee for these securities." The Dow's young utilities index, which had started in January of 1929 at 86, had been down to 19 in April of '33. Not many weeks after the TVA became law, Willkie had dined with a seemingly friendly Arthur Morgan in New York in a private room at the University Club. Willkie had interpreted the new law to say that while the TVA might generate power in the South and transmit it, private companies would distribute it and sell it. Morgan did not necessarily agree—but he assured Willkie that a solution could be found. Reassuringly, in the meantime, the TVA was a limited entity. It could make power, but had no means to distribute or sell it. Willkie wanted to lock in Commonwealth and Southern's future by writing a multiyear contract with the TVA under which C and S would deliver and sell power generated by Wilson Dam.

Later in the summer, it became Lilienthal alone who controlled negotiations with power companies. Lilienthal set up a meeting in early October 1933 at the Cosmos Club in Washington, the club being, in

Lilienthal's words, "about as neutral a ground as we could think of." A decade later, Lilienthal wrote about the meeting: "I recall that we were two exceedingly cagey fellows who met at lunch that noon. In appearance, Willkie was a much better looking article than he is now, not being heavy." Following the meal, the men repaired to an enormous dark-stained table in a corner of the Cosmos Club lounge.

Willkie played the Indiana connection for all it was worth. He, the man from Elwood, Indiana, should after all be able to get along with Lilienthal, who was from Michigan City. Indiana University and DePauw could see eye to eye. What, Willkie asked Lilienthal, could the TVA want with the sale of power? That work was the private sector's job. Why not stick to generating power, or perhaps generating and distributing, and leave the actual sale to Commonwealth and Southern? The TVA had its appropriation from Congress—$50 million. But it might not get more. Willkie would pay $500 million for this contract with the TVA. If it wrote contracts with Commonwealth and Southern, the TVA might be able to balance its budget.

Willkie was a formidable debate partner. He was not a rogue, like Insull. He was even friendly with progressives, just like Lilienthal; Willkie's own candidate for president, Newt Baker, had also talked about expanding the role for government in the power business. He was one of New York's most successful corporate lawyers. And he had the power companies up and running—six in the South, producing electricity from steam plants. There was even some good news, of a sort, for his industry; the Dow Jones utility index had hit the 30s over the summer of 1933 and was now trading in the mid-20s. This was still a far cry from its original high, but at least above the level of 19 that the index had seen in April.

All that Lilienthal had were two things: a portfolio full of blueprints for future dams, and Wilson Dam, for whose electricity Willkie's own Alabama Power already had a contract. In later meetings and perhaps in this one as well, Willkie tried yet another tack. No one, he argued to Lilienthal, went into government without the intention of going into the private sector later. The private sector, after all, was where the business lived. If Lilienthal was too nasty, then he was not likely to find work at private utilities companies. Lilienthal was, by his own admission, "pretty badly scared" by the time he left the Cosmos.

Yet after the meeting, Willkie found that he was not getting all he sought. A contract between the U.S. Army Engineers and Willkie's Alabama Power & Light was due to expire January 1, 1934, and Willkie wanted Lilienthal to see that that contract was renewed promptly. Lilienthal confidently dragged matters out, aware that control of timing was a big part of any such battle. He was emerging as a harder partner to negotiate with than Morgan. Lilienthal really believed something that Willkie thought improbable: the future of power might actually lie in the public sector, with the TVA, and Willkie, therefore, and not he, ought to be the mendicant in meetings. Lilienthal was forcefully playing the role Frankfurter had taught him a decade before, that of virtuous government's agent at the table wrangling with the less virtuous corporate attorney. And the play was unfolding precisely as Frankfurter had envisioned it would.

Willkie still couldn't believe Lilienthal's audacity. "I think he was about as shocked at my ideas, as in the next conference or two I began to explain them, as I was shaken by his assumption that the whole TVA would amount to nothing except selling a little power to his companies," Lilienthal later explained. Still, even as he stalled, Lilienthal was furiously building up the TVA. He was even creating an electric appliance retail program to finance the purchase of TVA power by poor southerners, a version of Insull's old idea of selling irons to Schenectady housewives; this also replicated an old habit of Willkie's, selling appliances to boost demand. Roosevelt backed Lilienthal up with an executive order on the matter. Late that fall Lilienthal also wrote and signed TVA's first contract with a town, Tupelo, thereby entering the distribution and sales market; a contract with Knoxville followed.

In the scheduled meetings, Willkie fought back, "with a good deal of bellowing," as Lilienthal later recalled. Lilienthal was taken aback to find Willkie trying histrionics: "I couldn't have the blood on my hands," Willkie replied when Lilienthal made clear that TVA wanted to move into much of Commonwealth and Southern's territory. The blood to which Willkie was referring was the blood of shareholders, who had already seen the Commonwealth and Southern share price come down. But also, perhaps, the blood of customers that C and S might have served.

But Willkie was not merely a good actor; he was also truly concerned. And though Lilienthal could not help but like Willkie, and even envy him his confidence, he did not see Willkie's concern as justified. To Lilienthal, as to Frankfurter, shareholders were misty background figures. And so Lilienthal ambivalently wrote off Willkie's outrage. "I always marked off this walking back and forth, arms swinging, etc., as just a part of Willkie's selling technique."

None of this was especially visible to those who were not C and S shareholders or customers. What they could see at this point was the hope that the TVA offered. Morgan imagined that the region would not only wisely develop power but also move away from the old concept of private property and generally foster "the development of ethical attitude and conduct." Lilienthal warned the power companies that when it came to the Tennessee Valley, the valley's interests would prevail over the interest of the private sector—and the TVA and its spokesman, Lilienthal, knew what that interest ought to be.

Meanwhile, what the rest of the country saw was the architectural reality of the rising dams. Like the Hoover Dam, they astounded. Much later, critics would write about the dams' effect on people—F. A. Gutheim would call the TVA an example of "the architecture of public relations" and say that from "the conception of the scheme to its final execution you feel that each decision has been made in the light of the fact that the public would come, look, and judge by what it saw"; Lewis Mumford would write in the *New Yorker* of TVA that "the pharaohs did not do any better." Visitors were beginning to come to see the new marvel. At some points during the Depression the number of visitors at Norris, Wilson, and Wheeler would rise to a thousand a day.

In Washington, the whole story seemed an early proof of the success of the New Deal experiment, and it showed that the TVA was beginning to put the United States back where it belonged in world competition. The Soviets—to be sure, with American advice and supervision—had ruled the world with Dnieprostroi when it was completed in 1932. But the TVA was building more than seven dams as large as Dnieprostroi. Thirty-five Boulder Dams, or ten Grand Coulee Dams—Grand Coulee was a New Deal project, but not a TVA one—could have been built with the amount of material used at the TVA, Lilienthal would note.

He grew more ambitious. In the coming years he would not only move his family to Norris but even acquire a horse, a bay named Mac, and ride it through the town's streets. Like the rivers in his projects, his thoughts cascaded forward. Flood control and dams were not enough; there had to be generation of power. Selling electricity should also be the TVA's job, no matter what Morgan said. Appliances were also good; if he really could convince southerners to buy them, they would use more electricity. Lilienthal negotiated with General Electric and the National Electrical Manufacturers Association for low prices. Lilienthal also tried his hand at designing a logo for his new appliance entity, the Electric Home and Farm Authority. To the modern eye, it looks close to something from the Europe of the 1930s: a black fist holding a blazing bolt, and the words "Electricity for All." He asked Eleanor Roosevelt to write the introduction to an EHFA flier.

But these projects were only the start. The TVA would address farm problems by rerouting waters. The South was especially poor in recreational bodies of water—nothing like Wisconsin, where Lilienthal had been fishing when the call had come to join the administration in the spring of 1933. But now the TVA was making lakes.

The soil was a problem, but here there was a TVA solution, too. TVA power could provide phosphate fertilizers, made with power from the dams. TVA engineers would discover a cheap way to deliver them in concentrated form. Farmers might not understand about phosphates, so the TVA created a demonstration farm project to teach them. Lilienthal and Harcourt Morgan were especially enthusiastic about the phosphates, describing them as having an "almost magic" effect. Treeless stretches made the land susceptible to floods. The TVA could plant new forests. In the first four and a half years of TVA's existence, Lilienthal gave forty major speeches.

Roosevelt was also inspired. The TVA seemed to be succeeding, and faster than he ever could have hoped in his days as a progressive governor in New York. Ickes at the Interior Department was busy facilitating matters by handing out cash to towns so that they might build their own power stations and no longer be dependent on the private sector companies. The same November that Cove Creek got going, his PWA gave $7.5 million to North Platte, Nebraska, for a diversion dam and reservoir. Roosevelt was especially interested in fostering the creation of rural pools of farmers to buy power, and was now considering writing legislation to that end. Already Paul Douglas of Chicago had brought Ickes together with others to help individual farmers form cooperatives so that they might buy power at a reasonable price. This was the beginning of what would be known as the Rural Electrification Program.

Tugwell too seemed to be doing splendidly. For weeks he had engaged in a tiff with the pragmatist George Peek, head of the AAA, over the administration of National Recovery Administration codes. Another thing that Peek objected to, probably all too loudly, was the left-leaning general counsel at his own AAA, Frankfurter protégé Jerome Frank. Tugwell the idealist won, and Peek was moved over to head a temporary committee to look into the creation of a new export agency. Influential columnist Arthur Krock trumpeted Tugwell's victory, writing that "the hero of the contest, Mr. Tugwell, is a brilliant, engaging, good looking man." Krock called him "the winnah and new champion." Rex had arrived.

As 1933 ended, it was clear that the administration was now thinking about more radical goals for power. In December, Arthur Morgan, whom Willkie had found easier than Lilienthal, also turned sharp. Morgan held a press conference to show off a ten-million-volt generator and talk about the future of power transmission. He also took the occasion to note that the TVA would supply Tupelo with power for $12,370 the next year, compared with the $31,144 the town had paid in 1933. Morgan warned, the paper reported, that in taking over private plants in its territory, the TVA did not plan to buy inflated shares of private companies. The government would not be bilked by Willkie's Commonwealth and Southern. The president, Frankfurter, Lilienthal—all were sanguine. Willkie realized he was up against no ordinary opponent.

7

A Year of Prosecutions

January 1934
Unemployment: 21.2 percent
Dow Jones Industrial Average: 100

At the new year the papers carried a report that the Justice Department was creating a division to look into civil and criminal infractions of the tax code. But taxes were not the only area where the Roosevelt administration was on the hunt. Both the administration and Congress would step up efforts to see that justice was carried out in the cases of all those who might have misstepped in the period of the crash. J. P. Morgan's executives, other Wall Streeters, the utilities would all be targets.

Clearing out the corrupt forces that had brought the crash in the first place seemed to the New Dealers the best way to recovery. Nineteen thirty-three had been a year of experiment; 1934 would be a year of prosecution. The Davids of the New Deal—not only the literal David, David Lilienthal, in this instance, but also Roosevelt's young prosecutors—would go up against the Goliaths of the old world.

It was obvious who the Goliaths would be: Sam Insull and Andrew Mellon. Both men were so old as to be from another era—Victorians: Insull was born in 1859, Mellon even before that in 1855. Both were anglophiles—Insull, of course, actually came from Britain. Shareholders in Insull utilities had lost millions when his empire collapsed in 1932. Investors had paid $88 million for Insull's special gold debentures; now, after all the sales, only $460,000 in cash stood behind the engraved certificates. *Time* magazine estimated that all told, Insull's losses had cost shareholders hundreds of millions. In Chicago, city coffers were so bare that teachers had gone many months without pay. The guilt of Insull seemed to mount as the downturn extended.

Mellon was an equally enticing target. As treasury secretary, he had presided over a market in which everyone else lost their cash. If Insull was a national target, then Mellon was both a national and an international one. In the Soviet Union, Madame Krupskaya, Lenin's widow, happened just then to be arguing for the publication of a series of didactic children's books: in one of them, a Marxist version of *Jack and the Beanstalk*, Jack the Giant Killer slew Andrew Mellon, caricature of capitalism. What the prosecutors hoped was to finish the work of the progressives of the 1920s—the attack on Mellon by Wright Patman in Congress, the attack on Insull that Paul Douglas had begun in Chicago, for instance. Ferdinand Pecora had made a stunning start on this as counsel to a subcommittee in charge of an inquiry into the stock market. Pecora had had J. P. Morgan on the stand in May of the previous year, and the country would not soon forget that. It had emerged that Morgan, the leading figure in the world of finance, had not paid any income tax in 1931 or 1932. This was the sort of revelation that warranted replicating.

But there was a hitch. Both Insull and Mellon were still popular in their towns, among the people who knew them well. In the context of the credit that Insull had built up with citizens, even the collapse of his companies might not convince Chicago of his evil. As for Mellon, many Pittsburghers were still loyal. Both observers and prosecutors understood therefore that the men might win court victories on home turf. Still, in an odd way acquittal at home might even strengthen the Feds' case—it would be a form of proof of the corruption of local politics and the need for national intervention.

Over at Commonwealth and Southern, Willkie felt the concern strengthen. Months had passed, and now the utilities index was still in the 20s. Would it go into the teens? He could see that the New Dealers were consolidating their power. Lilienthal was giving speeches in small towns about the electricity he would bring. And Roosevelt, the word was, was planning to nominate Rex Tugwell to the undersecretary slot at the Department of Agriculture, a promotion. The New Dealers were becoming stronger. Willkie, like Insull, was a utilities man, and in such a world, Commonwealth and Southern could also become a target. Willkie's wife, Edith, disapproved of the New Deal—so much that he teased her about it.

But Willkie reassured himself that there were still many reasons both he and Commonwealth and Southern would be all right. He still saw himself as more like Roosevelt than Insull. Willkie was still looking for a way to square his vision of reform, which was more Wilsonian, with FDR's protean version. A few years earlier, at a conference, Willkie had actually encountered Insull. The discussion had been about press criticism of the utilities industry. Insull had complained about it; Willkie had countered by telling the older man that he believed press criticism was healthy for enterprise. Insull had rebutted Willkie's rebuttal: "Mr. Willkie, when you are older you will know more." This disdain for free speech came from another world, a world that was not Willkie's.

As for the differences between Insull's empire and Commonwealth and Southern, there was evidence. In 1932, James Bonbright had published a book attacking holding companies while praising Commonwealth and Southern as a corporate model that functioned as a holding company should, making capital available to its subsidiaries in the most efficient fashion. What's more, Willkie was now in the process of preparing, rather publicly for that matter, to clean house further at Commonwealth and Southern. And of course Willkie had his old acquaintance with Newton Baker, one of the Democrats' leaders in the area of utility reform. He was also quite friendly with Oswald Ryan; they knew each other from Indiana. Now Ryan was general counsel to the Federal Power Commission, Willkie's principal regulator. Willkie supposed he might observe the attacks on figures like Insull and learn from them.

But he did not think for long, for he had his work cut out with Morgan and Lilienthal. On January 4, he finally concluded a geographic truce with the TVA. Commonwealth and Southern agreed to sell the TVA a transmission line and properties near Norris Dam. The agreement also divided up the South with the TVA—

until 1937, or the completion of the new Norris Dam Power House, whichever came first. Commonwealth and Southern could not sell power in TVA's new distribution area, and the TVA could not sell to Commonwealth and Southern customers outside certain counties.

The arrangement bought Willkie time. Like Lilienthal, he and the power companies in the South knew how to use that time to their advantage. They were now going to court and winning injunctions to halt TVA construction. The administration might think in Frankfurter terms, but the average southern judge did not—even if he was a Democrat. The fall's midterm elections could change matters as well. Roosevelt's Democrats might not do so well, Willkie thought. Then Lilienthal, who came across as so arrogant in meetings, might begin to understand that he had to be reasonable. The Supreme Court might find the TVA itself unconstitutional. In the end, Willkie told himself as he regrouped, Commonwealth and Southern could still win.

Meanwhile, the two Goliaths were distracting the country, taking the country's eyes off C & S. The beginning of the year found Insull ensconced in Athens, Greece, a fugitive from justice. Alone walking about Athens, Insull realized now that he regretted some of his past—especially the fact that his brother Martin had written negatively about Roosevelt. Insull's prophecy about the consequence of ad hominem attacks was coming true. As soon as FDR was elected, Insull had known he was in for trouble. Still he believed, legitimately enough after Roosevelt's "Ishmael and Insull" speech of the campaign, that an American trial would be political and unfair.

Insull was buying time, as well. He wanted to be in Britain, but Greece, unlike Britain, did not have an extradition treaty with the United States, so he was safe, at least for the moment. His wife joined him there. In Chicago, federal lawyers were using a young technology, the photostat, to work up a case so extensive that its documents, taken together, weighed two tons. The prosecutors wanted to get him on mail fraud, for sending out stock prospectuses that included false promises of profit. The State Department had begun pressuring the Greek government, and the Greeks for their part repeatedly assured the State Department that they would force the fugitive out and deliver him up. Yet the Greek government agonized—there was a domestic audience to consider, and it did not want to seem entirely under the American thumb. Insull's lawyers and doctors discovered they could win repeated stays and postponements for their client with the argument that Insull was in ill health. And so he was: Insull had diabetes. Insull's status became a national story in the

States, where papers reported that Greek police were guarding the invalid's door.

Mellon was also in the news. In February, Roosevelt nominated Robert Jackson, a young prosecutor from his own state, to become general counsel of the Bureau of Internal Revenue at Treasury. The idea was to beef up investigative and prosecutorial work. On March 11 came news that Homer Cummings, Roosevelt's attorney general, was preparing a tax suit against Mellon, as well as against T. S. Lamont of J. P. Morgan and Thomas Sidlo, a law partner of the reformer Newton Baker, of Cleveland. Seeing this last name may have troubled Willkie—Roosevelt was reaching dangerously near to his corner of the world. The attorney general would also look into the finances of Jimmy Walker, the New York mayor whose resignation Roosevelt had provoked in 1932 in the midst of a legislative inquiry into city corruption.

The question in the tax suits was whether it had been legal for Mellon, Lamont, and the others to use certain tax loopholes in preparing their returns. Lamont, the son of Thomas W. Lamont, another Wall Street giant, and the brother of the socialist Corliss, had sold stock at a loss, then deducted the loss on his 1930 return, and then permitted his wife to repurchase shares of stock in the same company. The investigation of Mellon was looking into whether he had underpaid taxes by a million dollars in 1931. It might not be as dramatic a charge as the one against Morgan; still, it made for good press.

But taxes were not the only front on which the attorney general attacked. Cummings was also announcing an investigation into the Aluminum Company of America, a Mellon company, to determine whether it broke antitrust laws.

Even as the country digested this news, Insull was again in the papers. On March 5 the Greek government gave Insull forty-eight hours to leave, but Insull won a medical stay. The U.S. consulate kept watch on his building from the balcony of the consulate, only a block away. Yet on March 15 Insull vanished, somehow managing to escape his apartment out the back way. He had evaded the Greek police, a national embarrassment of a magnitude that provoked a cabinet crisis. After several days came reports that the fugitive was aboard a tramp steamer, the *Maiotis,* heading toward Egypt. The State Department quickly asked Joseph Robinson, Senate majority leader, to put through special legislation that would enable U.S. officers to snatch up Insull in countries where the United States exercised extraterritorial powers, such as Egypt. The bill passed without debate, and Roosevelt signed it on March 23.

Meanwhile, however, both the mortified Greeks and others had lost track of the *Maiotis*, and in Alexandria crowds scanned the horizon for a glimpse of Insull. "Not since the passage through the Suez Canal of Mahatma Gandhi has any individual been so awaited," wrote the normally staid *New York Times.* Within a day or so Greek authorities were in wire contact with the owners of the *Maiotis*, and a few days later the *Maiotis* docked at Istanbul to pick up provisions—potatoes, macaroni, salad. Washington demanded that Turkish authorities arrest Insull, and unlike the Greeks they complied immediately, subjecting Insull to a mock trial. Soon the septuagenarian was traveling toward the United States in State Department custody; the name of the ship bearing the captive was, appropriately, *Exilona.*

In April, Roosevelt nominated Tugwell as undersecretary. By now the prominence that the journalist Arthur Krock had noted was beginning to show its painful side. Tugwell had spent the year fighting, with FDR's support, for a radical updating of the old Food and Drug legislation, the idea being to regulate more thoroughly from Washington "the purveyors of doubtful nostrums and unregulated foods," as he put it later. Others however saw his effort as an outrageous theft of a function normally provided by the private sector—quality control. At one point Eleanor Roosevelt, who herself had a sense of humor, invited Rex to lunch. The lady seated next to him, Tugwell would later report, "turned out to be one of the editors of *Good Housekeeping*, a magazine that offered to approved products something known as the *Good Housekeeping* Seal of Approval." Tugwell commented that "the lady in question was very high and mighty." The guest from the magazine spoke angrily to Tugwell—probably more so than Mrs. Roosevelt had intended. But "the situation was saved," Tugwell concluded later, "in a most unexpected way: an awkward waiter spilled a bowl of tomato soup in my lap and I was able to withdraw without dishonor." Nonetheless, the event stuck with Tugwell: still an idealist, he could not see why the *Good Housekeeping* lady had been so angry.

Now Tugwell did what he could before his confirmation hearings, to establish that he was a moderate. He delivered a mild and inspiring speech to Dartmouth students in which he told them that coming into the world, they would confront "not revolution but the same old system with some new changes." Tugwell spoke too to the American Society of Newspaper Editors, on the twenty-first. The journalist Frank Kent reported that Tugwell had succeeded in his mission of taming the press: "He buttered the editors until they glistened like greased poles in the sunshine." Tugwell's point, though, was a real one: there was no treason in being an economic planner.

Then suddenly it was May, and Mellon's turn again. Homer Cummings charged that Mellon had earned $9.2 million in 1931, not the $6.8 million that Mellon claimed. Sidlo settled. But Mellon, normally quiet, struck back, accusing Cummings of a "campaign of terror" designed to railroad the jury in Pittsburgh into indicting Mellon. Mellon had not underpaid his taxes, as Cummings was suggesting, Mellon's lawyers said. He had overpaid them, and would now appeal for a refund. On May 8, the grand jury in Pittsburgh refused to indict Mellon. "Not a true bill," someone in the jury room—probably a juror—wrote on the government's charges that were handed back to the judge. But for the administration, the event was not entirely a loss: prosecutors figured that in the eyes of the country, even publicizing Mellon's income would hurt him. The same day that Mellon went free, Insull gave himself up as a prisoner at Cook County Jail. He moved into the hospital ward, sharing the space with thirteen-year-old George Rogalski, who had confessed to kidnapping a two-and-a-half-year-old girl. The papers reported that a crowd of 3,000 had watched Insull enter.

Meanwhile, Morgenthau, whom Roosevelt had just made treasury secretary, was not planning to give up.

When newspapers criticized the assault on Mellon, Jackson had turned to Morgenthau to ask whether the prosecution was worth it. As Morgenthau recalled the exchange, he commented, "You can't be too tough in this trial to suit me." Jackson then jumped up, exclaiming, "Thank God I have that kind of boss." Morgenthau had then gone one better: "Wait a minute. I consider that Mr. Mellon is not on trial but Democracy and the privileged rich and I want to see who will win."

The fury surprised some of Roosevelt's earlier allies. James Warburg, the departed financial adviser, was in the process of preparing and publishing two books pointing out the economic errors he felt Roosevelt had made—*The Money Muddle*, which came out that May, about the gold standard, and another, *It's Up to Us*, a counterattack on New Deal economics. *The Money Muddle* quickly became a best seller. Now Warburg thought about broadening his arguments—would Roosevelt stop at nothing? Ray Moley had taken his distance and left the administration. He started his own weekly magazine, *Today*, financed by Vincent Astor and Averell Harriman. He still admired Roosevelt, but was baffled at the inconsistencies. He later would write his own, softer version of Warburg's analysis, saying that the surprise at beholding Roosevelt "arose chiefly from the wonder that one man could have been so flexible as to permit himself to believe so many things in so short a time. But to look upon these policies as the result of a unified plan was to believe that the accumulation of stuffed snakes, baseball pictures, school flags, old tennis shoes, carpenter's tools, geometry books and chemistry sets in a boy's bedroom could have been put there by an interior decorator."

Tugwell for his part was still in the game, anxious about his confirmation hearings for the undersecretary post, scheduled for June. In March, Congress had turned against Roosevelt for the first time and supported the American Legion, when it overrode his veto of legislation increasing government employees' pay and veteran pensions. Like all of the brain trusters in Washington, Tugwell had not appreciated the level of pressure in Washington until he experienced it. Keeping his family happy while he was always at work was a problem, especially since he was emerging as the favorite target. It didn't help matters that he had taken a salary cut to leave Columbia. As assistant secretary of agriculture, Tugwell was responsible for enforcing the Pure Food and Drug Act. In early June, the *New York Times* got a hold of a tiny but painful story: Sometime at the beginning of the year, Tugwell had been forced to fine his own father. The agriculture department had determined that the canning firm of Tugwell and

Wiseman—Tugwell's father, Charles, was a partner—had mislabeled cans of grapefruit and orange juice in a "false and misleading" way to disguise the fact that sugar had been added to a product labeled natural. For Tugwell, who adored his parent and knew so much about the man's financial struggles, the headline must have been bitter: "Tugwell Fines Father."

Tugwell's allies spent a weekend in the country prepping him for interrogation with mock questions: "Who are you, anyhow?" "Are you a Communist? "Did you ever spread manure?" Tugwell worried. It likely seemed ironic now that he had not even attended that interview with Stalin in 1927. For now, among all those on the junket, he stood a chance of losing something important because of it.

The preparation was worth it. One senator, L. J. Dickson of Iowa, read aloud Tugwell's "make over America" poem of 1915. Over and again, senators assailed Tugwell as a red and questioned him on his loyalty to the United States and the Constitution. Tugwell told them he was not interested in implementing Soviet ideas, but rather Roosevelt administration ideas. Tugwell was already sensing what would become a pattern—the frankest of the New Dealers, he said what the others dared not. "Tugwell had been unjustifiably used as a whipping boy," Douglas concluded from the sidelines. The *Spectator,* the student newspaper at Columbia, was busy criticizing him from the left. Columbia kept granting him leaves of absence, but Tugwell wondered if its president, Nicholas Murray Butler, would take him after it was all over.

In this instance, things worked out. Six of the seven Republicans on the committee reviewing his nomination voted for him in the end. The full Senate confirmed him 53–24; afterward, Tugwell celebrated at a party hosted by his friend *Babbitt* author Sinclair Lewis. Absorbing the hits as Roosevelt's Red seemed, for the moment, worth it. *Time* put him on its June 25 cover.

That same June Roosevelt took another series of steps in the name of helping the economy. One, hardly given its sufficient due at the time, was to sign a treaty that Cordell Hull had been working toward since the disastrous London conference the previous year. The Reciprocal Tariff Treaty, as it was then known, ended penalty duties. It also granted the president the authority to shift tariff rates. The new agreement was classic Roosevelt—and perhaps classic Hoover—in that it strengthened the power of the executive. But it also happened to be good for the economy; trade increased dramatically. Americans understood this, after the bitter Smoot-Hawley experience. The country had seen trade volume narrow by 40 percent. Hull later said that "in both House and

Senate we were aided by the severe reaction of public opinion against the Smoot-Hawley Act." Still, those who changed their views were mostly Democrats. George Peek, Tugwell's old nemesis, was now criticizing Hull. The Republicans, obstinate, clung to their old party position notwithstanding the mounting evidence.

That month Roosevelt also gave Wall Street its own policeman: the Securities and Exchange Commission. Publicly traded companies must register with the SEC, and its officers were free to investigate and punish them. In one of his cleverest and most cynical moves to date, Roosevelt named Joseph Kennedy the SEC chairman. People said he had picked the fox to guard the henhouse. Harold Ickes noted the event in his diary: "The president has great confidence in him because he has made his pile, has invested all his money in government securities, and knows all the tricks of the trade. Apparently he is going on the assumption that Kennedy would now like to make a name for himself for the sake of his family, but I have never known many of these cases to work out as expected." In the year and a half he was at the helm of the SEC, Kennedy would recommend hundreds of prosecutions. At *Today*, Moley devoted an article to praising Kennedy, writing hopefully that Kennedy's and the SEC's inauguration would bring for America a "reign of law in finance." The article did not spell it out, but it contained a hidden criticism: arbitrary prosecutions were subtly different from consistent enforcement of existing rules.

In late May, on the 28th, Keynes, the British economist, visited with the president. The meeting between the man who was becoming the world's most influential economist and the U.S. leader was not entirely successful, lasting fifty-eight minutes, and Frances Perkins, whom Keynes saw afterward, would later recall that Keynes told her the session did not go well. Roosevelt gave a similar report, telling Perkins that Keynes had left him, disappointingly, with a "rigmarole of figures. He must be a mathematician rather than a political economist." Still, Keynes told other New Dealers, including Perkins, that he thought the spending of all the New Deal programs was a good thing, because cash outlays gave the common man purchasing power: "With one dollar paid out for relief or public works or anything else you have created four dollars' worth of national income." Marriner Eccles, now at the Treasury with Morgenthau, was hearing intellectual justification for what he already believed instinctively. Whereas before there had been almost no framework to explain what Roosevelt was doing, now a respectable one was forming. Spending promoted growth, if government was big enough to spend enough.

All these events, but especially the public prosecutions, were sending business executives into new fits of housecleaning. Willkie busied himself making over Commonwealth and Southern. Late in June came Commonwealth and Southern's annual meeting; Willkie saw to it that the departure of four board directors was announced—men not engaged in actual operations. Officers of actual operating utilities were elected to take their place. Willkie also cut off local lawyers in the states who had served Commonwealth and Southern and who had been too easy to bribe with stock. Commonwealth and Southern would manage its own lawyers in-house now.

A few days later, on June 28, Roosevelt gave a Fireside Chat that likely did nothing to quell such executives' anxiety. The president posited that "much of our trouble today and in the past few years has been due to a lack of understanding of the elementary principles of justice and fairness." And people seemed to like such remarks. Harry Hopkins had hired Lorena Hickok, a journalist and friend of Eleanor Roosevelt, to report on the state of the country and New Deal projects. In 1934 she wrote, "I carry away the impression that all over the area, from Knoxville, Tennessee, to Tupelo, Mississippi, and on up to Memphis and Nashville, people are in a pretty contented, optimistic frame of mind.

They just aren't thinking about the Depression any more. They feel that we are on our way out and toward any problems that have to be solved before we get out their attitude seems to be 'Let Roosevelt do it.' " With Hickok some of the time was Grace Falke, Tugwell's assistant. Both Hickok and Falke believed that getting people off barren land into industry was the best antidote to Tennessee-level poverty. The fact that some of the radio power through which the citizens heard Roosevelt had come via Roosevelt's own TVA made him seem all the more impressive.

Come August, it was Lilienthal who was in the news, announcing that all of the TVA's power had found a market and that the construction of new dams was ahead of schedule. Lilienthal and the TVA's engineer, Llewellyn Evans, sailed to Britain to study something new that the United States might copy: what the Britons called their "grid," which used power from both private and public companies. "The problem of linking public and private companies is similar to that of the Tennessee Valley," Lilienthal told a reporter in New York at the Hotel Roosevelt before he sailed. In fact the UK grid owed some of its efficiencies to Sam Insull, who had advised Westminster repeatedly. The British press saw irony in the fact that Lilienthal of the TVA wanted to have a look at it. Lilienthal, the *Electrical Review* chuckled, said he was looking for a man of "seasoned specialized judgment, such as [Mr. Lilienthal] regards as essential for determining future policy in the States, but circumstances have deprived the American people of his experience and that of his associates."

The British were on to something. Another reality was becoming clear that summer of 1934: the drama of the prosecutions and the spring cleanups was hiding something. While the Norris Dam and other New Deal projects might be ahead of schedule, the economy was not, at least not in the sense of being where it had been before. The American Federation of Labor had reported in late spring that 780,000 workers who had been reemployed by the National Recovery Administration in the autumn had been unemployed again by the spring of 1934. William Green, the AFL's leader, began fighting with Richberg at the NRA over employment numbers: the AFL wanted industry, not relief agencies, to solve the economic problem. The job had to be done, and it had to be partly Richberg's, since, challenged Green, "We cannot indefinitely support one-sixth of our population on money borrowed against future taxes." There were still nearly eleven million unemployed. And the Dow was heading in the wrong direction; it would spend the summer below the 100 mark of the preceding winter. Looking up, the New Dealers were taken aback. "The Depression was refusing to disappear," Tugwell later wrote of the whole period.

Roosevelt now regrouped. Midterm elections were that November. The prosecutions had originally been about punishing financiers for the crash. If Insull's debentures had lost all their value, then Insull must be guilty. But now the prosecutions and investigations should also be about justifying the New Deal, economically and morally. The Fireside Chats would help. Roosevelt and others had noticed that the medium of radio really did seem to create a new reality, separate from the reality of old politics. If voters focused on the voice and the message, and not the tardy recovery, that might carry the Democrats forward.

It was a hope that the administration concentrated on—for even as government lawyers prosecuted, they were also finding themselves on the legal defensive. The National Recovery Administration, for instance, was under fire in Congress, and from business, as a bureaucracy out of control. A report submitted to the fifty-seventh annual meeting of the American Bar Association noted that by June 25 of 1934, some 485 codes and 95 supplements had been approved by the president and 242 more by the Administrator for

Industrial Recovery. In the period of a year, 10,000 pages of law had been created, a figure that one had to compare with the mere 2,735 pages that constituted federal statute law. In twelve months, the NRA had generated more paper than the entire legislative output of the federal government since 1789.

To survive, Richberg and the Justice Department warned—accurately—that the NRA must pass review by the Supreme Court, and soon. In any case, the law required renewal by Congress in 1935. Its constitutionality was a big question: could such a large program really be legal under the Constitution's commerce clause? Marking Constitution Day in September, the *New York Times* commented that in regard to such legislation, "the winds of controversy over this issue are already rising." With the election coming up, the New Deal had to score victories in national politics; if it couldn't win a genuine return to prosperity, then it might succeed with the negative victories of bringing down the villains.

Richberg and the Justice Department were on the hunt for a test case through which they might prove that the National Recovery Administration was constitutional. They had not yet settled on a case, or even an industry. One possibility that emerged over the course of 1934 had to do with the oil code, one of Ickes's babies.

Ickes, like Lilienthal and Frankfurter, tended to see life in legal terms. When markets didn't cooperate with legal precepts, that was because they were being insufficiently policed. He had discovered that at numerous points oil was being extracted clandestinely and illegally, outside his NIRA production quotas, and sold at prices that undercut the policy to force prices upward. He was outraged and opened a campaign against the oil bootleggers, describing them as possessed of a "sly animal cunning." He also sent investigators all over East Texas to catch the violators—the purveyors of illegal "hot oil." Hot oil, officials at the Justice Department believed, might be best positioned to win them a victory in the Supreme Court. One of the principal aims of Justice at this point was to prove that the NRA did not violate the commerce clause, which confined Washington's regulatory authority to interstate commerce. Oil traffic was clearly interstate activity.

And hot oil was not the only violation that officials were reviewing. There was the case of William Elbert Belcher, who had refused to enforce the timber code at his lumberyard in Centerville, Alabama. Belcher had paid his workers less than the 24 cents an hour that the code required—perhaps only 7 ½ cents.

Poultry was a sector that many of the New Dealers expected to be transferred to the public or cooperative sector. Tugwell had sought to distract Roosevelt from the burden of office by taking him out to an experimental government center at Beltsville, Maryland. There, only ten miles from the White House, the government maintained swine, cattle, and "a large poultry breeding operation." The poultry trade of course was known as a messy one—distasteful to the eye, smelly, and on the East Coast, often run by Jewish immigrants who ended up in court for corruption, crime, or worse. Animal slaughter was also a topic made for headlines. In these days, in which antibiotics were unknown, there always remained the possibility that a merchant who did not conduct his business properly might sell fatally tubercular meat or milk. And clean food was square in Tugwell's "bailiwick.

A young Harvard lawyer, Walter Lyman Rice, had already successfully prosecuted a member of the New York chicken industry for violating antitrust provisions. His case had made it to the Supreme Court, which had determined that the live poultry business fell under the jurisdiction of federal laws and was covered under the Commerce clause. The NRA lawyers began to strategize about chickens: if government attorneys could target a poultry dealer—and show, in effect, that he was endangering consumer health by violating NRA rules—that would put the virtuous qualities of the NRA in the best possible light.

Beginning in June, inspectors had begun visiting ALA Schechter poultry, a slaughterhouse in Brooklyn, looking for poultry code violations and taking notes. One of the inspectors later confirmed in testimony that the case had been set up in order to clean out the poultry trade, known for its corruption: "We are going to get an indictment and convict the Schechter brothers and that will be a whip over it," he had been told. Over the course of the summer inspectors swarmed the Schechters' business. They looked for, and found, evidence of violations of the 40–48-hour workweek mandated under the NRA; they also found that the Schechters, like Belcher, had not always paid the minimum wage mandated by the bill.

By July, or just a few weeks after the first visits from the code officials, a grand jury had indicted the Schechters on not one or two but sixty counts. Some of charges were criminal ones, which meant they would not only have to pay fines if convicted, but also might have to serve time in jail. Among the charges were that they had threatened violence against agents and inspectors; that through illegal transactions they had burdened the freedom of interstate commerce and were subject to federal regulation; and that they had also violated NRA code rules involving hours and pay. The prosecutors said that the Schechters had violated code rules about the selection of chickens and that they had, in the same weeks as the inspector's visits, "knowingly, willfully and unlawfully sold for human consumption an unfit chicken" to a buyer, a man named Harry Strauber. Finally—and this escalated matters—the Schechters were charged with conspiracy to flaunt the code. The case would come to be known as the Sick Chicken case, and would be heard now, in the autumn.

In order to investigate the way that cases such as *Belcher, Schechter,* or Andrew Mellon's demanded, government agencies needed staff. They also found that prosecution became easier if they revised the rules of the game. In the instance of Insull, that meant proving crime and fraud in his bankruptcies. Insull argued that this amounted to illegal retroactive action; prosecutors were imposing current law and sensibility on a different time. As he told his State Department escort on the *Exilona,* "What I did, when I did it, was honest; now, through changed conditions, what I did may or may not be called honest." He and Mellon had commiserated—Mellon had written him, he told the escort, to sympathize.

When it came to Mellon, the rule changing involved the definition of what constituted illegal tax behavior. It also meant creating staff to find such behavior. At the Treasury, therefore, Morgenthau acted, dramatically increasing the number of tax officials. Between 1934 and 1935, the staff at the Bureau of Internal Revenue rose to 16,000 from some 11,000.

The New Dealers were comfortable with all these changes. Morgenthau was beginning to feel at home in his office, and was thinking of putting his own stamp on Mellon's old department. In October 1934 he announced the establishment of a Treasury Department Section of Painting and Sculpture, its purpose to "secure suitable art of the best quality available for the embellishment of public buildings." Mellon collected classical art; Morgenthau would collect more modern art—and do it on behalf of the American people. Mellon had disliked tax loopholes. Morgenthau disliked them too, but wanted what Mellon had not—to punish taxpayers for using those loopholes, even when government had created them intentionally.

When it came to taxes, the law was shifting in Morgenthau's favor. In a case involving stock shares, *Helvering v. Gregory*, Judge Learned Hand that year found that it was illegal to use a corporate device for tax purposes for which it was not intended. Even Justice Sutherland was going along. He would affirm the finding, writing at the new year that "to hold otherwise would be to exalt artifice above reality."

Mellon was likely surprised at Morgenthau's tactics. His dislike of loopholes had been less on moral grounds and more on practical ones. As secretary he had preferred a broad tax base to one with numerous exceptions and had even asked the Bureau of Revenue to supply him with "a memorandum setting forth the various ways by which an individual may legally avoid tax." With each of his many rate cuts, he had tried also to reduce loopholes. A simple system had always seemed to him the best way for the government to get the most money. "A removal of the artificial value of tax exemption will restore all securities to natural conditions," he had written in *Taxation: The People's Business.*

That did not, however, exclude another point in which Mellon firmly believed: any man had the right to use legal loopholes. This was the traditional common-law distinction between avoidance, which was legal, and tax evasion, which was not. And Mellon, too, might quote Learned Hand, for in *Helvering,* the judge had also said that there was "nothing sinister in so arranging one's affairs as to keep taxes as low as possible. Everybody does so, rich or poor; and all do right for nobody owes any public duty to pay more than the law demands . . . to demand more in the name of morals is mere cant."

Mellon's argument was the same as Insull's: to prosecute a man for doing something that was legal was not acceptable. A number of people within the Treasury agreed, quietly. One, it later emerged, was Elmer Irey of the special intelligence unit of the Treasury. Irey had gotten to know Mellon while investigating the Capone brothers of Chicago; a civil servant for many decades, he would later publish a book about his targets, *The Tax Dodgers.*

Yet Irey had been reluctant in this instance. "The Roosevelt administration made me go after Andy Mellon," he told the co-author of his memoirs. Jackson, Irey recalled, had insisted on his help: "You are qualified and I need help." Morgenthau himself also called Irey after Irey hesitated: "Irey, you can't be 99⅔ percent on the job. Investigate Mellon. I order it." Irey demurred, explaining that he was friendly with Mellon, who was, after all, his former boss. He also added that from what he knew of Mellon, Mellon was innocent. But Morgenthau would not relent: "I'm directing you to go ahead, Irey." Irey sent one of his most skilled agents, Ralph Read of San Francisco, to help Jackson's prosecution team.

Shortly, the prosecuting team scored what they considered a coup. They got Mellon to confirm that he had used five of the loopholes on his old list. Morgenthau considered this admission, made under oath, as extremely damning. "Things that the courts approved outraged the Secretary's personal sense of justice," the coauthor of his memoirs, John Morton Blum, noted.

Mellon had the wherewithal to defend himself, and so did Insull. The power industry generally, if not the monolith the antitrusters depicted, was still organized enough to begin a countering action. In September a few shareholders of the Alabama Power Company went to court to protest Commonwealth and Southern's agreements with the TVA. Specifically, they asked that the court invalidate the sale of properties to the TVA. Loomis was perturbed at all the lawsuits; it was heartbreaking to see what had been a bold industry descend into perpetual litigation. He continued to withhold his talents therefore from the business sector: "He didn't want to have to fight the world," a relative noted. And this action seemed to work against Willkie. But while Willkie was a nominal defendant, via Commonwealth and Southern, the suit was also much in Willkie's interest, for it sought to curtail the TVA's authority. The Wilson Dam had been built in the name of national defense. The TVA had been permitted in the name of making a southern river navigable. Did their existence now give the government the right to litigate a takeover of the power business? The case, known later as *Ashwander*, might come before the Supreme Court, which could then invalidate the entire TVA.

On the last Sunday in September, Roosevelt delivered the sixth of his Fireside Chats. The talk summarized the administration's new attitude. First, it went back and forth on the correct policy to bring about recovery. On the one hand Roosevelt rallied the country in a collective campaign, calling for the "united action of management and labor" to bring recovery and proudly pronounced that "we are bringing order out of chaos." On the other hand he insisted that the country still counted on "the driving power of individual initiative." The president also complained about labor unrest. He spoke about a model overseas—Labourite Britain. "Did England let nature take her course? No." As the British press had already noted, Roosevelt said, much of the New Deal program was merely an effort to catch up to Britain's reforms.

In addition, Roosevelt asked for time: "There should be at least a full and fair trial given to these means of ending industrial warfare," he said in regard to the NRA. Finally, he railed against corruption—"thoroughly unwholesome conditions in the field of investment"—and made the correction of trouble a principal task of recovery.

In autumn came Insull's moment. After posting bail, and a long summer in the Seneca Hotel on Chestnut Street preparing his statements, Insull was ready. The federal prosecutors put forward eighty-three witnesses to show that Insull had criminally defrauded shareholders. The prosecutors told themselves that the sheer volume of their work ought to convince.

But Insull was about to strike back. Called to the stand, he began, not with the structure of his business as it stood in 1930 or 1932, but with his own story. It was an American classic: a poor childhood in Britain. A period as Thomas Edison's assistant, and then, the master's accountant. Opportunity in Chicago. The recent failures: "My judgment may be discredited, but certainly my honor will be vindicated," he said. The federal prosecutor, Leslie Salter, brought out Insull's salary to embarrass him—$500,000. Insull's attorney responded by showing tax returns that demonstrated Insull had given away more than his salary in charity in several years. When the business failed, Insull had not fled; on the contrary, he poured his own money into it. One of his problems was that he borrowed too much against his own name. The move to Europe had only come later. To the Chicago jury, which had known Insull longer than the prosecutors, the story was not one of simple crime and theft, but rather of the challenges of city building.

On a Saturday afternoon in late November, the jury reported out its verdict: not guilty. Insull celebrated—and was "showered with telegrams of congratulations," noted the *Nation* sourly. To the magazine, the case simply illustrated "the difficulty of sending a rich man to jail." Still, observers in the utilities world took note. Insull might have been a rogue. He might have handed out shares in his corporations too freely. He had lost the money of thousands of shareholders. And he had used aggressive accounting tactics. But in his day he had also achieved much: lighting up Chicago, earning enormous sums for some shareholders, building the opera. The British trade journal was in fact correct: Insull had done first many of the things that Willkie was now trying at Commonwealth and Southern and David Lilienthal and Arthur Morgan were trying at the TVA. To condemn Insull was to condemn enterprise.

Still, the New Dealers were not downcast, for sentiments such as those of the Chicago jurors toward Insull were not hurting the Democratic Party. On the contrary, the November election of a few weeks before had been a Democratic sweep, strengthening the number of Roosevelt's party allies in Congress. The days after the election found Democrats claiming 9 new senate seats, giving them a 69-seat majority. The crucial two-thirds majority was in Democratic hands. In the House, the early returns showed Democrats holding 318 seats compared with the Republicans' 99, and later it would emerge that the Democrats had done even better. In

Illinois, the land of Insull and Lincoln, Democrats had gained several congressional seats. Congressman Oscar De Priest, the sole black member of Congress and a Republican, was defeated. Voters selected to replace him with another black man, the Democrat and Harvard graduate Arthur Mitchell. The switch reflected a national trend: African Americans were becoming Democrats in great numbers for the first time since the Civil War. Pennsylvania elected a strong advocate of organized labor, Democrat Joseph Guffey, to the Senate, displacing Mellon's old spokesman, the pro-tariff Senator Reed. Where the New Deal was faltering economically, it was gaining politically. Roosevelt's radio voice was succeeding.

Relieved, the prosecutors regrouped. They noted that in the lower courts, NRA lawyers were winning some victories. The point at which the Supreme Court must affirm the NRA's constitutionality was drawing near. That month, just before the election, a federal jury at the courthouse in downtown Brooklyn had convicted the four Schechters. It was the first felony case that the Justice Department won under the NRA, and Walter Rice, the special federal prosecutor, was triumphant, calling it "a sweeping victory of immense importance." The brothers faced two-year sentences. *Schechter* was in the running to become the Supreme Court test case.

Tugwell relaxed with the president in Warm Springs, writing in his diary on November 23, "I spent nearly the whole day with FDR yesterday. Went to the pool and swam, he driving his own car. Went back and had lunch with him and talked until five in the afternoon." He was just back from a trip to Italy, and had got the chance with Mussolini that he had missed with Stalin. Like other Americans before him—Thomas W. Lamont, for example—Tugwell was impressed. Their meeting had been in a large rectangular room in which Mussolini sat at one end, positioned to intimidate. But, Tugwell noted to himself, the dictator had got up to meet him halfway. Now Tugwell and Roosevelt talked about international economics, both concluding that the United States could not be too optimistic about foreign trade.

"I felt again," Tugwell wrote of Roosevelt in his diary, "as I have before that my mind runs with him as with almost no one I ever knew before," adding: "I wish I had more wisdom. . . . I think he overestimates my intelligence; he couldn't overestimate my loyalty and affection." Even the small jokes that Roosevelt made at his expense only made him feel closer. Just before Christmas, cabinet members dining at the White House would comment on the poor quality of the champagne; the president later asked Harold Ickes whether he had "ever tasted worse champagne," then pointed out, laughing, that it was domestic champagne from New York, recommended to Mrs. Roosevelt by Rex.

In Washington several hundred social workers, labor leaders, and economists gathered at a conference on economic security. Frances Perkins declared that the country was being swept "by a wave of enthusiasm for the President's promised program of social security." Among those attending was Paul Douglas of Chicago, who presented a program for unemployment insurance that would pay benefits for twenty to twenty-six weeks. Roosevelt threw some cold water on the meeting when he rejected "fantastic" schemes and focused narrowly on plans for unemployment insurance such as Douglas was proposing. Still, the mood was more hopeful than it had been even in 1932.

On November 10, Roosevelt appointed Marriner Eccles, the Utah banker, as governor of the Federal Reserve Board, the top Washington position at the Fed. Accepting the job, Eccles, like so many Republican progressives before him, declared allegiance to Roosevelt and his party. "Previous to the last national election, I had always supported the Republican national ticket," Eccles said, "but I was not satisfied with their policies, which were not sufficiently liberal and progressive to meet changed conditions." "Governor Eccles will run the Fed as the White House wants it run," wrote *Time*'s editors, always to the point.

"Americans are following Roosevelt . . . as the Israelites followed Moses," concluded the London *Morning Post,* abashed to see such behavior in the world's most independent people. Eccles envisioned new Federal Reserve legislation, which indeed would become law the following year. Under the new Federal Reserve Act, monetary officials in Washington would control the money, supplanting New York. The law would give the Washington Fed the power to buy and sell bonds and therefore to control the quantity of money and credit in a more formal way. As chief in Washington, Eccles would also get a new title: he and subsequent Federal Reserve heads would not be governor. They would be chairman.

As for Willkie, he was now not sure whether to fight publicly. His company's stock was dropping at 1¼ On the one hand Roosevelt was making his threats more explicit than before. On November 18 the president traveled to Tupelo—the TVA's first municipal customer—to announce that what was being done here was "going to be copied in every state of the Union before we get through." Where was Roosevelt's mention of the fact that TVA was heavily subsidized? On the other hand the president was trying to reassure industry.

After the election—only days after—Roosevelt wrote to Newton Baker, "One of my principal tasks is to prevent bankers and businessmen from committing suicide!" The president issued a round of invitations to power executives, and there was one for Willkie.

The Commonwealth and Southern president duly attended his meeting, greeting the president with a reminder that Commonwealth and Southern was the president's power provider in Warm Springs: "We give you good service, don't we?" Roosevelt announced that his talks with the industry had been "entirely amicable." There was an element of satisfaction there. Now that Roosevelt had two more years, it was just as well that Willkie was dealing not only with Lilienthal but also the ultimate arbiter of utilities' future in the United States.

After his meeting, Willkie wired Edith, who was an old critic of FDR's: CHARM EXAGGERATED STOP I DIDN'T TELL HIM WHAT YOU THINK OF HIM.

As 1934 moved toward a close, Willkie confronted two realities. The first was that as much as he wanted to fight back more openly, he could not. It would be irresponsible toward his shareholders to declare war against Lilienthal; he would then have "blood on his hands," as he had told Lilienthal. Both Willkie and

Commonwealth and Southern simply had too much to lose. They had to work out more deals.

The second reality was that in the battle between David and Goliath in this particular industry—utilities—he was not, in the end, going to find a way to ally himself with the Davids. Uncomfortable as it made him feel, Willkie had to admit that he was finding himself siding with the monster.

8
The Chicken Versus the Eagle

November 1, 1934
Unemployment (November): 23.2 percent
Dow Jones Industrial Average: 93

There was one man who had nothing left to lose. He was Martin Schechter of the Brooklyn Schechters, the chicken butchers whom the Justice Department was prosecuting for violating the NRA. On November 1, the day they lost the first time in court, Martin and his brothers decided that they, the chicken, would fight back against the Blue Eagle. Within a year, their story would move from obscurity to the center of the debate about the New Deal.

More unlikely heroes than the Schechter brothers of East Fifty-second Street and Rockaway would have been hard to find. Unlike Willkie, they had neither a legal education nor corporate money behind them. Unlike Insull, they were unknown and unable to purchase city officials. Unlike the wealthy East Siders of their city, they had no social status to fall back on. Unlike their opponents in the Justice Department, the Schechters were inconsistent, almost comically ambivalent, about their case. These were not board members, not stock market players, but rather slaughterhouse men who served a market as humble as they were. Their English was laughable.

The name Schechter actually means "ritual butcher" in Yiddish. It derives from the Hebrew word for the same thing, *shochet.* Centuries before Martin and his brothers opened their business, both terms were used to describe those qualified to perform ritual slaughter of animals as required under Jewish dietary law. The brothers' parents, David and Molly, had brought them over from Europe. The 1930 census, by Herbert Hoover's old Commerce Department, says they were from Poland. Sam, Alex, and Aaron, who was also known as Abe, are on the list. The children were born in Hungary, while the family was in transit to the United States. There was also another brother, Martin, though he was not listed with the family in the census. David, the father, gave his profession as "rabbi" to the census worker, who in turn described his workplace as a "church"—this was the period when Americanization meant Christianization. In those years the term "rabbi" had a looser meaning than it does today; the fact that David was a rabbi could have meant that he was ordained, and it could have meant that he was merely a pious scholar. Still, the label meant that David Schechter ran a religious household.

For several years the brothers operated two firms: ALA Schechter Poultry Corp. and the Schechter Live Poultry Market. Their product spanned the range of poultry: leghorn fowl, colored fowl, heavy "fancy Indianas," and broilers that could cost as much as 27 cents a pound. Chickens were shipped to New York markets from across the country. There the Schechters bought them, then slaughtered and sold them, mostly to retailers. In short, the Schechters were middlemen who took their cut. This was just the sort of economic activity of which Tugwell disapproved and against which the Agricultural Adjustment Act tax had been directed.

The men operated a kosher business: the value in their product was that it conformed to Jewish dietary law. They employed rabbis and Jewish ritual slaughterers, *shochtim*. As was appropriate, these workers observed the Sabbath, halting work on Friday afternoon and returning only after sundown on Saturday.

Kashruth, the principle of keeping kosher, was about religion, but it had an earthly purpose as well. It had long served as a primitive but effective health code—one of the first health codes. Sorting out dangerously unhealthy animals of any sort was a core principle of *kashruth*; the Jewish proscription against eating pork probably came out of caution about trichinosis, a disease that often affected pigs in Europe and here.

The main fears were simple infection and tuberculosis. Everyone knew someone who had died of TB—from the children on the streets of Brownsville to Benjamin Strong, the New York Federal Reserve head whom the disease felled in the late 1920s. The word *glatt* in the phrase "glatt kosher" meant "smooth," referring to the desirable smoothness of the lung of the nontubercular animal. Tuberculosis was mainly passed through meat and milk, but businesses like the Schechters' were also cautious about poultry, which could pass along infection. Part of the value of their product was the health guarantee of the *shochtim.* Unhealthy birds the *shochtim* ceremoniously discarded. Customers, whether they were retailers or families, also had the right to choose their birds, and this in turn ensured that everyone involved had a chance to determine whether the product was as healthy as possible. *Kashruth* was not a modern health code, but it was a health code, a ghetto version of the Good Housekeeping Seal so angrily defended at the White House lunch by Tugwell's table neighbor.

Much in the Schechters' culture would have made it easier for them to sympathize with the New Deal than with its opponents. The Schechters were first-generation immigrants: in America, but not yet entirely of America. They were the kind of newcomers who had invested their money in the Bank of United States, the one that had failed. Their vocabulary was profane. Their intonation and syntax were the sort we today associate with a stand-up comic from the Catskills. It was the sort of speech high school teachers banned from their classrooms and judges banned in courtrooms.

To people like the Schechters, the New Deal sounded good, for the old deal was problematic: their industry had declined steadily since the crash of 1929; by 1933, nearly a thousand of the 2,000 or so jobs in their field in the New York area had evaporated. There were people in Brooklyn who would have gone straight over to the Soviet model, given the choice. One of the areas served by the Schechters, Brownsville, had elected Socialists to the State Assembly in the past. A Hebrew Butcher Workers Union Local 234 had existed in Brownsville since 1909; there was even, at the time of the NRA creation, or soon after, a group that proudly called itself the New Deal Kosher Butchers Association, Inc.

But this confidence in progressive thought antedated any New York experience. European immigrants as a collective had experienced life under a random ruler—a czar, a sadistic regional governor, a Hapsburg emperor. This left people like the Schechters with a natural sympathy for the underdog, and the New Deal was a project undertaken in the name of the underdog. One of the first New York projects to win PWA funding was the Triborough Bridge, whose construction the Schechters or their fellow tradesman witnessed as they carted their fowl about New York.

The NRA code that applied to the Schechters was the Code of Fair Competition for the Live Poultry Industry of the Metropolitan Area in and About the City of New York, a lengthy and forbidding document. Section 2 of article 7 declared it was prohibited "knowingly to purchase or sell for human consumption culls or other produce that is unfit for that purpose." Businesses could not sell unfit animals. Nor could they exchange in "destructive price cutting." The code prohibited "straight killing," defined it as "killing on the basis of official grade." The rule meant that customers might select a coop or a half coop of chickens for purchase, but they did not "have the right to make any selection of particular birds." This went directly against the old marketplace rule of customer choice.

Yet all these rules, the letter of transmittal assured, were "not designed to promote monopolies or eliminate or oppress small enterprises." The argument was that they would help small business by eliminating competition. FDR had personally signed this code into law by executive order just a few months before the first inspectors appeared at the Schechters' door. As for the Schechters, they had not signed on to it personally, but were bound to comply as members of the trade.

Even finding themselves under inspection over the summer of 1934, the Schechters had at first been inclined to go along. They knew of Walter Lyman Rice, the special federal prosecutor, and probably feared him. The young federal prosecutor had earlier pulled them in as witnesses—not necessarily willing ones—in his poultry corruption case of the late 1920s. Earlier in 1934 he had been in the news looking into poultry violations of the Sherman Anti-Trust Act. They had escaped that first case mostly unscathed, and sought to get through this action by keeping their heads low as well. They had tried to ingratiate themselves with investigators over the summer. Aaron Schechter reported later that he and an inspector, Philip Alampi, had gotten along "swell" except when Alampi was "insulting" the customers.

The brothers had also followed the rules involving "straight killing" as best they could. Aaron later told stories of customers who had asked to select their birds, and his own rule-abiding response: "I am sorry, but we are under the code, and that is straight killing." This, Aaron would point out, had caused him to lose customers. Brother Alexander offered to help an investigator interpret the company books, as he would later report. "Well, I helped him. In case he needed me or the bookkeeper, I always helped. If he does not know what is what, and I knew something about it, I would help him and show him and the other gentlemen would, too."

Joseph Schechter would later recall an inspector who had appeared "by the name of Bob." The inspector had been a "very nice boy. . . . He don't know from a chicken. And I started in to teach him what a chicken is, and my man and myself teach him what a chicken is, what a rooster is, and what a spring is."

But what the Schechters had kept coming back to in that first trial was that the inspectors from the government had been too hard to deal with. Aaron had recalled that Alampi had started arguing with customers and had also insulted one of them. "He told the customer that he is full of shit, and 'I am the Code Authority, and I got a right to do anything I want, and if you don't like it, get out.' " As Aaron had noted, this had seemed wrong: "Well, that hurted me a little bit, to hear that, so I called him aside and I said, 'Mr. Alampi; please try to restrain yourself from using language like that, because I will lose my trade, and the result will be that you will chase me out of business.' " Later, at another point, Aaron sent Alampi away. Alampi had returned with police. "Well, why don't you let him in? He is the Code Authority," Aaron remembered the police saying. "I said, 'I will not. I am not going to let anybody in here to ruin my business.' "

This little exchange, between a local man with a local policeman over an intruder from some distant office, had been a sort of parable of the offense to traditional federalism that the NRA and the codes represented. A local policeman was the locals' own guy. Washington's representative, or even an industry representative, was a trade enemy or an intrusive foreigner. In other words, the code inspectors had now indeed been doing something that did oppress small business, to repeat the language of the poultry code. They were busting in on an intimate private relationship: that of the small businessman with his customer. When the documents arrived, the Schechters' sense of resentment had only grown. The brothers had known that code authorities were after them, but learning of the sixty-count indictment, they realized the elaborate nature of the setup. They had been charged with selling unfit chickens to two men, Sam Tanowitz and Harry Strauber. This time, Walter Rice was not using them; he was targeting them. What's more, "the indictment was lengthy and vague, making it impossible to understand the nature and cause of the accusation against them," their attorney complained. The Schechters had trouble keeping up with their business while on trial; they realized they would have to spend the autumn downtown at the courthouse; they might go to jail. Their larger competitors could only profit from their misfortune.

The jury trial was presided over by the Honorable Marcus B. Campbell and took place in the District Court for the Eastern District of New York—just on the Brooklyn side of the Brooklyn Bridge. The scene in the courtroom was a culture clash. On the one side were the Schechters, clearly working people and clearly foreign born. Their lawyer was Joseph Heller, a fellow Jew who had attended Brooklyn Law School. On the other side were the gentlemen—gentlemen of the New Deal, and gentlemen of the New York legal establishment. The prosecutor was Rice. Judge Campbell was a Mason, a member of the fancy Brooklyn Club on Remsen Street, and a member of the Union League Club. He had been appointed to the federal bench by President Harding, and had been an active Republican in the Sheepshead Bay section of Brooklyn before that. He was of Brooklyn, like the Schechters and Heller, but his Brooklyn might have been the Brooklyn on another planet, it was so different.

The Schechters were especially frustrated that the government did not understand the consequences of its own "sick chicken" allegation. To sell a sick chicken broke the NRA code, and that was all the government lawyers understood. But to suggest, as they had, that Schechter chickens were unfit was also to suggest something that the Schechters viewed as far worse: that they were not good Jews. It was to suggest that their kosher slaughterhouse was not really kosher, and so unworthy of customers. In other words, the poultry code officers had done something worse than anger the Schechters. They had offended their dignity. These were things worth fighting over. The idea that a government regulation was higher than a religious precept was not something the Schechters were entirely ready to accept. Felix Frankfurter's quasi-religious feeling about Harvard Law School neither the Schechters nor their *shochtim* would have understood.

Because of their trade, the Schechters were also able to see something clearly, something that FDR had not seen. It was that the NRA code did not make sense. The clash came in several areas. The first was prices. The code forbade setting prices too low, in part to combat a general "low price problem"—deflation. But one could not drive all prices up generally by ordering a specific business to charge more. Something larger about the currency had to change.

And the NRA's price-fixing rule was destructive to smaller business. Their customer base was fragile: "Our business is not steady for every week," as one brother had testified in the fall. The market was fickle. "Every wholesaler," Martin Schechter explained to the Brooklyn Court, "has chicken dealers who go around, they are not steady customers." And: "Chicken dealers will walk into my place and wouldn't like my price, and he would go out again."

In fact, if there was anger in the city in those years about the price of meat, it was anger that the price was so high. Customers of kosher meat were among the most vociferous and organized of those angry people. The new A & P supermarket was selling fryers at 19 cents a pound, whereas the market price of live birds was something like 21 cents—competitive. This was the productivity the economists had studied in the 1920s. The next year, 1935, Jewish wives in Brighton Beach and Coney Island would boycott kosher meat with the demand that prices come down. Of course the live markets had to consider such facts.

Yet the prosecutors had not. "Your price is not very stable, is it?" a prosecutor asked Martin Schechter at one point, seeking to draw out evidence that Schechter was violating the code by undercutting with low prices. "The market isn't stable," Schechter replied. It might be 15 cents today, the market quotation, and tomorrow 18 cents. "We got our prices according to what the market might be," answered Martin.

The attorney, Walter Rice, accused the Schechters of competing too hard. And again, Martin Schechter struggled to grasp what to a businessman seemed an ungraspable concept.

RICE: You are in very keen competition with your competitors?

SCHECHTER: I do not understand that question.

RICE: Do you know what competition means?

SCHECHTER: I do.

RICE: There is a lot of competition between you and your competitors, is there not?

SCHECHTER: There is a lot of competition in every other business, the same thing.

RICE: Yes, and the competition in the whole slaughterhouse business is very keen, is it not?

SCHECHTER: Well it is keen in every other business in the same way.

It was not merely the word "keen" that Schechter had trouble with. It was also the difference between the language of law on the one hand and the language of economics on the other. Walter Rice was saying that the economy must operate one way because the law said so. Schechter was saying it could not. The market had its own natural laws, the laws of chicken blood, competition, and profits. It was neither good nor evil.

The Schechters especially resisted the notion that they were as immoral as the prosecution was trying to make them out to be. Their business was a small one: Martin was secretary; Alex, president; Aaron, treasurer. They were assisted by Shochet Gershon and Shochet Weisman. They paid themselves $35 a week, a wage that was less than that of the staff of the code authority who investigated them. The prosecutors repeatedly asked them about stockholders, and encountered incomprehension: there were no stockholders, other than the brothers. There was not even a formal business agreement.

And when the prosecutors ignored all this logic, it only made the Schechters and their friends move from resentment to outrage. They had seen that there was a kind of intellectual bigotry to the way the prosecutors were proceeding, and found it impossible to refrain from objecting. Walter Rice aimed to discredit as ignorant a long-standing member of the trade, a broker, Louis Spatz. He asked him: "You are against the code, aren't you?" "Certainly," Spatz replied. Next Rice moved onto the topic of Spatz's qualifications for such statements.

RICE: You are an expert?

SPATZ: I am experienced but not an expert.

RICE: You are not an expert on the effect of competitive conditions upon the prices of live poultry?

SPATZ: I am experienced—

RICE: Are you an expert? . . .

SPATZ: I am not an expert about anything.

Later Rice went on: "You have not studied agricultural economics." Spatz: "No, sir." Rice: "Or any sort of economics?" "No, sir." Rice: "What is your education?" Spatz: "None. Very little." But Spatz got past the intimidation, and struck back: "In my business, I am the best economist." Rice parried: "What is that?" And Spatz now qualified: "In my business I am the best economizer."

Rice thought he had gained a small advantage in the case by making Spatz betray his lack of education through the inappropriate use of a word. More generally, both Rice and the judge had sought to use the Schechters' social class against them. The prosecutor announced that he wished to "have that word spelled in the minutes, just as he stated it," so that the error might go on the record.

The uneducated Joseph had often digressed on the stand to talk about a broken leg he suffered at the time of the investigation—"the only thing I can move around is with sticks or crutches." Eventually, Rice grew impatient: "Now if your Honor please, I object to this constant repetition of his physical condition." Schechter: "Well it was that way." The Court: "Don't argue."

The four brothers could not have been entirely aware of it, but in other courts, far from Brooklyn, a similar debate over government regulation of the market was taking place. All across the country, the NRA was being litigated. The prosecution of the various illegal traders of "hot oil" was coming along. Ickes had assiduously tracked illegal sales of oil, but more had continued to pop up. Ickes told a lawmaker, "I have moved heaven and earth on the matter"—stopping the "hot oil" boys. The *Belcher* case was being prosecuted before a federal judge in Birmingham, though it was becoming clear that a fact about William Belcher—he was blind—made it hard to make him seem the exploiter.

Nonetheless, as the Schechter case moved forward, the scale of the prosecution's ambitions had come to look grotesquely large. The Schechters were accused of selling unfit chickens, but this accusation, in the end, was based on the selection of ten chickens, which was then reduced to three suspect chickens, which, upon autopsy by the health authorities, turned out to include only one unhealthy chicken. It was an "eggbound" chicken—

eggs, upon its slaughter, were discovered to have lodged inside it, something that would have been hard for the Schechters to detect before sale. That they had knowingly planned to sell an unfit chicken was hard to prove.

When it came time for judgment at the end of October, Judge Campbell sensed that the jurors, New Yorkers all, might be compelled by the Schechters' story and put off by the government's prosecutorial zeal. He warned them, "Decide it on the evidence, and not on some views you may have." He also did what he could to stop Heller, their lawyer, from making the jurors aware of the consequences of their action. Heller spelled it out in his summation: "Gentlemen of the jury, would you like to be put behind bars for a thing like this?" At which the judge intervened: "Now about the bars, they do not do that. The sentence rests upon my conscience, not theirs." He had warned, a few moments earlier, that "you are trying to tell them what punishment is attached to any one of these offenses; that is not the province of the jury."

Judge Campbell fined them $7,425, years of salary, and he sentenced them to jail: Joseph to three months, Alex to two, and the other brothers a month apiece. This last was particularly painful, since the Schechter brothers had families—Joseph's wife felt humiliated.

As they had testified in the case, they had never been in trouble before. The game of life seemed stacked against them. "First Felony Case Is Won under NRA," the *New York Times* trumpeted; Walter Rice was quoted in the paper speaking of "a sweeping victory of immense importance."

That winter, the circuit court rejected the Schechters' appeal. But they determined to fight on. Columnists Drew Pearson and Robert Allen would later mock the Schechters' persistence: Joe Heller, whom they described as speaking with a "Brooklyn Hebrew accent," and looking "hawk nosed," "labored over his lawbooks in Manhattan, determined to rank his name alongside of that of Daniel Webster." Still, though their case seemed improbable, they would go higher, even to the Supreme Court.

And perhaps the Schechters' move was not so improbable, for now there was a shift in the country, one that Roosevelt could not entirely overwhelm, even through his radio bond with the people.

The shift started with the NRA itself and the discovery of its punitive side. The Schechters were not the only NRA violators headed for prison. In York, Pennsylvania, a former Cornell fullback named Fred Perkins had paid 20 cents an hour to a staff of ten who built lighting batteries for him. NRA officials had insisted that he pay the code rate—40 cents an hour. Perkins had showed his books to the officials—his profits for 1933 were only $2,531—and offered to raise the wage to 25 cents. Then he had asked for an exemption to the relevant code—he was denied. The government prosecuted, *Time* magazine reported. Not able to pull together the $5,000 required for bail, Perkins spent weeks in prison even before his autumn trial. He lost.

In Alabama, there was another, similarly ugly story. The day before the Schechters had lost their first case, the federal judge in Birmingham, William I. Grubb, dismissed code violation indictments against an Alabama lumberman. The judge had agreed with Belcher that his decision to allow his employees to work more than forty hours a week was not something that federal law could address.

Many economists were disturbed. Irving Fisher, in a letter to his son, quoted another professor who said the letters "NRA" stood for "National Retardation Affair." Senator Huey Long of Louisiana, always eager to tilt with Roosevelt, was urging citizens to ignore the NRA codes. Long, more primitive than either Roosevelt or Hoover, saw similarities between the two presidents. At one point he would tell the Senate, "Hoover is a hoot owl. Roosevelt is a scrootch owl. A hoot owl bangs into the nest and knocks the hen clean out and catches her while she's falling. But a scrootch oil slips into the roost and scrootches up to the hen and talks softly to her. And the hen just falls in love with him. And the first thing you know—there ain't no hen." In the end, the Justice Department withdrew on *Belcher*, fearful of losing.

Some unions still liked the National Recovery Administration and hoped, of course, that its labor provisions might be preserved no matter what. That spring William Green of the AFL would call for a general strike by 18,000 New York clothing workers if Congress did not extend the NRA for two more years. But business was finding its voice on the topic. The man who sold that clothing, Percy Straus, the chairman of R. H. Macy, the department store, criticized the codes bitterly for the way they pushed up prices: at a time when no one had enough cash, "the consumer is forgotten." It was rather the employer who was forgotten, alleged the executive director of a retailers group, the National Dress Manufacturers Association. Mortimer Lanzit told a crowd at the Hotel Astor in January that the industry's code was hurting employers by giving them fewer rights than workers. Turning FDR's phrase against him, Lanzit argued that it was the employer who was the "Forgotten Man," and that the International Ladies Garment Workers

Union placed so many demands on companies that they "blocked reemployment."

In New York, Ray Moley was now playing broker between the administration and business, hosting what quickly became known as "Moley dinners." In late January, Tugwell attended one. He wrote in his diary that he "was, so to speak, put on the spot. I spoke for about twenty minutes or half an hour respecting the administration." Tugwell went on: "After I had spoken about fifteen of the businessmen spoke in turn. The general burden of their talks were rather querulous complaints about the unfriendliness of the administration and about the lack of confidence which businessmen have in the present administration. They rather indicated that until there was greater confidence, prosperity would not return." He added that "Willkie was present and did a good deal more than his share of the complaining."

Even within the NRA, there was disagreement. From the Consumers' Advisory Board of the NRA, Paul Douglas wrote dissents to the codes. He was, after all, the same old Douglas, happy to point out something that seemed wrong, even if he was going against the group in doing so. It seemed wiser to Douglas that the administration spend its energy on unemployment insurance and the broader concept of what was sometimes called "social security." Other critics on the left pointed out that the NRA helped big business at a cost to smaller businesses. This argument was valid. A price set to suit a big firm, with its economies of scale, was low enough to drive a smaller firm out of business; a wage set high enough to meet Washington's goals might be tolerated by a larger firm, but it killed off a smaller one. The NRA institutionalized cartels. And cartels were perceived by most citizens as one of the principal reasons the average fellow now had so much trouble.

General Johnson and the NRA now occupied Hoover's new Commerce Department. The task had expanded to fit the building. Nineteen thirty-five was the year when a new board game invented by a Philadelphia man was becoming a surprise best seller: its name was Monopoly.

A woman from Connersville, Indiana, wrote to President Roosevelt:

> *I have been employed as a clerk at E.J. Schlichte Company this city for seven years five months until the N.R.A. went into effect. They let me out said they coulden' pay me $14 a week. When the NRA Went into effect, I was so happy I had planned to lay in some coal and pay on some bills I owe, I guess I was too happy.*

Yet other critics focused on the sheer unreality of the top-down culture of the NRA. Its demands for synchronization of wages and salaries were, when one thought of it, worthy of parody. As the humorist Ogden Nash put it in a poem:

Mumbledy pumbledy, my red cow,
She's cooperating now.
At first she didn't understand
That milk production must be planned . . .
. . . But now the government reports
She's giving pints instead of quarts.

In words less succinct but equally authoritative, the scholars at the Brookings Institution came to the same conclusion. The codes didn't work. "In trying to raise the real purchasing power the NRA put the cart before the horse," wrote the Brookings authors. The idea of increasing national income—getting true growth—by "a general increase in nominal wages," was, the authors said, a doubtful one. "The conclusion indicated by this résumé is that NRA on the whole retarded recovery." In Congress, Roosevelt managed to win a commitment of a one-year extension for the NRA. Still, he now knew that the Supreme Court test would be decisive.

Roosevelt and the New Dealers figured that if they pulled in their horns a bit, the assaults on the NRA might abate. At the AAA, that meant the firing of Jerome Frank, the left-leaning general counsel. The occasion was a contract the former clerk of Oliver Wendell Holmes, Alger Hiss, had written requiring landlords to keep tenant farmers for the length of their cotton contracts. Frank was out, as well as Lee Pressman, with whom Tugwell would work again. But Hiss, lower ranking, stayed on for a time. Still, as Tugwell, who supported Frank, wrote in his diary, the issue was really part of a new plan "to rid the Department of all liberals."

Frances Perkins likewise proceeded cautiously. In February, the giant Labor Department building whose cornerstone Hoover had laid was finally finished, and Frances Perkins and her department moved in—the department men's chorus, Crescendo, sang at the occasion. But Perkins and her staff were not entirely comfortable with the grandiose feel. They had come into office claiming differences with Hoover, and yet the building was a daily reminder of the similarity of the two administrations' work. She and Douglas had their plans for unemployment insurance and pensions for senior citizens in draft bill form. At a tea at his house the year before, Perkins had sat beside Justice Harlan Stone, and he gave her a tip. She had confided her fears that any great social insurance system would be rejected by his court. Not so, he said, and whispered back the solution: "The taxing power of the federal government my dear; the taxing power is sufficient for everything you want and need." If the Social Security Act was formulated as a tax, and not a government insurance, it could get through.

Still the legislation she developed did not speak of taxes or entitlements. It spoke of an old-age reserve "account," the implication being that an account with the government was safer than one at a local bank. This was a powerful argument, given the rate at which banks had so recently collapsed. The administration depicted the whole program as government insurance for senior citizens, insurance that would come in tandem with unemployment insurance for workers and aid for the indigent.

Even this model, however, was encountering resistance. Roosevelt himself saw that while the program's revenues might cover its costs now, the numbers from the actuaries suggested that there would not be enough money for old-age pensions for future generations. Morgenthau too made it clear to Perkins that he disapproved of any demands on the Treasury, even in future.

"Ah," Perkins reports Roosevelt saying, "but this is the same old dole under another name. It is almost dishonest to build up an accumulated deficit for the

Congress of the United States to meet in 1980. We can't do that. We can't see the United States short in 1980 any more than in 1935." Perkins noted—as others before her had in similar situations—that the president was "in the midst of one of the minor conflicts of logic and feeling which so often beset him but kept him flexible." Roosevelt's opponents were firmer, especially Bennett "Champ" Clark, a Democratic senator from Missouri. If the Social Security program was entirely about social welfare, he said, then why not allow private companies with pension programs already in place to choose to stay out of the government program? This would allow a genuine private-sector counterpart against which to measure the government program. Clark argued hotly that adding the Social Security levy to the costs of supplying the private pension would be onerous for some employers, and prohibitive for others. Without the opt-out of the Clark Amendment, companies would give up supplying private pensions. Why should they pay double when the government would do their work for them?

Outside Congress, there were also challenges. On January 2, 1935, the old chairman of the Democratic National Committee, Jouett Shouse, published a major article pointing out that the New Deal had created thirty agencies, nearly all close to the executive, leaving "the average citizen bewildered." Shouse was joining up with Al Smith, Roosevelt's old ally, to form a group called the Liberty League, whose goal was defeating him in the 1936 contest. James Warburg, angrier as the months passed, was giving speeches against Roosevelt's "planned economy." He was preparing to publish yet another book about Roosevelt. This time his title was *Hell Bent for Election.* In his introduction, Warburg would describe a party program that sounded just like Roosevelt's—one that planned public pensions, more rights for organized labor, and so on. Then he would point out that that agenda had been not Roosevelt's in 1932 but rather that of Norman Thomas, Jim Maurer, and the Socialist ticket. It was a rant, but a rant that would sell nearly a million copies.

In this, the sixth year of the Depression, people were developing new responses to adversity, some of which competed. Francis Townsend, a doctor, was selling his own pension scheme, at campfire hearths and in meeting halls across the land. Father Coughlin was on the radio, railing about the evil dole and "relief that has failed to relieve." Huey Long had his new post in the Senate.

But there were also those whose responses existed on an entirely different plane—the spiritual, or the personal. One was Father Divine. More aggressive than the Schechters, the cult leader sought out chances to provoke—his own form of civil disobedience. In

New York City, his Peace Restaurant mounted an anti-regulatory battle of its own, refusing to post hours of labor as required by New York State law, provoking action by state authorities. He operated a boarding-house without a license, so that officials pulled him into court. He allowed his truck drivers to drive without licenses—another clash. Father Divine's followers sought to register to vote under their quirky names; a judge rejected the application by sisters Truth Delight, Charity Star, Mary Magdalene Love, and Joy Praise to vote under those names. On the names, city officials were as intolerant as the Schechter prosecution: "We do not intend to let them make this department ridiculous," the chief clerk said.

Father Divine's power was impressive. By now, a letter addressed simply "God, Harlem, USA" would reach him. On Easter Day of 1935, ten thousand of Father Divine's followers, mostly black, a few white, mounted a four-hour parade along Harlem's Seventh Avenue, singing "The world in a jug, and the stopper in his hand." Father Divine himself rode, the paper reported, in a blue Rolls-Royce Victoria, "a white policeman standing like a footman on the right running board, and a negro policeman poised likewise on the other." Father Divine wore Easter gray. Faithful Mary, his chief aide, followed him.

Father Divine was also developing a new plan—to make a statement through property purchases. He was buying up cheap farmland along the west side of the Hudson, widening his empire. He was also considering whether he might lead his flock into politics.

Bill Wilson, the hard-drinking stock analyst, for his part, was also developing a new response to adversity. He didn't like the New Deal—a Green Mountain man like Coolidge, he found that his instincts ran against it—and he wrote letters to Roosevelt, longhand, as he drank. People should find a way to solve their problems closer to home. His wife, Lois, collected the little pieces of paper—she didn't know why, but she later said she had thought they might be of value some day.

In May, Wilson headed out to Akron to see if he could gain control of a company through a proxy battle. The deal failed, humiliatingly, and he found himself standing in the Mayflower Hotel bar with only $10. Frantically he phoned a local minister, Reverend Walter Tunks, who gave him the name of Henrietta Seiberling, a woman who had tried to help another alcoholic, a doctor named Robert Smith.

The next day Wilson and Smith met, in the afternoon, after Smith had slept off a binge. The meeting changed their lives. "Dr. Bob," as he was known, was also from Vermont—St. Johnsbury, where Coolidge had attended school for a time. The two alcoholics decided that by sharing their challenge, they could help one another. Then and later, they talked out a few precepts: professionals weren't necessary; alcoholism was a sickness, not a moral weakness; alcoholics were never cured, but a group might keep them on the wagon. Their insight was in part medical, in part psychological, and in part sociological. Part of the problem of the alcoholic was loneliness, especially nowadays—there was no longer the sort of New England village green where the men had grown up, to find consolation. Yet the traditional answer—going to a clergyman, or a doctor—did not seem to work for them. The two could not retrieve the old Vermont village, but they could build a new village, a community of alcoholics.

The Roosevelt entourage was working on another plane entirely, busy consolidating after the midterm election results. Chase was writing an article for the magazine *Current History* declaring that the New Deal was a victory for collectivization. Tugwell entertained the president with work on his favorite topic, agriculture. The descendant of fruit farmers reconnected with the descendant of Dutchmen on a topic dear to both their hearts: trees. Carrying mounted specimens of insects and leaves over to the White House from the Agriculture Department, Tugwell showed the president the damage that a new tree blight—Dutch elm disease—was doing to trees in the fifty-mile radius of Manhattan. As undersecretary of agriculture, Tugwell was spending $560,000 to forestall further devastation. Roosevelt was in the process of giving Tugwell his own agency, a resettlement administration, where he would be less likely to quarrel with agriculture secretary Henry Wallace. There he would rehabilitate poor people and poor land together—the unity of it pleased Tugwell. Roosevelt established the Resettlement Administration by executive order, thereby avoiding the need for congressional approval. The RA would be funded with emergency funds in another bill, one that could not be labeled "Tugwell." Roosevelt admitted to himself that he was finding it costly to defend Tugwell. Frankfurter was back in the country after his own pilgrimage, to Britain, and two of his protégés, Tom Corcoran and Ben Cohen, had jobs in Washington. Already, the year before, the pair had written to Frankfurter in Britain that "[t]he Tugwell crowd has been pushed by its enemies—and its own loose talk—away over to the left. Ray is vacillating considerably toward the right"—the Ray was Moley.

There was one place where resistance built that the administration could not ignore: the courts. On January 7, Supreme Court justices had rejected the government's defense of presidential authority in the "hot oil" cases. This was an affirmation, at least in part, of the public's sense that the NRA's activities could go too far.

The Justice Department took consolation in the fact that the opinion left unaddressed whether Congress had the power to regulate industry—and fought its other battles. The very next day, January 8, Attorney General Homer Cummings donned a pair of gold cuff links that had been given to him by the president and began the oral arguments in the gold case. Cummings reminded the justices of the emergency of the Depression, and railed against the "supposed sanctity and inviolability of contractual obligations."

Roosevelt, uneasy, sought to determine what his options would be if the Supreme Court ruled that his gold policy was unconstitutional. Days after the oral argument began, he told Secretary Henry Morgenthau and Homer Cummings at lunch that he hoped to keep the bond market in confusion until the Supreme Court decided the gold-clause issue. Then, if the Court decided against the administration, things would still be so rough that the people would turn to the president and say: "For God's sake, Mr. President, do something."

Cummings liked the tactic. Morgenthau was horrified. "Mr. President," he told Roosevelt, "you know how difficult it is to get this country out of a depression and if we let the financial markets of this country become frightened for the next month it may take us eight months to recover the lost ground." Morgenthau might be Roosevelt's yes man, but he had already learned a few things at Treasury. Roosevelt indicated to Morgenthau the next night that he had been kidding, but Morgenthau believed the reality might also be that the president had simply, upon consideration, changed his mind.

As it turned out, the Court did not overturn the administration's gold policy. After all, Congress had the power to regulate the currency, and it, in turn, had given Roosevelt the authority to manage the money. Roosevelt was satisfied, writing to Joe Kennedy, "With you I think Monday, February eighteenth was a historic day. As a lawyer, it seems to me that the Supreme Court has at last definitely put human values ahead of the 'pound of flesh' called for by a contract." But Justice James McReynolds delivered a soliloquy: the New Deal's "flippant approach to currency manipulation" was dangerous. Congress had no power to destroy the gold-clause commitment. Roosevelt was like a tyrant. "This is Nero at his worst. As for the Constitution, it does not seem too much to say that it is gone." The meaning of the news was something the country found hard to grasp. Clearly it affected all private contracts. The *Christian Science Monitor* noted that it cut the value of "$75,000,000,000" in contracts—nine consecutive zeros being, at that time, something Americans were not accustomed to seeing. Railroad bonds were hurt, but so were Liberty Bonds, which seemed, to McReynolds, especially bitter. Utility bond holders' losses were also now assured. The Dow moved up to 107—it had been 105 the day before—in the hours after the news, but the movement this time was more one of relief that a decision had been made than cheer at the news. The next day the index settled down and hung around 100 for the better part of a month. The gold issue, at heart, was one that was applicable to all contracts and property. Clearly McReynolds would react the same way if it came to the Schechters—to their right of contract with employees, to the right of the consumer to pick his chicken, to the general intrusiveness of the NRA.

The cases against Mellon—there were three now—were also not progressing quite the way the government had hoped. Robert Jackson, the general counsel for the Bureau of Internal Revenue, spent the spring before the Board of Tax Appeals seeking to prove that Mellon wrongly claimed deductions for losses on his 1931 returns, committing tax fraud.

The preceding year, Mellon had fought back with a letter. Now the seventy-nine-year-old fought back from the witness stand, and, like Insull, made his impression. He repeated the points that were now becoming familiar in such defenses: It was wrong to assail a man for doing what was legal. When it came to his own case, Mellon had not claimed too many deductions, he had claimed too few.

Mellon's son and daughter had grown up watching their parents experience legal trouble from time to time—legal trouble was something that affected all wealthy families. Still, Mellon's son Paul was shocked at the persistence and aggressiveness of Morgenthau and Jackson. The week that Paul Mellon planned to marry and head off for a honeymoon in Egypt, he received a subpoena requiring that he appear in court. Rather than allow the prosecution to interrupt his honeymoon, Paul climbed out a window of the Mellon Bank that led to an inner courtyard, climbed in another window, and escaped in an unobserved elevator. The younger Mellon and his bride made away safely without being disturbed by Treasury lawyers again.

In late March, Mellon turned eighty, and he took the opportunity to speak out. The Dow stood about where it had in the fall of 1921. When reporters approached him for birthday thoughts on the state of the world, he did not argue his own case but rather chose to make a philosophical point that was both devastating and optimistic. Mellon commented that "present conditions, however distressing, especially in terms of human suffering, reflect on a passing phase in our history." In the context of American progress, Mellon said, the Depression was a "bad quarter of an hour." The summary shocked not so much because it seemed so wrong, though it did—it was the equivalent of predicting a stock's rise at its lowest point, going long at the bottom of a seemingly bottomless market. But the larger audacity in the statement was that it dared to question the premise of the New Deal—that crisis was somehow permanent.

In April, Jackson persisted, attacking a trust that Mellon had created for his paintings. The charity did not meet the conditions it claimed to meet and therefore represented illegal tax evasion; Mellon's charity was a "mask" for selfishness. Mellon's art was becoming a theme in the prosecutions, evidence of wealth and greed. His attorney, Frank Hogan, was mocked as a criminal lawyer representing one more criminal. At the annual banquet of the Washington press, the Gridiron dinner, on April 13, one of the skits mocked Mellon and Hogan. The Hogan character, played by a journalist, announced the opening of the Andrew W. Mellon

National Gallery—and displayed an art collection that included an image of the district jail, with the title *Right up Hogan's Alley.*

On April 29, Jackson suffered an embarrassment when the Board of Appeals rejected a 4,500-word document he had submitted asking for access to certain Mellon letters. The board judges called his memorandum "false, ill-tempered and not useful." Jackson, insulted, moved to withdraw himself from the case—and win time for the government lawyers to regroup in Washington. As for the matter of tax rates themselves, some in Washington and on Wall Street believed that they could not go higher. The U.S. tax schedule had high thresholds—the top rate began to be applied at an income of $5 million. But federal rates were high and there were state taxes. Benjamin Anderson of Chase noted that the total rate for the wealthy New Yorker, 69.9 percent, was the highest in the world. Why would capital want to come to the United States when it could invest elsewhere? In order to come back, Anderson would argue that summer, America had to remember the importance of being relatively competitive in tax terms. Death duties on large estates were especially important, and here the United States was losing out to Britain.

Lilienthal for his part was feeling more confident. Early that year he had had a chance to show up Willkie in public when James Warburg's New York Economic Club hosted them both at the Hotel Astor. Warburg presided, and the guest of honor was James Bryant Conant, Harvard's president. Lilienthal called the holding company structure a "financial tapeworm. The patient always seems hungry and the more he eats the thinner he seems to get. The patient thinks he ought to have more food." Lilienthal's conclusion: "The doctor may decide that what the patient needs most of all is to get rid of the tapeworm."

Later that spring, testifying before Congress, Willkie would come up with a successful counterargument. He read aloud from a letter that FDR had written in the 1920s around the time of his purchase of his center at Warm Springs. Roosevelt wanted to link Warm Springs to a high-tension power line nearby, and complained that Warm Springs suffered from the "high cost and inefficient service of small local power plants." Holding companies were the remedy to that, yet now, for some reason, Roosevelt's team was vilifying them.

That spring, Willkie and other power executives pointed out three angles from which the Roosevelt administration was attacking them. The first was now old history: the government was spending public capital to enable the TVA to outcompete the private companies. It was also, through Ickes, wooing towns by subsidizing construction of their own power plants on the understanding that they would then purchase directly from TVA. The second was now old history too—the prosecution of figures like Insull. Though Insull had won his case, his business was annihilated. From time to time the grand old man made statements about a new chapter in his life, but there was little reality to the boast.

But it was now a third angle of attack that was the greatest threat. Frankfurter had assigned Corcoran and Cohen to the New Dealers. Corcoran had joined the White House staff in March. There had long been public utilities legislation in the works, but the pair were now drafting an aggressive version, and Roosevelt's strengthened majority in Congress meant that it might become law. This version planned to kill off all but a few holding companies by 1940. The act's official name was the Public Utilities Holding Company Act. Willkie called it the "death sentence act." The government's argument that only it or the TVA could provide modern-scale power had been wrong. "Think it over, Uncle Sam," a desperate advertisement by United Gas Improvement Co., the oldest holding company, read.

On Commonwealth and Southern letterhead, Willkie penned a letter to all 200,000 shareholders of Commonwealth and Southern, pointing out that the PUHCA endangered their holdings mortally. Much of the destruction, he noted, had already happened. The value of utilities securities had dropped $3.5 billion since Roosevelt gave his inaugural address. Utility holding companies were near death, and PUHCA would finish off the job.

Another power executive, F. S. Burroughs, vice president of the Associated Gas and Electric Company, told Congress in April that his firm's securities had lost more than $500 million in market value due to earnings lost "from acts of governmental agencies during the past six years." That was a figure that matched the scale of destruction by Insull's empire that the anti-Insull crowd always cited. The new Liberty League added its voice on the TVA, stating: "Never have the dreams of bureaucrats flowered so perfectly as in the Tennessee Valley."

The most important suit for Lilienthal and Willkie was *Ashwander,* brought by shareowners in Alabama Power, a subsidiary of Commonwealth and Southern. The shareholders argued that Commonwealth and Southern had no right to contract with the TVA. In February, a federal district judge found that the plaintiffs were right. "Because there are generators at Wilson Dam that doesn't give the TVA or the U.S. government the right to sell as a private agency in Alabama," Judge W. I. Grubb—the same justice who had found for

Belcher—ruled in Birmingham. Judge Grubb now told the towns that taking government loans in exchange for giving its business to the TVA—Ickes's program—was illegal. Across the country various lower courts were blocking David Lilienthal's projects on similar grounds. Willkie was not visibly pushing *Ashwander*—indeed, he was actually, as party to the contract, still on the defendants' side. But news of *Ashwander* was heartening, for it might mean that the Supreme Court might find the TVA unconstitutional.

At the Justice Department, the lawyers who originally wanted to use *Belcher* to test the NRA were now casting about for a new case. Seeing the Schechters defeated twice, they decided this would be the case they would ask the Supreme Court to hear. Felix Frankfurter, who had opposed *Belcher* as a test case, also argued against taking on *Schechter.* As a student and friend of Louis Brandeis, he knew that the author of *The Curse of Bigness* would not be sympathetic to so big a project as the NRA. ff suggests most impolitic and dangerous, Tommy Corcoran telegrammed to President Roosevelt of *Schechter* after speaking with Frankfurter. Frankfurter further suggested that Roosevelt "hold whole situation on NRA appeals in abeyance."

But events did not go Frankfurter's way. Many of the other New Dealers, especially Donald Richberg, believed that with *Schechter,* unlike *Belcher,* they stood a good chance of ramming the case through and vindicating the New Deal. After all, the Schechters had lost in the lower court. And while the New Deal might not have the support of all the country, or many of its judges, it still had powerful allies. It also appealed to nationwide prejudices.

Evidence of this showed up in the work of Washington's leading columnist team, Drew Pearson and Robert Allen, who tended also to emphasize Rice's unarticulated theme—that ghetto traders could not be in the right. The pro-Roosevelt columnists would later title their lengthy account of the Schechter case, "Joseph and His Brethren." Playing the Jewish aspect for all it was worth, they would write of the Schechters that "where the kosher butchers of the city work in filth, blood and chicken feathers, they operated jointly a prosperous pair of smelly chicken companies."

The oral arguments of *Schechter Poultry Corp. v. United States,* docket number 854, began at 3:47 p.m. on May 2, 1935, in the old Supreme Court at the Capitol. Donald Richberg, the NRA's lawyer, presented the argument that this, like other NRA cases, was a case about the national emergency of the Depression and therefore required a special kind of law. The NRA, he said, was part of a "national problem, which cannot be considered wholly dissociated from the condition which brought about the act." The government had been asked to "protect against the evils of this unparalleled depression." If it didn't, and defended liberty in the abstract, it was merely protecting "the liberty to starve." The case as Richberg and the new solicitor general, Stanley Reed, presented it was about grander things: the desolation of the Depression generally, the authority of government to reach down and determine little outcomes. Their deductive argument had the advantage of conveying the urgency of crisis, but it also risked appearing grandiose or vague. Brandeis asked a sharp question in regard to the Commerce Clause matter: "From where do the slaughterers buy their chickens?"

Later came Joseph Heller's turn. And he argued inductively, from the bottom up. The Schechters' business, he began by pointing out, was indeed not really interstate commerce, for the Schechters purchased almost all their chickens in the New York area, from "commission men" who in turn purchased them as they were shipped in. In the language of the Commerce Clause debate, the chickens "came to rest" in New York; they were no longer part of the greater chain of interstate commerce, at least not at the stage in which the Schechter brothers were actors.

Then Heller moved on to the issue that had generated ten of the seventeen sustained counts in the indictments against the Schechters': "straight killing." Here he argued not the legality of the code but its practicality, making an effort to translate the Jewish poultry market into the general American culture. "And let us say that John Jones"—not Sam Tanowitz—"for example comes in and says, 'I would like to buy three chickens weighing five pounds apiece.' Under this straight killing provision of the Code, we could not sell them to him in that way."

Heller also emphasized what he and the Schechters had come to recognize as an important part of the case, the insult that the code delivered to small business: "My client has never assented to this code, and he was put out of business by this code." Heller, finally, pointed out the cost that the code exacted from the consumer: his essential right to choose what he was buying. "He would be prohibited from making that selection. We would have to pick out the first three chickens that came to our hands."

The Schechter lawyers expected sympathy from James Clark McReynolds, the judge who had declared the Constitution "gone" in the gold case. And they were not disappointed. McReynolds wanted to probe the meaning of straight killing, and he started with the chickens.

"How many are there in a coop?" There were thirty to forty, according to the size of the coops. "Then when the commission man delivers them to the slaughterhouse, they are in coops?" They were in coops. "And if he undertakes to sell them, he must have straight killing?" He must have straight killing, yes. As Heller put it: "His customer is not permitted to select the ones he wants. He must put his hand in the coop when he buys from the slaughterhouse and take the first chicken that comes to hand. He has to take that."

At this point there was laughter in the court.

Then Justice McReynolds asked: "Irrespective of the quality of the chicken?"

Irrespective of the quality of the chicken, Heller replied.

Later on, Justice Sutherland asked, "Well suppose however that all the chickens have gone over to one end of the coop?" (More laughter.)

Late in the game a big Wall Street law firm, Cravath, DeGersdorff, Swaine and Wood had joined Heller as the Schechters' counsel. Now Frederick Wood tried to point out the gravity of the widening of the government's powers. He argued that it might be all right to go the way of Mussolini or Hitler, but a constitutional amendment was necessary for that, not merely an act of Congress.

But it was the merriment in the courtroom over Heller and his chicken crates, and not the Mussolini analogy, that seemed to matter. The laughter showed that the

NRA was a shaky house and, as Heller then put it, "The whole Code must fall."

Throughout May the criticism of New Deal agencies grew. In the middle of the month another Liberty League leader went on to assail the New Deal, saying that the country was now wrongly "mesmerized by alphabetical white rabbits."

Sensing potential for a show, crowds packed the old Capitol courtroom May 27, the day the opinion was to be delivered. Two other cases were announced before *Schechter.* The first was a unanimous opinion finding that Roosevelt had acted wrongly when removing William Humphrey, the FTC commissioner. Next, the Court overruled the Frazier-Lemke Act. This was a blow for property rights—the act had limited the ability of banks to repossess property. The Court ruled that this violated the takings clause of the Constitution. Contracts between private individuals were important after all, the majority opinion said. Even a contract between a starving farmer and a nasty bank had to be honored, and the government did not have the power to intervene.

It was Justice Hughes who read the *Schechter* finding. It too was unanimous. "Defendants do not sell poultry in interstate commerce," he said early on, thereby rejecting the authority of the NRA. "Extraordinary conditions may call for extraordinary remedies. But the argument necessarily stops short of an attempt to justify action which lies outside the sphere of constitutional authority. Extraordinary conditions do not create or enlarge constitutional power." The NRA had abused the Schechters, and other businesses, through unconstitutional "coercive exercise of the law-making power."

In a separate opinion, Justice Cardozo used language more biting to speak about the doctrine of delegation, the Constitution's limitation of Congress's power to let regulators write law. "Here, in the case before us, is an attempted delegation not confined to any single act nor to any class or group of acts identified or described by reference to a standard. Here, in effect is a roving commission to inquire into evils and upon discovery correct them." This, he summed up that day, was "delegation running riot." Cardozo concluded that the wage and hours provisions of the codes were not legal if the industries they regulated were local in character; the codes must be thrown out. In a final—and perhaps unconscious—reference to the work of the Schechters, Justice Cardozo concluded with a butcher-block metaphor: "Wages and the hours of labor are essential features of the plan, its very bone and sinew. There is no opportunity in such circumstances for the severance of the infected parts in the hope of saving the remainder. A code collapses utterly with bone and sinew gone." The Supreme Court justices were sending a message to business. McReynolds believed that an unmistakable signal such as *Schechter* would hearten investors and employers.

But more important was the message they were sending to the White House. Later that day, Justice Brandeis collared the two lawyers who had advised the New Dealers so closely, Tommy Corcoran and Ben Cohen, in the justices' robing room. Their teacher Frankfurter's suspicion had been correct. The justice told Corcoran: "This is the end of this business of centralization, and I want you to go back and tell the president that we're not going to let this government centralize everything. It's come to an end." Brandeis also added a second comment: "As for your young men, you call them together and tell them to get out of Washington—tell them to go home, back to the states. That is where they must do their work." On the surface, it seemed a near irrelevant aside from an angry older man to a young one. In fact, though, Brandeis's second comment fit in clearly with his first. There was something that he just couldn't stand about the New Deal itself, with its new laws and offices. The country might heal itself better if it stayed at home, cultivating its own garden. Revolutions were dangerous, and the best way to prevent them was to deprive them of personnel. Later, Senator Borah, who knew Justice Van Devanter personally, delivered a defense of the Court from a different angle on the radio: "We live under a written Constitution . . . fortunate or unfortunate, it is a fact."

The American papers seemed to draw their collective breath. They did not want to write too much, at least not until Roosevelt spoke. The UK press, with less at stake, blared its instant conclusion at the news of *Schechter*: "America Stunned! Roosevelt's Work Killed in 20 Minutes," read the headline on the London *Express.*

The tabloid was correct. The case did indeed mean death for the NRA. By mid-June, thousands of employees in Washington received their last pay. The codes began to fade, even though there were some vehement protestors, mostly among larger firms. David Lilienthal's appliance sales arm, the EHFA, could no longer operate under Lilienthal at the TVA; the executive order was null. For Lilienthal, *Schechter* was a signal of a tough road ahead. The TVA was already beset by dozens of lawsuits and injunctions; more were to come, and Lilienthal was not sure how he could handle them all. Willkie and his fellow power executives hoped that now they might manage to kill the dread utilities legislation, or at least alter it so that it was no longer a death sentence.

Roosevelt, who knew Brandeis less well than Frankfurter did, was surprised that the justice had gone along: "What about old Isaiah?" he asked, using his nickname for Brandeis. The president was furious. In a press conference a few days later at Hyde Park, he and Eleanor sat together before reporters. Eleanor was knitting a blue sock. Marion Frankfurter, Felix's wife, was also in the room. Roosevelt castigated the press and the court. The NRA and the Humphrey case, as well as Frazier-Lemke's repudiation, were all getting in the way of a change that must happen in the United States. What were the justices thinking, interpreting the Commerce Clause this way? They were, he told the reporters, going back to "the horse and buggy age."

Still, the hour was that of the victors, who now knew that while Roosevelt's voice mattered, theirs did too. At 15 Broad Street in New York, Frederick Wood celebrated—his wife had given him a miniature chicken coop housing two cotton hens and six chicks, which stood on his desk. Wood declared that with this case the justices had finally been "going at the fundamentals" of New Deal law. Heller celebrated too. He told the papers that *Schechter* showed that "the humblest individual receives the utmost protection under our form of government." "Nine to nothing," the Schechter brothers were heard to be repeating to the media at Heller's offices at 51 Chambers Street. "We always claimed that the code authority attempted to make us the 'goat,' " they announced in a statement to the press. Meanwhile, the papers reported that some 500 cases against people charged with breaking NRA codes were now to be dropped.

The Schechters were concerned about the cost of the suit. But gratification was also theirs. Mrs. Joseph Schechter of 257 Brighton Beach Avenue displayed to the press a poem, titled "Now That It's Over."

No More excuses
To hide our disgrace
With pride and satisfaction
I'm showing my face.
For a long long time
To be kept in suspense
Sarcastic remarks made
At our expense.
I'm through with that experience
I hope for all my life,
And proud again to be,
Joseph Schechter's wife.

Her cheerful mood accorded with that of the country. The Dow now staged its longest rally since Hoover had first lifted the beneficent hand.

9

Roosevelt's Wager

July 1935
Unemployment (July): 21.3 percent
Dow Jones Industrial Average: 119

Shortly after *Schechter*, around the time of the "horse and buggy" press conference, a little-noticed event took place at the White House. Felix Frankfurter moved in.

The arrival of Frankfurter signaled a shift in Roosevelt's outlook. He was tired of utopias, he now decided. They had not necessarily helped the economy. The hope that experiments like the NRA would bring full recovery had not proven valid. Roosevelt had played around with economics, and economics hadn't served him very well. He would therefore give up on the discipline and concentrate on an area he knew better, politics.

The president formulated a bet. If he followed his political instincts, furiously converting ephemeral bits of legislation into solid law for specific groups of voters, then he would win reelection. He would focus on farmers, big labor, pensioners, veterans, perhaps women and blacks. He would get through a law for pensioners, and one for organized labor, with the aid of Frances Perkins and Robert Wagner. Rex Tugwell would take care of the poor and homeless of the countryside—Tugwell was to have a staff of more than 6,000, $91 million, and options on ten million acres of land, all to try out suburban and rural resettlement. There was also $2.75 million for Dutch elm disease in New York, New Jersey, and Connecticut.

If the politics was right, the wager said, the economy would follow suit and he could take credit for the rally himself. Indeed, he would *deserve* the credit. This attitude was the sort that William James, the philosopher at Roosevelt's own Harvard, had written about in his famous essay "The Will to Believe"—if you had faith in an outcome, you could help to make that outcome occur. Frankfurter made a good audience, and suddenly, with Frankfurter as his partner, he had a surer sense of the way to go. The next two years would yield the results of the first part of that bet. The outcome in the second question would emerge definitively only later in the 1930s.

For a man eighteen months away from an election, the plan made sense. Roosevelt had Congress behind him, and he had his agencies, even if he did not have the courts. He could use his authority to win the votes—or lose the authority. After all, he had challengers advocating more radical programs than his. Huey Long in the Senate had his evangelizer, Gerald L. K. Smith. Long's "Share Our Wealth" program promised senior pensions, free higher education, and employment for all. Francis Townsend, a doctor, had built a national movement with his own pension blueprint, the Townsend Plan. And Father Coughlin, the radio voice, had turned against the president. On May 22, just before the *Schechter* opinion came down, Roosevelt received what must have seemed a nearly equal blow in the news that Coughlin pulled 23,000 into Madison Square Garden to attack FDR. Coughlin assailed the "Morgans, Baruchs and Warburgs," little caring—perhaps it didn't matter—for Warburg's battle with Roosevelt. Coughlin argued that capitalism should now be "constitutionally voted out of existence." Roosevelt thought a lot about Coughlin, for Coughlin, with ten million listeners, was the superstar of Roosevelt's own medium, the airwaves: *Fortune* in 1934 had said that Coughlin was "just about the biggest thing that ever happened to radio."

To preempt the demagogues, Roosevelt had to prepare his new legislation so that it pleased the same force that sealed the fate of the NRA: Brandeis. This was one place where Frankfurter came in. Frankfurter knew how to craft a law that could please Old Isaiah. Through the summer months, Frankfurter would see the president nearly daily, recording in a letter to Marion: "FD wants me really around—so that I've not dined out of the White House once." The new White House policy became that the president must not repeat his angry press conference about the Supreme Court, or even mention it. Rage must wait.

Tommy Corcoran and Ben Cohen were also helping. The president had David Lilienthal at the TVA, another Frankfurter "hot dog." With the legal talent Roosevelt was now marshalling, the shift he sought became feasible.

To be sure, there were economic justifications for the new policy. Helping the worker with his pension made his family happier and more productive. Bringing down big enterprises and wealthy families liberated smaller companies and strivers to thrive—this was Brandeis's thesis. Giving cash to new constituents meant that they would spend and strengthen the economy—that was what Marriner Eccles, now governor at the Federal Reserve Board, was still telling the president. Taxing big business might also balance the budget, just as Roosevelt had learned as a young man. The president relished squeezing cash for the poor out of the well-to-do, especially after Tommy Corcoran and Ben Cohen or Robert Jackson had worked him up, regaling him with tales of wrongdoing by the rich.

But the emphasis remained political. "He illuminated objectives—even fantastically unrealizable objectives. These excited and inspired," Ray Moley would later write of Roosevelt, only slightly bitterly. "When one set of these objectives—FDR loved the word—faded, he provided another." The fact that he shifted did not have to matter.

Frankfurter, now closer to Roosevelt than ever, noted all this, and also understood that the president was making history, turning away not only from utopians but also from the moderates in his own party. "Last night," he wrote in a memorandum that summer, "after a very delightful dinner on the South Porch, the President asked Ferdinand Pecora and me into his study in the Oval Room." Thinking aloud, Roosevelt told his guests about Democrats in Congress, "at bottom, the leaders like Joe Robinson, though he has been loyal, and Pat Harrison are troubled about the whole New Deal. They just wonder where the man in the White House is taking the old Democratic Party." The Democratic lawmakers feared, Roosevelt concluded, "that it is going to be a new Democratic Party which they will not like." Still, Roosevelt was resolute: "I know the problem inside my party but I intend to appeal from it to the American people and to go steadily forward with all that I have." As Roosevelt in 1936 would freely acknowledge to another adviser, the election was about a single issue—Roosevelt. The country had come so far from Coolidge, who had sought to remove the "me" from every scenario he evaluated.

In advancing this plan, Roosevelt was also refining his definition of his forgotten man. Before, the forgotten man had been something of a general personality—albeit always a poor one. In projects like the NRA, and with grand planners like Arthur Morgan, there had at least been the attitude that the country was all in it together. Now, by defining his forgotten man as the specific groups he would help, the president was in effect forgetting the rest—creating a new forgotten man. The country was splitting into those who were Roosevelt favorites and everyone else. The division started at the top. The president pulled increasingly close to legal pragmatists and yes men—Frankfurter and his entourage, Lilienthal at the TVA, Henry Morgenthau, even Henry Wallace at Agriculture. (Moley, increasingly on the outs, had especially little regard for Wallace: "His oratorical support of his boss" in campaigns, Moley would write later, "was as intolerably partisan as a paid party *Spieler,* and though he could find no wrong with his own party, he routinely compared the opposition to 'Nazis and blackguards.' ") He was also pushing away those old allies whose views were too idealistic or simply inconvenient—Tugwell, Moley, Arthur Morgan. Jim Farley, his postmaster and national chairman of the Democratic Party, would encourage this change. It was not good to look too radical. To Tugwell, it was bitter: "I had worked hard and felt I was entitled to speak." Yet he found that "Roosevelt agreed with Farley to keep me quiet and hidden."

Lower down, the constituencies were also of the president's choosing. Roosevelt rejected, for example, the Bonus Army marchers who had helped him win his first election, refusing to sign a bonus into law. To his mind his plan for Social Security would take care of the marchers; they could receive theirs with all the rest. Congress in any case would do his work for him, restoring programs that had been in place before he cut them back in 1933. He also created new constituent groups. Roosevelt disliked handing out money to the poor. He wanted, as he said, to "quit this business of relief." Instead he would create work now in other ways. That summer—the summer of 1935—Hopkins was spending the first dollars in the Works Progress Administration, a program that would, the papers said, start 100,000 projects and hire by the millions over the coming months. General Johnson would be the administrator, a job to replace his old post at the NRA. Here, Hopkins and Ickes, always competitive, were going head-to-head in an alphabet competition: Ickes had his PWA, and now Hopkins had the WPA. The WPA work was project-oriented: WPA staffers ran hospitals and dug ditches, opened libraries and served a million school lunches a day. But there was also a financial distinction: Ickes' projects generally were the ones that cost over $25,000; Hopkins's ran under the $25,000 line.

Hopkins established the Federal Writers' Project to employ unemployed writers, and gave them the legitimately useful task of writing travel guides to towns and regions across the country. Among the hires were 150 jobless newspapermen, whose new jobs had a circular aspect: they were to chronicle the advances of the WPA. Hopkins picked Hallie Flanagan of Vassar to create a theater that would air plays about the social conditions in the country—and again, spotlight New Deal progress.

This was a chance to hire new tiers of intellectuals. The Federal Writers' project engaged not only John Cheever and Ralph Ellison but also Anzia Yezierska, John Dewey's old friend. A young black writer named Richard Wright repeated a few lines of a song he had heard: "Roosevelt! You're my man! / When the times come / I ain't got a cent / You buy my groceries / And pay my rent!"

A National Youth Administration would provide work and education for thousands of college-age young people and high schoolers. The NYA had its own vast bureaucracy; one worker would be a young Texan, Lyndon Johnson. Consumers were Roosevelt's people too, for the fact that consumers were voters was Roosevelt's central epiphany. It was what made the spending that Eccles advised so very useful.

Then there were the blacks. The Roosevelt camp conducted an intensive outreach to all black groups. William Andrews, one of two African Americans in the New York State Assembly, likened FDR to Lincoln before his colleagues in the assembly; Irwin Steingut, also a Democrat although not black, added that the comparison was apt because Roosevelt was "a great emancipator of his time." Harold Ickes would spend the year serving as Roosevelt's emissary to this group. And many responded—black registration to vote rose. Ickes's projects gave blacks a greater share of construction work than they had ever been allotted; a million blacks would take literary classes funded in some way by Washington that decade.

But there was also new hostility to the enemies Roosevelt had chosen: big companies, employers, the wealthy, those shadows that Moley had described as inhabiting the background of Frankfurter's life. Utilities were clearly the enemy now as well. The skirmishes were over; the class war was out in the open.

His plan in place, Roosevelt opened fire. He hadn't highlighted taxes in a while; they were seen now as merely a part of the greater 1930s story. Visiting her old school, Todhunter, in May, Eleanor had inspected state and federal tax returns prepared by the girls as a study project. Now, however, FDR too turned to taxes. In a message sent to be read aloud to Congress, he railed against the "great accumulation of wealth" and called for a tax bill to change society. The Mellon prosecutions of the spring came to mind. The president wanted rich families to pay an estate tax when they died, but he also wanted their children to pay a second levy, a new inheritance tax, when they inherited the money. There would be a graduated corporate income tax, a shift from the old flat rate for companies, in accordance with Brandeis's philosophy that big was bad. There would be a sharp increase at the top of the rate schedule for earners above $50,000. And there would be a tax on intercorporate dividends. Roosevelt relished the suddenness of his surprise. Speaking of the chairman of the Senate Finance Committee, he told Ray Moley, "Pat Harrison's going to be so surprised he's going to have kittens on the spot." Moley disapproved again. The "proposals ran counter to the New Deal's most elementary objectives," he said; limiting corporate surpluses would prevent companies, especially small ones, from using their cash to keep workers in downturns, and deepen depressions by leaving companies nothing with which to pay dividends in hard times. "It will aggravate fear," summed up the *Boston Herald.*

Roosevelt was unfazed. Indeed, rather than invite a full-fledged review, he boldly proposed the change as a rider to other legislation. Some members of Congress—those to Roosevelt's left, pushing for even more punishment for the rich—needed little convincing. Hearing the clerk read Roosevelt's words, Huey Long cried out from the floor, "Mr. President, before the President's message is referred to the Committee on Finance, I wish to make one comment. I just wish to say 'Amen.' " But others were disturbed that the president would try to slip such major legislation in as a rider before the July 4 break. The president shortly canceled the rush order but persisted with the plan.

Among the first to pick up on what Roosevelt was doing was Lilienthal. Arthur Morgan still backed the idea of sharing a grid with the power companies;

striking such a deal, after all, would give him the time and space to build up his utopia. Lilienthal always said he was above politics—"a river has no politics," as he had said. Still, this was disingenuous, for his actions were all about gaining power. Now Roosevelt had given him new ammunition. Lilienthal therefore busied himself trying to write contracts with municipalities to squeeze Willkie out of the towns. Where he failed—and he was still, mostly, failing—he compensated by delivering his speeches. Earlier in the year, the town of Norris had been finished; both Lilienthal and Arthur Morgan were settling in—their homes were within five minutes of each other. "Boys and Girls Have Been Making Money in Norris" would read a headline in the educational trade press. The Norris School Cooperative enabled children to pool their labor, and even sell insurance and make loans. Nancy Lilienthal, the daughter of director David Lilienthal, would be one of the child leaders in the cooperative.

In June, the *New York Times* announced that the TVA would have a distinguished summer worker—John Roosevelt, the president's son and a freshman at Harvard, who was enlisting as a volunteer to chop and heave along with other TVA workers. The younger Roosevelt, it was promised, "would displace no" laborer. On July 4—to underscore that the TVA was a patriotic project—Lilienthal staged a rally of 30,000, some 7,000 more than Coughlin had drawn, for "TVA appreciation day" in Tuscumbia, Alabama. Two governors were there to survey the majestic train of fifty-seven floats, and the president sent a message to be read aloud.

Meanwhile, Roosevelt was preparing his pen to sign Bob Wagner's labor legislation. The act gave workers the same right to organize and to bargain collectively that Roosevelt had hoped to secure through the NRA. In the spirit of other 1935 legislation, the Wagner Act included an economic justification: labor had not been sufficiently organized heretofore, and that itself had caused downturns. The inequality of bargaining power between employees and employers, the act said, tended to "aggravate recurrent business depressions." Ignoring the importance of productivity, the economics of the law were lopsided.

Under Wagner Act rules, a union, once in place at a company, might keep out workers who did not join—the so-called closed shop. The same union need not ever again be subject to election by ratification but could represent the workers more or less in perpetuity. The effect was the most coercive of any law passed in the New Deal. Roosevelt had second thoughts about it. Especially disturbing was the act's clear warning that refusals by employers to allow workers to organize "lead to strikes and other forms of industrial strife." The suggestion was that employers who did not interpret the new law generously could expect to pay for that with strife—violence—and might even be subject to such violence as a result of pre–Wagner Act refusals. Again, Lilienthal acted with alacrity: he began readying a statement noting that 17,000 TVA workers were entitled to rights and collective bargaining under the new Wagner Act.

The Wagner Act contained news enough for a year. But there was also the Social Security Act, Perkins's and Douglas's plan to provide pensions for senior citizens. Here lawmakers had given Perkins, in particular, a hard time. Senator Thomas Gore of Oklahoma—a Democrat, though no relation to the Gores of Tennessee—had been blind since childhood. Though presumably precisely the sort of person whom acts like Social Security might protect, Gore was sarcastic. "Isn't this socialism?" Why no, Secretary Perkins replied. "Isn't this just a teeny weeny bit of socialism?"

Now Champ Clark staved everyone off, blocking agreement on the legislation through the heat of July. Roosevelt, however, did not relent, and finally he prevailed: the Social Security Act would be voted on without Clark's amendment, which had supplied the private companies with an opt-out. There was a promise to study Clark's concept for consideration in later legislation, but the chance to continue a form of American pension that would show workers there was an alternative to the government provision was fading.

The motion cameras were ready when Roosevelt entered the Cabinet Room on a mid-August Wednesday. Perkins was there, along with Bob Wagner and Joseph Guffey of Pennsylvania—the Guffey who had replaced Mellon's old ally. There were no newspapermen and the print reporters, humbled, had to ask photographers for details for their story. The participants pulled close to the table and signed what was to be the most famous legislation of Roosevelt's presidency: the Social Security Act. "The civilization of the past hundred years with its startling industrial changes has tended more and more to make life insecure," Roosevelt said in a statement. Now, with this pension bill for older citizens, that insecurity would be reduced. Government would begin to provide "a law to flatten out the peaks and valleys of deflation and inflation—a law that will take care of human needs." Roosevelt used different pens for different parts of his signature, so that a number of those present—especially Perkins—would have a keepsake from that day.

Social Security seemed a gift on a scale most Americans would never have expected a president to be able to offer. At a time of—still—so much need, the idea of help seemed in itself a blessing. Even though the first Social Security payment, check number 00-000-001, would not be issued until 1940, people knew that the money was coming.

To many of the progressives the news that their ideas were finally becoming law was intensely gratifying. Roosevelt hoped the program would make older workers comfortable with the idea of retiring earlier, leaving more work for younger people. Perkins and the progressives liked the unemployment insurance as much as the senior citizen pensions. Paul Douglas would head for Italy that fall of 1935—he had a new wife, the daughter of the sculptor Lorado Taft. From Siena, he wrote: "One who for fifteen years has worked for such legislation may perhaps derive a pardonable sense of satisfaction in the fact that the American public has finally realized that it needs the greater protection against unemployment and old age which pooled insurance gives."

That same August, the TVA had more good news: in the first fifteen days of the month the directors had hosted 30,000 tourists, there to inspect the rising Norris Dam. The dam was 253 feet—like a seventeen-story building. And at the end of August, on the twenty-sixth, Roosevelt made the utilities' nightmare into law: he signed the utilities act. The language of the death sentence was softened, but it was still dire. Holding companies had to limit themselves, *Time* explained, to a "single integrated system, and" multilevel companies must be reconfigured. The law still restricted private companies in a way that gave significant new advantage to the TVA. On the last day of August, Roosevelt also signed into law his new tax bill, increasing the top tax to 79 percent, increasing estate taxes, and lowering tax-rate thresholds so that more families would pay higher taxes. As if the new holding company act were not enough punishment, the tax bill included a graduated corporation tax—to punish big business—and a dividend tax designed to weigh down holding companies. The summer was ending, but Roosevelt told himself he had reduced, for the moment, the chance of losing constituents on the left.

In September, news brought further confirmation of that assessment. An assassin felled Huey Long, perhaps the greatest single political threat to Roosevelt's political coalition. Roosevelt now made a promise to a prominent journalist of what he called a breathing spell—a ceasefire—in his war against business. Stocks, which had been rising, promptly marched up some more. But a breathing spell indeed was all it was. The class war was far from over.

Tugwell, perhaps watching Lilienthal, now sought to dramatize his successes. On the last day of June he and Eleanor Roosevelt had hosted a conference on the future of housing and resettlement at Buck Hill Falls, Pennsylvania. Stuart Chase had attended. There they had talked about the importance of government's role in developing communities for migrants and others: "A community does not consist of houses alone," said Tugwell. "You cannot just build houses and tell people to go and live in them. They must be taught how to live," Mrs. Roosevelt echoed. Mrs. Roosevelt liked Tugwell because he said what other reformers only thought; "My hat is always off to your courage," she wrote when he got in one of his tangles.

Determined to publicize his work further, Tugwell thought of his old friend Roy Stryker, one of the coauthors of his economics textbook back in the 1920s. Through photos and drawings, Tugwell believed, that textbook had triggered more thought than any words-alone text could have. More recently Stryker had suggested to Tugwell that they do a "picture" book together.

Now Tugwell had an idea. He had seen the modern murals that Henry Morgenthau's Treasury had commissioned across the country—so many that, that year, a newspaper critic would note in wonder that the New Deal had functioned "more lavishly as an art patron than all the previous administrations lumped together." Could not he also convince minds through art? Tugwell asked Stryker to work at the Resettlement Administration, to gather evidence that made the RA's cause and work understandable to the public. Stryker had a good eye and began putting together a staff of photographers—then near-unknowns named Arthur Rothstein, Carl Mydans, Walker Evans, Ben Shahn, and Dorothea Lange. Some were already working for government; others were new.

When Stryker started out, he had trouble conceiving exactly what sort of pictures to commission. Tugwell coached him: "One day he brought me into his office and said to me 'Roy, a man may have holes in his shoes, and you may see the holes when you take the picture. But maybe your sense of the human being will teach you there's a lot more to that man than the holes in his shoes, and you ought to try to get that idea across.' "

If he could picture these forgotten men, Tugwell reckoned, the programs undertaken in their name would be allowed to continue. Indeed his whole department counted on that: the RA, Stryker later remembered, "simply could not afford to hammer home anything except their message that federal money was desperately needed for major relief programs. Most of what the photographers had to do to stay on the payroll was routine stuff to show what a good job the agencies were doing out in the field." Stryker channeled Tugwell's faith that agriculture was crucial to the future of the country, and would ask for a photo from Kansas sending the message that "there is nothing in the world that matters very much but wheat."

As summer 1935 moved into fall, organized labor meanwhile took in the meaning of its legislative victory. The most excited—logically enough—was John L. Lewis of the United Mine Workers. He realized that with a single blow, Roosevelt had created one of the greatest power blocs ever to appear on the American stage—and one that he, rather than the American Federation of Labor's Will Green, might lead. The moderate old AFL was not the right union for the wage worker. To allow Green to lead was to squander advantage.

On September 1, Lewis staged a giant outdoor meeting in Fairmont, West Virginia, with forty thousand coal miners and their families. The event celebrated not only the Wagner Act but also the just-signed Guffey Coal Act, which created a National Recovery Administration for coal mining. Lewis predicted Roosevelt's reelection—a year early—and announced happily that "the era of privilege and predatory individuals is over." The next month in Atlantic City, at the convention of the American Federation of Labor, Lewis walked along the boardwalk and, during a rain shower, ran into a union acquaintance. The acquaintance asked him whether he had been thinking of his old dream of a new kind of industrial unionism. "I have been thinking of nothing else for a year," Lewis replied, grabbing the friend's forearm. The same week, at the convention, the majority rejected a proposal led by Lewis to authorize a campaign to organize industrial workers, and Lewis decided to lead the workers out into a new institution, the Committee for Industrial Organization. It was time to part with the Will Greens and the Matthew Wolls.

Lewis was a theatrical man of the tough variety. To make the breakup more vivid, he provoked a quarrel with the president of the conservative Carpenters' Union. The men began to shout, and Lewis walked over and punched the fellow in the nose. The man collapsed, and all eyes followed as Lewis left the hall, a symbol of the strength of the new more aggressive industrial unionism. The CIO was launched. The morning after the AFL convention adjourned, UMW leaders and others met for breakfast to plan. Among the planners was John Brophy, now working as Lewis's arm at the UMW.

On the last day of September, Roosevelt traveled out to Hoover's dam, now called the Boulder Dam. The project was finally—and splendidly—ready for dedication. Roosevelt noted that the dam's construction created jobs for 4,000 men in hard-up years. Government had turned "unpeopled, forbidding desert" into something useful; the power generated here would increase the welfare of all, for, as he put it, "use begets use." Hoover retaliated with a nasty speech in Oakland, California, before western Republicans a few days later, charging that Roosevelt was "joyriding to bankruptcy." He also noted, accurately, that the Roosevelt administration had not brought unemployment back to anything near precrash levels. There had been make-work, such as the PWA or the WPA, but only 700,000 jobs had been created since 1932, leaving unemployment at one in five men. Hoover's points were valid, but what came through was the stridency. He spoke past the end of the appointed hour, forcing the embarrassed radio network, the National Broadcasting Company, to switch him off before he finished.

In October, the WPA announced it would employ 26,000 idle artists, musicians, and actors—20,000 by November. At her theater, Hallie Flanagan would shortly produce *Triple A Plowed Under,* a play about the problems and courage of the farmer under the New Deal. As at the TVA or at Tugwell's RA, responsibility within the WPA for supplying the jobs and overseeing the programs would be assigned regionally. The head of the program for New York City—the theater center—was to be Elmer Rice, a well-known and left-leaning playwright. The signal was clear: WPA product in the coming year would not necessarily be all pro-Roosevelt, but it certainly would be anti-Republican.

By November, the new CIO had opened an office in Washington. Its goal was "to foster recognition and acceptance of collective bargaining" in mass production industries. Lewis named John Brophy to head the office. The dreamer who had placed the last question to Stalin in Moscow had completed his transformation: he was now a lobbyist on K Street.

At the Supreme Court, the justices began to contemplate what they would do when the Wagner Act came before them, or if Roosevelt won a second term. After paying a call on FDR in November, Justice Ben Cardozo wrote to Felix Frankfurter of the president, "He seemed strong and happy. To have a picture of him talking with McReynolds would be precious." Frankfurter carried this bit of humor to the President like a gift. In a Thanksgiving note to Roosevelt, he wrote: "Can we not have such a photograph!? It would be a superb campaign poster—or might McReynolds enjoin you from exhibiting it!"

At the TVA, things were also moving into high gear. Work on the Norris Dam was now scheduled to be completed in January, months ahead of the original schedule. More than a million cubic yards of concrete would be poured into the dam by New Year's Day. Some 1,770 men worked in four 6-hour shifts. The TVA had bought out 3,000 families who lived in the basin behind the dam, now scheduled to become Norris Lake. Even the dead in the Norris basin were being removed; the bones of some 2,500 people, many early American settlers, were disinterred and reburied at a cemetery on higher ground. It was a gruesome thought, but one that seemed to underscore the inexorability of the institution.

Even those who did not like the new projects and new buildings often supported them. As the country told itself, there could always be something worse: if not Huey Long, then Father Coughlin, or the communists. That year Hitler was staging Germany's first war games since its defeat in World War I and calling up its first draftees. And many of the writers of the period did their best to reinforce the impression that the United States was in a fragile state. Late in 1935, Sinclair Lewis, the author of *Babbitt,* produced a novel on the question. The premise of *It Can't Happen Here* was that it could: Lewis pictured the nightmare of an America gone fascist. The country was run by "Corpos"; the thuggish police were called Minute Men. The point was not subtle: an America in trouble was readily capable of producing its own version of fascism.

The novel was not wholly respectful of Roosevelt—a fictional leader in the novel appointed Roosevelt ambassador to Liberia, and Hoover ambassador to Brazil. But the novel's effect was to make it seem as if Roosevelt stood for stability.

Also in December, a National Resources Committee put together by Roosevelt and chaired by Harold Ickes laid out the new plan for TVAs across the country. Roosevelt was finding Ickes remarkably efficient, and continued to reward him with projects. The plan suggested dividing the United States into twelve regions, with regional capitals: Philadelphia, Detroit, Baltimore, Cleveland, Cincinnati, Knoxville, Nashville, Atlanta, Chicago, St. Louis, New Orleans, Dallas, St. Paul, Denver, Salt Lake City, Los Angeles, San Francisco, Duluth, and Portland, Oregon were mentioned. Each capital would then be able to serve its region with ample managers. After all, as Ickes had already said recently, the national capital was so full of New Deal projects that new offices would be necessary in Baltimore and elsewhere. The presenters of the program for capitals took care to assure the public that their project would not infringe on the sovereignty of the states. But the implication was still clear.

Tugwell was finding his Resettlement Administration busy across the land. The goal of the administration was to help needy families and to improve land use across the country. If farmers lived on poor "submarginal" soil, they would be permanently in need, for they could not compete. Tugwell's job was to retire land that needed retiring, move those who needed moving, and build new communities for those who were moved. All in all, the RA had four areas: suburban resettlement, rural resettlement, rural rehabilitation, and land utilization generally. Where Arthur Morgan and David Lilienthal had Norris, Tugwell had "greenbelt towns," planned suburbs for workers near centers of industry: Greenbelt, in Maryland, between Washington and Baltimore; Greenhill, near Cincinnati; and Greendale, in Wisconsin. Tugwell liked the greenbelt towns, but he was anxious about rural resettlement projects, for they gave the opportunity at once for the greatest experiments and the greatest failures. In his view the Roosevelts were being too romantic when they imagined successful little communities in the countryside. Where would the people work? Industry was more the future.

Overall for 1935, Washington's spending was $5.6 billion, double the level for 1930. In the coming election year, 1936, Washington would spend yet more. In many towns, Roosevelt's presence was signaled by construction work—local contractors and local labor pounding away at the town square or at a new school.

From the swimming pool and pavilion in La Grange, Georgia—under thirty miles from Warm Springs—to a public library made of limestone in little De Pere, Wisconsin, to an eleven-story courthouse in Alameda County, to the already complete city hall of Brentwood, Missouri, the Public Works Administration seemed to be everywhere. There were also PWA zoos, PWA bridges, and PWA museums. The PWA buildings were for the most part good-hearted structures, often made of the slate or stone local to the area. They sent a message to the towns: Washington is there to help when the town is in trouble, and yet will not intrude on the community to do it. It was a relationship that the illustrator Norman Rockwell would depict and also sanction with his *Saturday Evening Post* covers. The WPA spirit was patriotic, and catching. In October, the artist gave a lecture at the Otis Art Institute in Los Angeles. Its title was "What Is Required of an Illustrator Today." The tiny Washington of just a few decades ago, the Washington whose anxious politicians had placed height limits on construction in the District of Columbia to keep out the skyscrapers, now began to dominate the national landscape.

Still, that winter, there was a familiar feeling in the country: the scenario of the midterms in 1934 was repeating itself. The politics was exciting, and the Dow was heading up again. But the index was not near old levels, and jobs were not materializing. Around the new year, Will Green took time off from his battles with John Lewis to make something clear to the press. People might be speaking about recovery, but business activity was still far below 1929 levels—and it was a jobless recovery: While "business has recovered half its Depression loss, only 30 percent of the Depression unemployed have been put to work." Some 11.7 million Americans still had no job in November 1935.

An expert from a culture distant from that of organized labor—Benjamin Anderson at the Chase Bank—warned of trouble too. The new law raised taxes on several classes of taxpayer. But it targeted the rich. It might sound amusing to impose high rates on wealthy people. But such taxes also caused enormous damage. The thing to focus on was not that the economy might be improving a little bit, but rather that the country was not getting the strong recovery that it should expect. The New Deal was causing the country to forgo prosperity, if not recovery. The wealthy, after all, were in a position to take risks with new ventures precisely because they were wealthy—they could invest in several projects at once. Under the new 1935 law, a very wealthy man would see more than three-quarters of any profits from new ventures taken by income tax. Any loss, however, would be the same man's to bear. This man would try to hoard his capital and wait—thus coming to fit the very stereotype of the idle rich man the New Dealers were hoping to propagate.

Father Divine, for his part, was still hoping for a chance to move the country into a new sphere, and influence Roosevelt as well. His modus operandi paralleled Coughlin's: At a convention at Harlem's Rockland Palace come January 1936—an event attended by representatives of the Communist Party, the Republican Party, and the Democratic Party—an audience of a thousand or so would chant and speak about a day when he would be a power behind the U.S. president. Divine's eclectic platform included an end to installment purchases for consumers, and a ban on tariffs. He also sought a ban on the salutation "Hello"; this should be replaced by "Peace," and the phone company should be forced to comply. Most telling, perhaps, was the plank that called for "Enactment of laws against newspapers and publications which employ words designating the differences in creeds, races, and conditions of peoples." Father Divine was an angler, just like Huey Long and Coughlin, and he had legitimate additional goals: an end to lynching, for example. He, unlike Norman Rockwell, feared that government help might make citizens more dependent. In 1936 he would write directly to Harold Ickes about a government law that required people to apply for relief before they could be hired for public works positions. Divine argued that the law damaged self-esteem and the will to work, and thus "lowers the standard of a person for the present, and for his future generation."

Willkie for his part was aware that others in his industry had given up. Figures like Alfred Lee Loomis were long gone from the scene: Loomis was busy publishing articles on sleep patterns in the cerebral cortex in *Science* magazine. But Willkie persevered. Under the new utilities law, Willkie had a new, second regulator: Joseph Kennedy, the chairman of the young SEC. Within a month of the Public Utilities Holding Company Act's passage, Willkie had been visiting with Kennedy. He still had hope for Arthur Morgan at the TVA. He thought Morgan retained some support from the president on a cooperative concept: the TVA and others, including Commonwealth and Southern, might create "power pools" and sell and buy electricity together. Willkie told reporters he wished he had the $150 back that he had given to the original Roosevelt campaign. Taking him up on his challenge, Akron Democrats wired an offer to pay up, writing: "Before you became a plutocrat you were a good Democrat."

Willkie also exhorted other "disgruntled Democrats" to speak out. Yet later in the year, in December, Willkie had been found rallying members of the Bond Club of New York to recognize the new Roosevelt campaigns for what they were, hate campaigns: "Surely," he said, "the haters have occupied the stage long enough."

Willkie's battle was making him nationally famous for the first time, and that part of the story he enjoyed. He spent days and hours selling appliances to Commonwealth and Southern customers all across the South. He dined out often in New York, and read history, enjoying his encounters with other Wall Streeters and the editors of the *Nation* or *Partisan Review* with equal gusto. And for the first time he had a sense that he was being watched in a new way, both in New York and nationally. In the coming year, the Willkies would be listed in the *Social Register* for the first time.

People noticed that Willkie also loved helping his subordinates. One of them, an executive at the Georgia Power Company named John Marsh, was married to a Smith College dropout. The wife had written a 1,037-page romance novel about the Civil War. Willkie liked the book and the author so much he sent out a memo to Commonwealth and Southern employees, plugging it. Her name was Margaret Mitchell, and her book, *Gone with the Wind,* would go on to sell four million copies.

Willkie believed such success created an unhealthy gap between man and wife. To make Marsh feel manly, Willkie promoted him at the power company.

In January at the Mayflower Hotel in Washington, the Liberty League hosted an event that, it hoped, would reset the course of the country. The dining room seated 1,200, but the dinner sold out; some 800 guests would have to content themselves with sitting in the next room. Republicans expected that they now might turn the tide. The guest was Al Smith. Smith argued fiercely against Roosevelt's "arraignment of class against class"; of the brain trust he said, "the young Brain Trusters caught the socialists in swimming and ran away with their clothes." Most outrageous of all to Smith was the rise of professors, the way Roosevelt had ignored others—himself, especially included—and constructed such a revolution with the brain trusters. Smith worked himself into a fury, and his rhetoric later would be even wilder. "Who is Ickes?" Smith would ask. "Who is Wallace? Who is Hopkins? And in the name of all that is good and holy, who is Tugwell and where did he blow from?"

The Republican Party was also in action. In February, Arthur Vandenberg of Michigan—the state where the banks had failed so badly—tried to articulate what was wrong at the annual Lincoln Dinner, held at the

Waldorf-Astoria in New York. Vandenberg said he was a liberal constitutionalist, something close to Willkie's vision of himself. Roosevelt, however, Vandenberg said, was something else. A thousand Republicans listened as he laid out his arguments: Roosevelt was leading "a government by executive decree." He was rejecting the old federalism and making the states his pawns. The country was ready for "restoratives rather than narcotics." It was possible to be a liberal and not go the way Roosevelt had. Roosevelt was a hypocrite, a "Dr. Jekyll and Mr. Hyde Party."

Nonetheless the anti-Roosevelt arguments were not taking. Partly this was because of their shrillness. The microphone had taught Roosevelt that moderate tones could be more convincing than loud ones, but his Republican counterparts still bellowed as they had in the days before electricity. They made their opponents sound reasonable—"No democratic European, whatever his party, can sympathize with the ear-splitting clamors of Tory Americans about measures most of us put through thirty-five or forty years ago," the French writer Odette Keun would comment. In their frustration Republicans, especially the wealthy, were now becoming parodies of themselves. It was that year, 1936, that Peter Arno of the *New Yorker* published his famous cartoon of rich people in evening dress telling one another: "Come along. We're going to the Trans-Lux to hiss Roosevelt."

But the critics had another reason to be loud—their own frustration at the genius of Roosevelt's wager. Roosevelt, they saw, had understood something that the Republicans had not. The contest now was not Democrat versus Republican but rather the classical republic versus the classical democracy. Government was less a representative republic than it had once been, more directly controlled by the people. The change had started back in the 1910s with the constitutional amendment to permit the electorate to pick senators directly, rather than through their state legislatures. Suffrage for women had accelerated it. And the Depression had accelerated it again—people who might not have had an interest in government before now found that hunger concentrated their minds. Instead of asking what government was doing on behalf of the general welfare, voters were asking in a very democratic way what Roosevelt was doing for them.

And as 1936 unfolded, they could see that Roosevelt was doing more for them than any president had done for the country in history. Washington continued to spend: as a share of the economy, the government was expanding to 9 percent from the 6 percent that had obtained when the New Dealers first rode with Roosevelt down to Washington. For the first year in peacetime America, federal spending would outpace that of the states and the localities. The spending was so dramatic that, finally, it functioned as Keynes and Waddill Catchings had hoped it would. Within a year unemployment would drop from 22 percent to 14 percent. Fourteen percent was still higher than the level that had been the peak in the early 1920s, but the Roosevelt team made the case that it was the rate and direction of change that mattered. The first campaign of 1932 had promised the repeal of Prohibition. This time there was the promise of Ickes's giant projects, not alcohol. But the effect was the same: like whiskey going down after so many years of tea.

The Republicans and the Liberty Leaguers could not compete with Roosevelt's new philosophy; to do so would be against *their* philosophy. The fact that they were trapped drove them crazy. In their rage they simply shrieked louder: "Let Tugwell get one of the raccoon coats that the college boys wear at a football game and let him go to Russia, sit on a cake of ice and plan all he wants," Smith would yell.

But there was another reason the critics' arguments were not penetrating, one that also addressed William Green's point. It was that the New Dealers' economic failures were working to their own political advantage. The country was now entering its seventh year of depression. The sense of futility was stronger than it had been in the early 1930s. Roosevelt's talk had had an aspect of self-fulfilling prophecy: because the first New Deal had not succeeded, many in the country believed that the United States was actually becoming the society of social classes that Roosevelt now described in his speeches. And they responded accordingly.

Whereas in the old America of the 1920s the sight of so many jobless men had provoked shock and alarm, now people accepted it, telling themselves that at least things were better than before. The same held for stock traders, who had stopped measuring success against the marker of 1929. People told themselves that the fact that the stock average was moving upward was the best they could hope for, even though it remained so far from the 1929 high. Recovery was supplanting prosperity as the goal.

Many Americans had recently seen a new film called *Alice Adams,* starring Katharine Hepburn. In the film, a girl from fine but simple people—Hepburn—is reduced to unseemly social striving in order to make it in a world where the gap between the higher-ups and the rest seems to widen. Meanwhile her family turns to questionable behavior—the appropriation of a company-developed glue recipe, and borrowing without asking from the till at work. In the end Alice prevails as the wealthy members of her town, led by her beau, recognize the error of their own ways. The employer of Alice's father shares the wealth—Huey Long sprang to mind—by making Alice's father more of a partner and less a wage slave.

The Republicans, comprehending at least some of their own failures, were seriously considering as presidential candidate a governor from the middle of the country—Alf Landon of Kansas—to help them reconnect with citizens. But even as they planned, they doubted whether Landon was a match for the incumbent.

Meanwhile, the administration continued to craft the recovery story line. That winter Roy Stryker, Tugwell's staffer, sent his photographer Dorothea Lange on her first RA assignment, to photograph the before-and-after experiences of families at a federal camp in Marysville, California. The camp manager at Marysville was Tom Collins, upon whom John Steinbeck would later model the camp manager in *The Grapes of Wrath*. Lange was now on Stryker's staff at a salary of $2,300 a year, and had a title: "photographer-investigator." Around the new year Stryker approved expenses of $600 for a trip in California, New Mexico, and Arizona. Through Stryker, Tugwell asked specifically that there be pictures of farm labor.

At the TVA, Lilienthal raced against time, building up his projects. The water behind the Norris Dam was rising. There was some trouble at the TVA. A family in the Norris basin—James Randolph, his wife, and seven children—was refusing to leave. Even as the waters touched the foundations of their house, the Randolphs would not move. "TVA Evicts Family As Waters Lap Cabin," the *New York Times* headline read. The evictors collected sixty chickens, one pig, and other animals and moved them to Jacksboro. But Lilienthal was already moving past the bad news, planning to announce in February that the TVA was now at work on a total of five dams.

Roosevelt was negotiating larger stumbling blocks than the hapless Randolphs. In January, the Supreme Court invalidated the Agricultural Adjustment Act, the NRA's twin. The excesses were the same ones again, involving the Commerce Clause and delegation. The AAA levied its processing tax nationally. In this instance, the receivers of a bankrupt Massachusetts cotton-processing company in Hoosac Mills argued that the AAA had no authority to levy the tax in the first place since it used the money for the sort of regulation that only states might impose. Agriculture was still a local activity. Stanley Reed, the lawyer for the government—the same one who had argued

Schechter—became ill and had to stop his argument and sit down.

This time, there was no horse-and-buggy explosion. The administration merely focused on dealing with the new challenge. Morgenthau alerted the president to the immediate problem that the Court's act caused: an enormous shortfall in tax revenues. The year before, the government had had $3.7 billion in receipts; losing the AAA's processing tax revenue meant losing $500 million, some one-seventh of the money. Though the president had just promised three days earlier that there would be no new taxes, now he had to consider reversing himself. Congress made the budget challenge harder by overriding the president's veto of cash for the Bonus Army veterans, at a cost of an additional $2 billion, or something like half of the prior year's revenues.

Morgenthau and his aides began an intense review of the tax problem. The revenues from business were disappointing, in part because corporations were not earning as much as they had, and in part because the companies were not distributing their cash in taxable dividends. Morgenthau and his advisers therefore came up with a novel plan to choke the money out of companies: an undistributed profits tax. If they could squeeze hard enough, the Treasury men posited, the companies would issue dividends or otherwise spend.

This in turn would put cash in the hands of the consumer-voter in an election year—exactly what Keynes and Eccles were telling the administration was important. Eccles was especially vehement—cash in hand for the visible consumer was important. Morgenthau estimated that about $4.5 billion in profits would not be distributed for 1936. He would write a law that would get at that money. Some of his advisers liked the idea because, in addition, it especially punished big companies, performing an antitrust function, something to please Brandeis. Morgenthau brought the whole concept to Roosevelt, arguing that the undistributed profits tax ought to replace the corporate income tax.

At the White House and in Congress, Roosevelt's advisers worked on the plan. For a business earning $10,000 a year, the tax on savings in the plan was 42 percent. For those with higher incomes it could be as much as 74 percent. But that was not all. The president was also talking about using the income tax in a new way—not just as a tax for revenue, but also as a means of social reform.

Mellon, the old tax hand, did not want to watch. Though robbing the corporate nest sounded amusing, to take the cash away was like taking the egg away from the bird, the offspring that was the insurance for future growth. But he was a private citizen now. So he worked harder on a new project, a more formal home for his own egg warmer, the idea incubator at the Mellon Institute. The institute had been a roaring success, producing research permitting sponsors to take out many hundreds of patents. In 1936, companies would give $816,000 to create sixty-nine fellowships. The final structure, costing $6 million and trimmed in aluminum, of course, would not be ready until the following year.

In February 1936, while everyone was still digesting the tax news, there came an important—albeit imperfect—victory for Roosevelt. In the *Ashwander* case, a 5–4 majority of Supreme Court justices—Brandeis, Stone, Cardozo, and Roberts dissented—found that the shareholders of Alabama Power had the right to their original suit against the TVA contract with Commonwealth and Southern. The court, however, also decided 8–1 that the TVA had the right to sell surplus electricity and operate in the marketplace. Whether the TVA itself in all its grandeur was constitutional the Court did not directly take up. In his diary, Homer Cummings noted that stockbrokers listening to Justice Hughes read aloud the Ashwander opinion first thought the opinion was pro-private sector, and so bought utilities. Midway, still unclear, they stopped buying. Toward the end, they sold wildly. *Ashwander* helped the TVA, by buying it time and dimming prospects for future challenge.

McReynolds, the dissenter, found his fellow justices to be disingenuously narrow. He argued in the dissent that "we should consider the truth of the petitioners' charge that, while pretending to act within their powers to improve navigation, the United States, through corporate agencies, are really seeking to accomplish what they have no right to undertake—the business of developing, distributing, and selling electric power." The case was really about whether government could "destroy every public service corporation within the confines of the United States." His vehemence may have come in part from the fact that McReynolds had lived and worked in Tennessee—he had attended Vanderbilt in Knoxville, at TVA headquarters.

"The tension of the last year is over and we can look ahead with fewer uncertainties!" exulted David Lilienthal, who had been listening to the news over radio and telephone. "For the last few weeks I have been on the point of calling Mr. Willkie and saying that however the case goes, we ought to plan on an early meeting after it to see what should be done. I am glad now that I didn't do that." Lilienthal noted in his diary that utility stocks had gone "up and up" recently, proving that "speculators are just as poor guessers as everyone else."

In fact, of course, even after the recent rises, the Dow's utility index was still breathtakingly low, at 32 or 33, less than one-third of its level in 1929. A whole sector of the economy, the one that had excited observers so much the previous decade, was being wiped out. Now Willkie and his industry understood that it might be a long time before it recovered. Within two days the National Resources Board sent Roosevelt a proposal to build another TVA in the Pacific Northwest. Willkie pointed out that between its power subsidies to municipalities and its other plans, the TVA was going far beyond its original legal pretext. Some 300,000 investors had invested $650 million in utilities in the South, and what the TVA was really responsible for was eroding that $650 million. The TVA paid no taxes, he noted. Let it pay taxes like other power companies; that would be a fair yardstick. He began to prepare another lawsuit, to try again to test the constitutionality of the TVA. His time was running out; Norris Lake was filling with water; C & S was making a profit, and he was more impatient than before.

Others, aware that they were beginning to sound like a broken record, argued generally that the president and the administration would hurt the economy with such projects. Ray Moley deplored the idea of "reform through taxation" and charged that it would send businessmen into "paroxysms of fright." At Chase, Benjamin Anderson was preparing a bulletin that tried to capture the longer-term economic damage that could result from Morgenthau's undistributed profits tax. The idea that corporate surpluses were bad, Anderson would write later in the spring, was a sheer fallacy that came down from Marx to Catchings and Keynes. Whatever recoveries the market and the economy were making, both were still behind. How would the Henry Fords of the 1930s succeed if they were not permitted to plow their profits back into the business? That had been the key to Ford's rapid growth several decades earlier.

But the administration, now eight months from the election and in full campaign mode, treated the pleas as so much background noise. Roosevelt himself, following Frankfurter's advice, still did not touch the issue of the recalcitrant Supreme Court. His allies, however, were jumping into action. Drew Pearson and Robert Allen, Washington's star syndicated columnists, were preparing a book about the Supreme Court. The title, chosen early, was *The Nine Old Men.* The new Supreme Court Building was a "mausoleum of justice." The justice who often provided the swing vote, Roberts, was the "the biggest joke ever played upon the fighting liberals of the U.S. Senate," the "foremost meat-axer of their cause." The section on Roberts was titled "The Philadelphia Lawyer"—a play on the old American pejorative referring to corrupt attorneys.

Cardozo, the book reminded, was the son of a corrupt Tweed Ring judge who had to be forced off the bench. Harlan Stone, the man who had passed along the tax secret to Frances Perkins, was "Hoover's Pal."

Such swipes, however, did not compare to the authors' attack on the Four Horsemen, as the anti–New Deal justices were known. The reference was simultaneously biblical and current, both to the Four Horsemen of the Apocalypse in the Revelation of John—war, famine, pestilence, and death—and to the members of the defense on Notre Dame's football team. Butler, a Democrat, the authors sought to attack for his Catholicism: all his life Pierce Butler had "striven zealously to promote the power and glory of the Holy Roman Church and the power and profits of big business," they wrote.

To smear Butler was to smear his fellow Catholic Al Smith. And the book also did that: "And as Al's hatred for Roosevelt has deepened, so also has Butler's, a hatred not merely against the President, as is Al's, but against all things for which the president stands." Willis Van Devanter was "The Dummy Director," who suffered from "literary constipation." But Van Devanter came off well next to Sutherland. The authors said of the justice and former head of the American Bar Association: "Van Devanter has brains. Sutherland has not." Treated worst of all, perhaps, was McReynolds, who often led the way when it came to reinforcing the traditional concept of "liberty of contract"—and who had snubbed Frankfurter's sociological arguments as poor logic years ago. Pearson and Allen titled their chapter on McReynolds "Scrooge." They also reported that court insiders had long ago tried to decide whether he was "chiefly stupid or lazy"—and then concluded he was both. The aim of the book was not so much to attack the Four Horsemen as to shame or intimidate Justice Roberts into switching sides and tipping the balance.

Tugwell was still under attack, but he tried to concentrate on his work. That same month—March—a draft proposal for one of the many new settlements came across the rural administrator's desk. This one was a cooperative farm for poor families to be built in a far-off, almost hidden place: an area called Casa Grande in Pinal County, Arizona. The land around the area had only lately become arable, after the construction of the Coolidge Dam, and seemed like a good prospect as a Resettlement Administration project. The condition of the people in the area was simply miserable. "Eight families occupied a shed, divided by chicken-wire into compartments measuring 18 by 24 feet, with dirt floors." Other families "lived in sheds made of box wood and cardboard, tin cans flattened." Now, on 3,000 or 5,000 acres, the RA would attempt to build a model farm community.

Tugwell's life was changing—he was spending time with his assistant, Grace Falke, and still wondering if Columbia might welcome him back. There had been a nibble—more than—from Yale Law School, but the job hadn't worked out. But he took his time over Casa Grande. He didn't like rural resettlement; still, this was the sort of experiment he had been dreaming of even back in the days of his Russian trip. Now he and his team "did all we could," as he would later recall. The project envisioned eighty individual farm units of forty acres each; the government bought the land and would supply the new farmers with everything from loans to get started to seeds to toilets and running water; the individual farmer-owners would eventually pay off their loans. Tugwell signed off, making a very small change—he increased the allowance for household equipment and furniture to $400 from $200. But he was not yet pleased—a life of experience in agriculture, his abiding instinct for efficiency, and his own advisers all told him that forty acres per family would yield only a bare living. The builders began the homes, but Tugwell sent his experts back to study whether the farm might work better as a large cooperative.

That same March, Tugwell and Stryker's photographer Dorothea Lange was returning from her field trip. As she would later recall, she was driving sixty-five miles an hour, tired and cold, when she saw a sign at Nipomo, California: "Pea-Pickers Camp." Later she remembered an "inner argument": "Dorothea, how about that camp there? What is the situation back there? Are you going back?" After twenty miles had passed, she did a U-turn, and found a thirty-two-year-old mother in a lean-to nursing a baby. There were older children; the mother "said that they had been living on frozen vegetables from the surrounding fields and birds that the children killed." Lange picked up her camera.

Lange did not have her pictures alone in mind when she left the camp; her first move was to alert the newspeople she knew that the people there were starving. As her biographer Milton Meltzer reports, this did much good: officials rushed 20,000 pounds of food to feed the California migrants. Beside the story about this, the local paper published two of Lange's photos of the thirty-two-year-old mother in her lean-to. A third photo—from the same shoot, but not in the paper at that time—would be the one later called *Migrant Mother.* It depicted a thin woman, almost recalling Mary Magdalene, holding her baby, with two others behind. The photo captured the despair of the Depression more than any Lange had taken.

In April, the unions paid Roosevelt back in a more formal fashion. George Berry of the printing pressmen's union and Sidney Hillman created the Non-Partisan League, which was dedicated to electing Roosevelt, the unions' answer to the Liberty League. That same month brought another achievement for the New Deal. Norris Dam's powerhouse was up and running. Roosevelt himself, pressing a golden telegraph key in his office, sent the two dam gates down to hold back the water. Willkie had lost *his* bet. Alabama Power's old contract with the TVA was now in jeopardy: from this date on, the TVA had leeway to sell power and find markets where it liked.

Meanwhile, Hopkins's WPA was now operating all across the country. Within nine months of its establishment, it had increased its rolls to over three million. The Federal Writers' Project in April employed a total of 6,686 writers. That February, Henry Alsberg, the director of the Writers' Project, announced that his authors would produce a new guide to America, a giant project. Authors would résumé cultural activities and geography of each state, from the festival of Los Hermanos Penitentes in New Mexico to the islands off Georgia, where Norsemen were believed to have settled a millenmium prior.

Roosevelt for his part was watching the Court, but still silently. That spring, the justices rejected a New York State minimum wage law—a law that Frankfurter had had a personal hand in drafting. The case, *Tipaldo,* was remarkably reminiscent of *Schechter.* A Brooklyn businessman—this time John Tipaldo, a laundry operator—had paid his laundresses less than the minimum wage. Would the Court stop nowhere? the president and his allies wondered. Did the states have rights at all to pass social legislation? *Tipaldo* seemed to say that they did not. Only 10 of 344 newspapers liked the *Tipaldo* decision. This time, the nation seemed to share the White House's sense of impatience. Hoover sided with Roosevelt, saying, "Something should be done to give the states back the powers they thought they already had."

Roosevelt was also waging battles on the tax front, for Congress rejected the first version of the undistributed profits tax. Morgenthau weakened: "I have come to the decision that I cannot take the risk of giving up something that I have in hand, namely $1.13 billion in revenue, for the possibility of getting roughly $1.7 billion." The administration then cobbled together, with lawmakers, a new plan: a graduated surtax on the corporate income tax—again, an antitrust measure—plus an increase on the intercorporate tax rate, as well as a new undistributed profits tax, albeit at lower rates than originally planned. Morgenthau later recalled: "Nobody in the Treasury wanted to testify. Everybody was frightened except Herbert Gaston, who wrote the statements I needed. I had to stand up like a column of concrete, but I had the backing of FDR. He wanted to wipe out special privilege." In the end the tax did make it into a new bill, in a watered-down version.

These little details did not really do much damage to the Democrats, for the Republicans were flailing. At a fretful convention in mid-June, uncertain party leaders selected Kansas governor Alfred M. Landon as their candidate. But Landon failed to distinguish himself from Roosevelt. The telegram on policy he sent to the Republican convention before the roll call vote of his nomination differed, as he would put it much later, "not too much" from the Democrats'. In the nominating speech, John Hamilton of the national Republican committee spoke of the importance of combating "great combinations of wealth"—a Democratic theme.

To add to its woes, the Grand Old Party was also stuck in an awkward place on foreign policy. The old "stay out" position, which had seemed reasonable in the 1920s, was looking increasingly questionable in the context of the news reports from Europe. Mussolini had occupied Ethiopia, and that same month—June—Haile

Selassie was begging the League of Nations for help in Geneva: "It is us today. It is you tomorrow." Left-leaning magazines were carrying reports of the torture and murder in Germany's early concentration camps, reports that were increasingly hard to discount as Marxist propaganda. Yet leading Republicans—Herbert Hoover being the prime example—were still interested in working with the Germans. Perhaps because Hoover himself had created what were called concentration camps in the United States during the 1927 flood, he could not fathom the German version.

Roosevelt, smelling victory, did not let up. He traveled to the Democratic convention in Philadelphia to attack the "economic royalists" of American business for bringing down the economy. The government, Roosevelt said, had to help citizens "against economic tyranny such as this"—there was no other power. As for taxation, it was now crucial: Ickes would remark in his diary that "the fundamental political issue today is taxation." A number of former Roosevelt allies spluttered with rage, including Warburg, whom Coughlin had so misleadingly assigned to the list of Roosevelt's current allies. Warburg, still fuming, published yet another book since his breakup with Roosevelt, *Still Hell Bent*, a follow-up to *Hell Bent for Election* of the year before. But the Warburg household was not content to stop there. Warburg's wife, Phyllis Baldwin Warburg, published *New Deal Noodles,* a book of alphabet rhymes about the New Dealers: "the whole new deal is full of hickies. One of these is Mr. Ickes"; " 'T' stands for Tugwell and dear TVA; 'T' stands for the Taxes we'll all have to pay." Ray Moley was still moving from critic to opponent. The president did call on him from time to time, but he was refusing, as he later put it, to be a "jester" in Roosevelt's court. Instead of talking with one another, the parties were now talking past each other. Moley warned of a devastating counterreaction.

To the incumbents that risk seemed small. Over the course of the summer, the effect of Roosevelt's spending was still strong, and the jobs materializing seemed a wonder after so many years. The same Benjamin Anderson who had tracked earlier damage now recorded the upturn in his office at Chase in New York. Together he and Colonel Leonard P. Ayres of the Cleveland Trust Company visited Landon in Topeka. They told him that the economy was moving up, and Landon later recalled, "I knew then that I was beaten."

Even as Landon doubted, a new factor began to work in the Democrats' favor. A drought worse than that of 1934 hit the land. Families in a thousand counties were affected, one-third of the nation.

The drought supplied Tugwell's RA with new purpose. That summer the RA provided aid—from cash to short-term jobs—for 400,000 families. The RA had an enormous loan component, and it declared a one-year moratorium on payments and allotted millions in new loans. It was a legitimate high point for the RA. What's more, the drought validated Roosevelt's allegation that the country was still in "emergency." It made a mockery of the "self-government" argument for localities by Republicans at the Cleveland convention, and it made the need for the New Deal seem permanent. Tugwell himself was to travel 2,000 miles, and to make a showing with Roosevelt in Bismarck, North Dakota.

Over the drought summer the writer John Steinbeck, already well known for *Tortilla Flats,* was traveling among the RA's dusty demonstration camps for migrants. He made the acquaintance of Tom Collins, the RA employee who designed camp operations. Collins was creating a women's club that would be a model for such a club in Steinbeck's *Grapes of Wrath.* Collins likely also introduced Steinbeck to Sherm Eastom, whose family was one of the models for the Joad family in the novel. Collins wrote lengthy reports to his employers—one appeared that summer in the *San Francisco News*—that later served as additional material for Steinbeck. In the September 12, 1936, issue of the *Nation,* Steinbeck wrote a column that served as a nonfiction outline for his book. He commented that people like A. J. Chandler, publisher of the *Los Angeles Times,* or William Randolph Hearst, or Herbert Hoover, operated big farms. These farms, he said, were proliferating at the expense of the disappearing medium-sized farm. Then he went on to report, of the arriving Okie, that "in the state and federal camps he will find sanitary arrangements and a place to pitch his tent." Conditions in the privately owned camps of farmers, Steinbeck reported, were by contrast horrific.

Meanwhile, too, Tugwell's buildings, his towns and settlements, were coming along—both under his direct supervision and via his influence. Beltsville, Maryland, saw the erection of a structure dedicated to animal husbandry, one of his old favorite subjects. Tugwell's greatest pride was Arthurdale in West Virginia: it had a vacuum cleaner assembly plant. The town also had a chicken farm run by a cooperative, a small-scale rebuttal to *Schechter.* Mrs. Roosevelt liked it, especially; later in the year she would note in her syndicated column that the chicken farm was "doing very well."

Casa Grande would also have chickens. The builders had worked ahead, laying out two-acre plots. But as the year advanced, Tugwell and his advisers were still turning over the format of the farm in their minds. The individual farms would simply yield too little. What were title and ownership worth to a man just scraping by? On paper a co-op or collective looked much more efficient, yielding $19 an acre instead of $14. Assailing "bigness" was fashionable now—that is what they were doing at the White House, at the Treasury, and in Congress. John Steinbeck might also find large-scale projects evil; the author's *Nation* piece in September criticized some of them. But election year or not, everything in Tugwell's philosophy and experience told him that a large cooperative taking advantage of the economy of scale would make more sense: one tractor for all. There was a sense of urgency now; the gossip against him was hard to ignore. For as long as he stayed in government, Tugwell determined, he had to do what he believed. He made Casa Grande a cooperative. The farmers would share the land.

In September, Roosevelt spoke at Harvard, Felix's home, his own alma mater, but also the institution of a man Roosevelt disliked—A. Lawrence Lowell, president emeritus, the man who had opposed Frankfurter on Sacco and Vanzetti. Bolder than ever, Roosevelt chose to omit the traditional acknowledgment of the host in his salutation. The audience was shocked, but Roosevelt enjoyed himself. Frankfurter telegrammed the president, IT WAS REALLY A GREAT TRIUMPH. YOU FURNISHED A STRIKING

EXAMPLE OF THE CIVILIZED GENTLEMAN AND ALSO OF THE IMPORTANCE OF WISE SAUCINESS. Roosevelt, anxious for praise from his adviser, wrote back, "Did you really and truly like it—more important still, did Marion really and truly like it? Your expression of the 'importance of wise sauciness' is perhaps better than mine. I told the boys afterwards that I had stuck my chin out and said 'hit me'—and nobody dared!" Even as Roosevelt drew closer to Frankfurter, he distanced himself from the others. Though Tugwell and Moley were doing productive things—Tugwell was at a high point—the change cast a shadow. Moley later wrote of Roosevelt: "He closed, one by one, the windows of his mind." Perhaps, Moley went on, "this is a disease that haunts the White House. In any case, Roosevelt developed pernicious attacks of it, and this lessened his capacity as a political leader and statesman."

That month, *Migrant Mother* was published for the first time. At the Tennessee Valley Authority, Arthur Morgan was still seeking a territorial truce with the private companies. To him the war with them was a distraction, perhaps even unnecessary. In August, however, the other two members of the TVA board had moved against him, resolving that "in future contracts the Authority will not agree to territorial restrictions on the sale of the Tennessee Valley Authority power to public agencies." It was a public victory for Lilienthal.

Roosevelt seemed to side with Morgan—at least until the election. Repeating his "breathing spell" action of a year earlier, he invited Willkie, the TVA, and a few others, including Thomas Lamont of J. P. Morgan, to a conference to talk about a power pool grid system.

Willkie, though suspicious, felt he had to go along with the administration, at least this last time. Whatever the status of corporate stock, there was plenty of hope for operating utilities companies. Commonwealth and Southern net earnings had headed up from 1935 to 1936 and they would be even higher in 1937. Roosevelt wanted his own breathing spell; he wanted the utilities businesses to end their interminable lawsuits. The utilities knew they could not win if the government continued to buy off towns by helping them to build their own distribution plants for TVA power. Perhaps Roosevelt would, after all, allow competition to challenge the TVA. Willkie felt, the *New York Times* wrote on October 11, that the White House willingness to talk about power pools was "an act of political statesmanship calling for an equal degree of business statesmanship on his part." In that original first meeting, Roosevelt may not have charmed Willkie. But this time it seems the president succeeded. And it was a ruse, if Robert Jackson, the attorney, is to be believed. For, as Jackson later recalled in a memorandum in his papers, Roosevelt around that period "had a profound dislike for Willkie."

Other utility executives, like Ferguson, did not even try to be agreeable. They had tired of Roosevelt's cynicism, which was in evidence even in the context of the power pool negotiations: as his representative for these government-to-utility talks Roosevelt had chosen Louis Wehle, the nephew of Louis Brandeis, author of *The Curse of Bigness.* What's more, Harold Ickes at the Department of the Interior was continuing to subsidize municipalities to build city-owned utilities to take power directly from the TVA—in October 1936 he announced a $3 million grant to the city of Memphis for precisely this purpose. This they regarded as a bribe. And of course, this was not all that Ickes was giving the towns; there were yet more pillared town halls and libraries. When it came to convincing towns that the federal government or the TVA belonged in their town, Ickes's helpful buildings had more authority than any politician.

A speech that the president delivered just weeks later seemed to validate these executives' fears—and reveal Willkie as naive. At Madison Square Garden, where Coughlin had stood, it was now Roosevelt's turn to let the invective fly. "I should like to have it said of my first administration," he told the crowds, "that in it the forces of selfishness and of lust for power met their match. I should like to have it said of my second administration that in it these forces met their master."

Now Wall Streeters had indeed become like the plutocrats featured in *New Yorker* cartoons—a few small men, isolated in an outsize ballroom. They knew that their fewness worked against them, especially as Roosevelt courted great swaths of society. Roosevelt had not pushed the antilynching legislation that Father Divine hoped for—it was the sort that southern lawmakers would filibuster. But that election autumn he dedicated one of three new buildings—"with more to come"—at Howard University in Washington, telling an audience there were "no forgotten men and no forgotten races." Blacks were moving into Roosevelt's fold. Father Divine would not be a power behind the presidency.

That autumn as well, something good was happening across the nation: Social Security was beginning to seem real. Sometime in the months before the 1936 election, millions of Americans found in their hands a small but riveting document known as an ISC 9. "There is now a law in this country," the pamphlet from a new office in Washington, the Social Security Board, instructed, "which will give about 26 million working people something to live on when they are old and have stopped working." The government would, it told the reader, "set up a Social Security Account for you, if you are eligible," and into this account "You and your employer will each pay three cents on each dollar you earn, up to $3,000 a year." That amount, the circular added, "is the most you will ever have to pay." Last came the promise: "From the time you are old and stop working, you will get a government check every month of your life. This check will come to you as your right."

The message could not be clearer: people who opposed Roosevelt might stand in the way of American rights. Landon was fading as a candidate. Jobs seemed to be materializing at a heartening rate. Data compiled later would show that in November 1936, unemployment hit 13.9 percent, the lowest level since 1931. But while yet more hiring might materialize at some point, voters believed that at this moment their choice was between the gifts at hand and the uncertain possibility of prosperity.

There were those who still questioned the terms of that choice. In Harlem, at his headquarters at 20 West 115th Street, Father Divine issued a message in his typically meandering prose: "Not one of the major parties, officially and nationally, or conventionally, has come to me and accepted of my righteous government platform." Father Divine therefore ordered his flock to "stay our hands," not to vote, and reporters noted that the polls of Harlem were deserted. Father Divine told his followers the movement would wait until 1940, when, presumably, they would be even greater in number. But his boycott amounted to a footnote in the 1936 story.

On the eve of the election, Frankfurter wrote to Roosevelt of his campaign "the Nation will crown it with victory." Many of the president's opponents voted for him. In an era where nothing was easy, the helmsman who tacked left, and then right, and then left still seemed the better choice. Roosevelt took forty-six of forty-eight states. His was the wager of the century, and he had won it.

10
Mellon's Gift

December 1936
Unemployment (December): 15.3 percent
Dow Jones Industrial Average: 182

One day seven weeks after the election, Andrew Mellon wrote to the White House. The octogenarian's letter was not about taxes, or his troubles with the treasury prosecutor, Robert Jackson, nor even deposit insurance for banks, the topic of his preinaugural conversation with Roosevelt four years and so many days in court earlier. Mellon wanted to tell Roosevelt a secret—a secret about his paintings.

My dear Mr. President,
Over a period of many years I have been acquiring important and rare paintings and sculpture with

> *the idea that ultimately they would become the property of the people of the United States . . .*

Mellon wondered if such a gift, and a building to go along with it, might meet presidential approval.

By this time the gallery idea was not truly a secret. Roosevelt had been at a Gridiron dinner at which Mellon's gallery had been spoofed in a skit earlier in the year. Treasury's lawyers had forced Mellon's attorneys to detail his holdings. And Mellon's critics had long ago begun to suggest that the gallery offer was an open effort to bribe the White House into desisting in its tax war against him. After all, some of the fiercest of Republicans were now seeking to smooth things over with Roosevelt. Even Hoover, who had been so ferocious during the campaign, was coming up with praise for the president: "The President is right," he would write from Palo Alto, California, in a January statement supporting the president's effort to see ratified a constitutional amendment banning child labor.

But what was new was something that Mellon was only now revealing—what had been on his mind all these years, the philosophy behind the donation of the collection. Mellon was not trying to bribe the government, or even placate it. He was trying to outclass it. For years he had tried to show, through business, that the private sector could give to the people, just as government could, and sometimes more. Then he had tried to demonstrate the same thing from his post at Treasury, through his tax cuts. Now, pleased but still not satisfied with his work through the first two methods, he was trying a third: charity.

In Mellon's head, the plan was entirely clear. By giving largely, generously, completely, and entirely, he would demonstrate that the private man could be as good a servant to the public as the government official was—certainly, he was ahead of Morgenthau. What's more, he would make his gift selflessly. It would not bear his name: "It shall be known as the National Gallery of Art or by such other name as may appropriately identify it . . ." Even the display of the paintings would be unselfish—they would be arranged by period and style, not by collector or collection.

Mellon's paintings must be spared the fate of those of his old collecting companion, Frick. Mellon's collection would not be eroded by taxes or prosecutions, as long as he could defend it. Mellon's gift would show the value of leaving art—or capital—to accumulate and compound in the shadows, untaxed. The National Gallery would be an object lesson that the high taxers could not forget.

The critics, even Robert Jackson, might say what they liked about the timing of his gift, but the whole idea was one that had come to Mellon years ago, while he was still at the Treasury. The Office of the Supervising Architect for the Capitol had reported to the treasury secretary; Hoover and Mellon had discussed and planned the development of the Federal Triangle together. David Edward Finley, Mellon's aide and a connoisseur of art himself, later recalled that "he had been embarrassed when representatives of foreign countries had come to the Treasury in Washington for debt settlements and other matters and had asked to be taken to 'the National Gallery' where they could see some of the great paintings they knew were in this country. Mr. Mellon would reply that there was no such National Gallery of Old Masters, but that he had a few paintings in his apartment which he would be glad to show his visitors."

By 1927, Finley later remembered, "Mr. Mellon was revolving in his mind plans for a National Gallery of Art, which he felt to be a necessity in the capital of a great country such as America." Never mind that a subsection of the National History Building of the Smithsonian was called the "National Gallery"; the new building would supplant it, and he would clear any legislative or bureaucratic obstacles that stood in the way of his plan. Andrew had not been the only Mellon dreaming of buildings. At Yale, well before his escape from the tax prosecutors, Paul had written a poem:

I built a temple in my inmost mind
Of pure white marble;
Its stern symmetry
Became the symbol of tranquillity.

After a cabinet meeting, Hoover for his part recalled, "he came to me and asked that that particular site [the gallery site] be kept vacant. He disclosed to me his purpose to build a great national art gallery in Washington, to present to it his own collection which was to include the large number of old masters he was then purchasing from the Soviet Government." Mellon, Hoover also remembered, "said he would amply endow it and thought it might altogether amount to $75,000,000. I urged that he announce it at once, and have the pleasure of seeing it built in his lifetime." But Mellon demurred. "He asked me to keep it in confidence," Hoover remembered, regretfully. "Had he made this magnificent benefaction public at that time, public opinion would have protected him." Still, Mellon had kept silent.

In 1930, while still at Hoover's Treasury, Mellon had established the A. W. Mellon Educational and Charitable Trust, and transferred his whole collection to it the following year, with the idea that the art would go into the National Gallery. Paul had been nonetheless surprised at the sweep of the gift, and was concerned about what it meant for him and his sister. A Joshua Reynolds painting, *Lady Caroline Howard,* had always hung in his sister Ailsa's bedroom at the Mellon apartment on Massachusetts Avenue in Washington. Yet, wrote Paul Mellon later, "we suddenly found he had put everything into the trust. . . . Father even put the Reynolds of Lady Caroline Howard, which I am sure he knew Ailsa loved, into the trust." Mellon the father "was terribly surprised when we said that we would have liked just one picture, one favorite picture. But it was too late." Secrecy was important, but so was the discipline of family philanthropy.

There had also been that letter to Paul in 1931, who by then had moved from Harvard to Cambridge for study in Britain. "I hope you are having some time to spend at the National Gallery." In the letter Mellon added: "I have gone deeper into the Russian purchases—perhaps further than I should in view of the hard times and shrinkage of values, but as such an opportunity is not likely to again occur and I feel so interested in the ultimate purpose I have made quite a large investment . . . The whole affair is being conducted privately."

As the Depression deepened, and Mellon moved over to London and back again, his purchasing had not ceased. Some of his paintings he put in a storeroom at the Corcoran Gallery, visiting them from time to time. But the secrecy remained important. Mellon savored it, confident that the full drama of his message would become clear only if the scale of his gift emerged suddenly, all at once. That was what the hideaway room at the Corcoran had been about. Not selfishness—unless the enthusiasm of the giver arranging his surprise could be called selfishness. Hence Mellon's irritation when anyone in his family, even his children, talked about his collection.

And hence his impatience at Jackson's tax prosecutions. Jackson had believed that through reference to Mellon's paintings he was exposing Mellon's wealth and vulnerability; what Mellon especially resented, though, was that Jackson was eroding his surprise. The etiolated Mellon rarely complained in public, but at the tax prosecutions he complained. He even rambled about "accursed publicity."

Mellon's unhappiness had only grown when Jackson forced him to tip his hand more formally in the spring of 1935. By claiming that Mellon had failed to declare certain income, Jackson was charging that Mellon had failed to pay sufficient taxes. Since the prosecution was a criminal one, Mellon's only recourse had been to find all means to defend himself. Logically enough, therefore, he and his lawyers pointed out that it was fine for the Treasury to allege that he had had larger income in 1931 than declared. But he had also had larger deductions than declared. And those deductions would include, of course, the paintings that he had purchased for the trust or the gallery.

What Jackson expected was that this would embarrass Mellon by betraying his wealth, by showing, as he put it, that there was something to "doubt" in Mellon's $200 million net worth. Instead, however, the very wonder of the list was itself a distraction. The gift collection as then planned did not contain very many paintings, only seventy, but the collection was so broad and of such high quality that it truly could claim to be America's foremost. It started with that Pocahontas, the only known one, painted from life during her visit to Britain in 1616, shortly before her death. And it moved forward through van Eyck, Botticelli, and Raphael's *Madonna of the House of Alba*—the last called the "million dollar picture," as Mellon's purchase had made it the first known seven-figure portrait purchase. And there was Rembrandt's brooding self-portrait, completed in his early fifties, the face of a genius whose lack of luck had made him look like a failure.

In October 1935 the story had even become semiofficial when Mellon filed a deed revealing that he had set aside $10 million for the construction of "a national public art gallery" in Washington. Two art experts—although one was Mellon's own broker, Lord Duveen—put the value of the Mellon collection now at $40 million.

Jackson's interrogation of Duveen in court had underscored the problem with the direction of the government's prosecution. Jackson sought to show Duveen as the rich man's servant, which Duveen was. But Duveen was also Baron Duveen of Milbank, a formidable presence of his own and a man who had logged many previous days in court as an expert witness and target himself. At the Bureau of Internal Revenue courtroom in early May 1935, he dismissed Jackson's assertion that the art was for Mellon's own uses—Mellon had talked about a gallery "five or six years ago." When Jackson tried to assail the scale of the deductions, Duveen had been equally scornful. Mellon's collection was "the finest in the universe" (he knew; he had a part in making it). Mellon's *Alba Madonna* was worth all three of the Raphaels in Britain's National Gallery.

Jackson, not giving an inch, revealed that he had had a look at Duveen's tax returns as well as Mellon's—one of Duveen's biographers, S. N. Behrman, reports that he then asked Duveen whether it was not true that he himself had lost millions in the early 1930s. Duveen had been undismayed. Jackson also questioned the stated value of van Eyck's *Annunciation* panel. "Perhaps you don't realize that there are only three small van Eycks in America," Duveen said. "And they cannot compare with Mr. Mellon's van Eyck." The *New York Times* report conveyed not the scene of shame that Jackson intended but rather one of fun and curiosity: "girl clerks" on their break crowding near the hearing room in the hopes of catching Mellon smoking one of his little cigars.

Duveen's very disregard for Washington seemed to put the government crowd back in its place—much in the same way that Mellon's comment about the dark quarter-hour had put the Depression in its place. Duveen cared so little for the Capitol that when interrogators asked where the Mellon gallery was to be situated he had said of the Washington Monument and the reflecting pool—"by the obelisk, near the pond."

Mellon's point, though not entirely articulated, was obvious to many of the observers. The only reason his art collection was so great was that he was supremely wealthy. And the only reason he was so wealthy was that he and his father before him had been allowed to invest and save. He himself had established charitable deductions in the United States for estates in order to ensure that rich people give to the public. His art collection was so large and had so much integrity precisely because it was a private collection.

As the details of the project seeped out, the argument against the gallery had become harder to make. Over the course of 1935 and 1936, the public learned that Mellon's gallery would not be a small thing. It would be different from the Corcoran Gallery or the Phillips Collection, already in Washington. Those collections were open to the public, but they were not national galleries. They were shrines to their philanthropic creators, William W. Corcoran, a cofounder of the Riggs Bank, and Duncan Phillips, a neighbor from Pittsburgh and heir to the Laughlin steel fortune.

Mellon, by contrast, truly was insisting that the fact of his giving—or that of any other donor to the museum—be pushed into the background. The gifts would be bequeathed directly to the people's representative, the federal government, just as he had planned. There would be no middle ground between public and private—no semiproprietary right that gave Mellon or those he designated a guarantee they could continue to manage the museum. Mellon, or another wealthy man, might give another Vermeer or a Giotto. But that art would be—just as he had conceived—integrated, by school and chronologically, into the general collection. And it would become public property as surely as the Capitol itself was public property, and even the fact that the paintings had once belonged to a captain of industry, say, one from Pittsburgh, would recede.

In 1936 Mellon, now past eighty, intensified his focus on the gallery. He traveled to Britain with his aide Finley and visited Duveen, who suggested he had many works in New York that might fit in with the gallery. Back in Pittsburgh Finley and Mellon walked the garden of his Woodland Road house, and, Finley later recalled, "his eyes would brighten as he talked about his last great project." Returning to Washington and down with a cold, Finley received a call from Mellon: he must go to New York and select paintings from Duveen that would suit the gallery. Finley took the midnight train and looked over the works with Lord Duveen in his velvet-lined chambers at 720 Fifth Avenue. "We were there for the greater part of three days," Finley remembered. Then Finley, thirty paintings, and twenty-one pieces of sculpture all traveled south to Mellon. The sculptures meant that Mellon, after so much study, would in the end be one-upping the British: their National Gallery, as fine as it was, did not have sculpture.

Some of those involved in the story later told it a different way—as a successful lure by Duveen. Sensing that Mellon was ready to give, Duveen had rented the apartment below that of his most important client on Massachusetts Avenue. He told Mellon, "You and I are getting on. We don't want to run around. I have some beautiful things for you, things you ought to have. I have gathered them specially for you." He then gave Mellon a key to his own apartment and invited him to visit the paintings when he was away. The biographer Behrman reports that Mellon, as it turns out, did indeed visit—in dressing gown and slippers. Eventually he bought up the art in the apartment.

Both accounts are probably true. Mellon always hesitated before he bought—Duveen's success with him lay in his ability to tolerate lengthy periods of indecision. The purchase when it finally came was in any case a giant one, outdoing the Hermitage acquisition by far. But Mellon, though fading, was still a formidable bargainer. Duveen had showed Finley a painting of Saint Paul. The identity of the painter was uncertain—Bernard Berenson, the critic, had categorized the large figure "as by Giotto." Duveen insisted it was a Giotto: "I say it is by Giotto and it will be by Giotto."

Finley reports countering that if Mellon bought it, it would be at the price of a painting by "A Follower of Giotto." Mellon, a good boss, backed him up: "I will buy the Saint Paul painting as 'A Follower of Giotto' and at a suitable price, not the price of a Giotto, and it will so hang in the National Gallery." The transactions were concluded December 15.

The number of artworks that Mellon had to offer was still not so very large. As John Walker, who would become collection curator, would note, in the end Mellon gave 125 paintings and 23 sculptures, nowhere near enough for the building he was beginning to envision. "The Mellon works of art would seem as scattered as sheep on a Scotch moor." But Mellon, again true to his old concept of seed capital, was reckoning that other philanthropists would fill a space if he set the model by providing it. It was now mid-December, and he was ready to write Roosevelt.

The president replied to Mellon's proposition the same week, on the day after Christmas.

> *My dear Mr. Mellon,*
> *When my uncle handed me your letter of December 22 I was not only completely taken by surprise but was delighted by your very wonderful offer to the people of the United States. This was especially so because for many years I have felt the need of a national gallery of art . . .*

Within days, Mellon and Roosevelt were taking their tea in the upstairs library on the second floor of the White House. They sat down at five o'clock, but what they said precisely is unknown: "He and Mr. Mellon were deep in conversation for some time," David Finley would write. The president "tossed" Mellon's offer to his attorney general for management. Last of all the president's personal secretary, Missy LeHand, "came in to pour tea, with some of the Roosevelt grandchildren to look on."

Roosevelt was in a grateful mood in these post-election days—the same week, mindful of all Cordell Hull's persistent free trade policy had done for recovery, he also wrote to the acting secretary of state to ask whether he might recommend Hull for the Nobel Peace Prize. Clearly he also wanted to show gratitude to Mellon. But this was still an anxious time for Mellon. There had to be more exchanges of letters and an act of Congress. Paul Mellon and his wife, Mary, were expecting the birth of their first child. But, the papers reported, Mellon was working hard on the art project. The grandchild, Catherine Conover Mellon, was born on the second-to-last day of 1936, but Paul convinced his father to stay in Washington until another round of letters with the White House was concluded. As it turned out, Mellon's concerns were overdone. The same Congress that would find other questions more challenging—most especially a plan of Roosevelt's about the Supreme Court—would find it easy to support the National Gallery project later that spring.

It may be that Mellon's impatience merely had to do with his age, and his sense of time passing. His son Paul later wrote of his father in this period that there were moments "during which his words came out helter skelter." He thought about buying art—in January he would buy a Hans Memling and a painting by a Frenchman, Jean-Baptiste-Siméon Chardin. He would buy these from Knoedler and Co., not Duveen. But Duveen was still close, along with the architect, pushing Mellon on the gallery. Mellon thought buildings should be limestone—the Mellon National Bank was limestone, as was the Mellon Institute, a trapezoid in Pittsburgh already well under construction. And so were a number of Washington structures Mellon had built for Coolidge. But limestone, Duveen was convinced, was not good enough for the gallery. Duveen would let Mellon pick which sort of marble to use in construction, but marble it had to be, Duveen insisted—just as in Paul's student poem. Duveen took Mellon on a car ride of Washington to remind him of the grimy look that limestone took on as the years passed.

Marble it would be, Mellon conceded. Like his son, he dreamed of bright structures. "Thanks for the ride," Mellon said. "It has been the most expensive ride of my life." Though we do not know all the details of Duveen's arguments, one that would have appealed to Mellon was that a grander gallery was likely to attract more "capital investment"—more art gifts—from other collectors. Mellon chose pink Tennessee marble and—as Duveen's biographer Behrman reports—did much for the marble industry by placing the largest single marble order in that state in history. More than a hundred carloads of light pink Tennessee marble would be quarried in the following year for the Mellon project alone. In a letter to Congress on his commitment to the gallery, Mellon reported that the marble increase alone involved an extra outlay of $1 to $2 million.

Mellon took care with the other details. Architect John Russell Pope had drawn a model—based on George Hadfield's 1820 courthouse building in Judiciary Square, some said. Others thought Pope was influenced by the German architect Schinkel. Mellon and his architect wanted a long, low structure because they wanted to spare Americans museum fatigue and that bane of septuagenarians, museum stairs. Mellon was not pleased with the number of columns—too many. But, he would tell Finley, "I would not want to hurt his feelings." In the end the message was communicated to Pope, who agreed that the columns on the north and south ends distracted, and made the change. At 782 feet, the gallery would be 36 feet longer than the Capitol itself. In New York, the *Herald Tribune* carried a three-part series detailing every aspect of the gift collection. Mellon was not merely giving, he was cataloging what he was doing for the recipients.

The full import of the secret was coming out. Mellon the miser was giving one of the greatest gifts a man had ever given to a country: a classical building with modern comforts, including upholstered sofas for the guests, a special smoking room, and ample light in its courtyards. It was a project that measurably boosted the economy in places such as Tennessee. A museum that would contain the world's best paintings. A collection whose ultimate curator—Mellon—had given unstinting attention to quality and shown utter disregard for nationality or provenance.

This was the opposite of Morgenthau's murals, of Dorothea Lange's migrant photography, which were both political art, and indeed, art whose representations of struggling industrial workers or farm workers aligned with the specific purposes of the Roosevelt administration. Watching the partisan art go up, a disconcerting critic had chastised Roosevelt's political opposition for failing to complain. "The losers prefer to await 'Landon' art," he had written mournfully.

Mellon tended the project carefully; construction would start in summer. And in summer, too, he would go to visit his daughter Ailsa in Southampton. Now that the gallery news was out, he could act openly. Perhaps inspired by the thought of his activity, the Dow climbed, moving close to 190.

At the time of these meetings and decisions, all the men involved knew that they might not see the gallery completed. Roosevelt suffered from sinus infections and atherosclerosis, and always, the increasingly debilitating effects of his paralysis by polio. Duveen was ailing. Pope was not well. Yet all four men recognized the challenge of Mellon's gift. It started on the aesthetic plane—the challenge to Morgenthau, and to the modern art of the 1930s generally. It continued on the architectural plane, with the scale of the building. But the challenge was also economic. What Mellon was saying was that the public sector could erect its structures—the Norris Dam at TVA; the new Supreme Court; a center for animal husbandry, Tugwell's area, going up in Maryland. But even in the fifth year of the New Deal, the private sector too could claim a proud place on the Mall, and occupy that place with its own structure of virtue. Mellon might be correct about the Depression being a bad quarter hour. History alone would tell whose edifice had the more enduring power.

11
Roosevelt's Revolution

January 1937
Unemployment: 15 percent
Dow Jones Industrial Average: 179

It was the wettest of possible inaugurations—a wild storm dumping a mixture of hail and rain onto the streets of Washington. Yet the four septuagenarians made their way through it on January 20, 1937, wading across soggy red carpet to wet seats. Pierce Butler, Willis Van Devanter, George Sutherland, and James McReynolds joined two others and listened as their colleague, Chief Justice Charles Evans Hughes, administered the oath of office.

But the justices were waiting for the two thousand words or so that would come next—the inaugural speech. The temperature hovered above freezing, and those in the crowds who had not brought umbrellas hid under blankets. Sutherland shivered visibly. John Knox, the clerk to McReynolds, later recalled his shock at seeing the old men up there. He was surprised to see Van Devanter, especially "since Van Devanter had told me once that he had been afraid even to take his hat off at Justice Holmes's funeral for fear of catching cold. And Cardozo was too frail a man to risk sitting out long in such weather."

Perhaps adrenaline protected the judges—or sheer resolve. They were bent on proving themselves as hardy as Roosevelt, who was braving the storm despite his own infirmity. The justices were showing they could stand up in the face of criticism. The Four Horsemen were also were making another, cold calculation. Roosevelt wanted to stop them, that they knew. And he probably had the power to do it. Still, he might now overreach.

All that month, there had been quiet signals that the White House was sorting out its options on the Court. The 1936 convention platform had suggested that the Democrats would offer a constitutional amendment—what the Democrats labeled a "clarifying amendment"—to allow Congress greater power. Roosevelt might now try to go further.

Certainly the administration seemed to be in action mode generally. On January 3, a Treasury spokesman announced that Mellon's gift would not affect the prosecution of the lawsuit against him in any way. The gift would certainly be accepted as a charitable deduction, Arthur H. Kent, assistant general counsel to the Treasury, said, but not one for 1931, the year at issue in the case. That same week, the Securities and Exchange Commission made targeting of the utilities executives easier by releasing the salaries of some executives to the papers. Willkie's 1935 salary stood out like an archery bull's-eye—$75,690, an enormous figure for the period. The average wage for all employed in Willkie's industry was much lower—less than $2,000 a year. This disparity could clearly be turned against Willkie.

Within a few days Hatton Sumners, chairman of the House Judiciary Committee, had revived a bill that guaranteed the income of retiring justices as long as they provided certain minor judicial functions. It looked like a proactive gesture to encourage retirement and, thereby, fend off a dramatic assault on the Court from Roosevelt. In January too McReynolds's clerk had opened the mail to discover a gift from Homer Cummings, the attorney general, Cummings's new book on the law. "To Mr. Justice Reynolds with the best wishes of Homer Cummings, Jan 11/37," the ink inscription read. Finding his name in the index, McReynolds discovered that Cummings had, on page 531, quoted an argument on "the problem of age" the justice had made more than two decades before, when he himself was attorney general. If a superannuated judge refused to resign, the younger McReynolds had argued, then the president should appoint an extra judge "to insure at all times the presence of a judge sufficiently active to discharge promptly and adequately the duties of the court." Upon reading this, the clerk recalled, "McReynolds' eyes narrowed." The justice gave the clerk the book, telling him, "I wouldn't have it around the house. Take it away!" Cummings's name became "unmentionable" in the apartment, Knox wrote.

Now, when it came, the inaugural address was every bit as expansive as the justices had imagined. "Our progress out of the Depression is obvious," the president declared; the achievement of recovery was as good as accomplished. This seemed a stretch, since with one or two men in ten still unemployed, the country was scarcely back. (Later data showed that joblessness had risen since the November 1936 low.) But the argument was expected. Roosevelt proceeded. It was high time that the old style of economics be buried: "We have always known that heedless self-interest was bad morals; we know now that it is bad economics." Those who forgot that did not recognize that in the long run, "economic morality pays." Wealth was simply not a sign of virtue. The country had developed a new understanding of life in the Depression. "This new understanding," Roosevelt said, "undermines the old admiration of worldly success as such." A decade had passed since Coolidge was quoted as saying, "The business of America is business." But Roosevelt made it seem like a millennium.

Next Roosevelt turned the weakness of recovery to his advantage. There were still millions in America who did not enjoy even the absolute necessities of life. Roosevelt dropped a line he and Sam Rosenman had crafted: "I see one third of a nation ill-housed, ill-clad, ill-nourished." This gave him, he said, a mandate to establish a "new order of things." The abiding downturn demanded a new government of unprecedented boldness. And next came a phrase that took even many of Roosevelt's allies aback: "We are beginning to wipe out the line that divides the practical from the ideal; and in so doing we are fashioning an instrument of unimagined power for the establishment of a morally better world."

Having learned the importance of the interest group in his first term, Roosevelt was now announcing that he would use the second term to make a perpetual interest group of that one-third of voters. But what stuck out was the phrase "unimagined power." The country itself remembered and knew, Roosevelt said, from the experience of wars, when it was time to move "beyond individual and local solutions." This was one of those times.

The old intellectuals had not done so well toward the end of the first term; many were already out of government. Tugwell was going. Lilienthal was slowly gaining ground on Arthur Morgan. Yet suddenly, in this address, FDR was talking just as expansively as he had when he inspired the old pilgrims at the outset. "Have we reached the goal of our vision of the fourth day of March 1933?" he asked. "Have we found our happy valley?" The difference was that this time the utopia would be more Frankfurter's than Tugwell's—one arrived at through crafted legal moves rather than bold programs, that emphasized the business of getting a law through over economics. This was the era of democracy; the era of the republic was passing. "In fact," Roosevelt said, "in these last four years, we have made the exercise of all power more democratic; for we have begun to bring private autocratic powers into their proper subordination to the public's government."

And, even in the rain, the president also made it clear that he was planning woe or worse for anyone who did not sanction the utter primacy of that new relationship. For, as Roosevelt put it, "evil things formerly accepted will not be so easily condoned."

The president also spoke of groups whose cooperation would be mandatory if his vision was to succeed. The first was the Court itself. Roosevelt carefully read a line about how men and women in the American republic would "insist that every agency of popular government"—the court included, the suggestion was—"use effective instruments to carry out their will." The second group whose support he demanded was wealthy taxpayers, who must be encouraged or coerced into going along with funding the new enterprise. This was what Roosevelt was getting at when he said that the "test of our progress is progress, not whether we add abundance to those who have much; it is whether we provide enough for those who have little." The last group whom Roosevelt called to task were employers. The sentence about "evil things formerly accepted" suggested that Roosevelt had yet more plans for utilities, even beyond the holding companies.

The vigor of the speech was big news. Frankfurter would write to Roosevelt that the "kids of 2036" would still be "reading and reciting your Second Inaugural." But the justices made a show of calm departing, just as they had with their arrival. Van Devanter could be seen taking his wife to lunch in the new Supreme Court Building's cafeteria. The court was in recess until February 1.

Frankfurter wanted the president to hold back; the aging of the Court meant that matters would go Roosevelt's way in any case. Frankfurter also had his mind on something else; the right outcome in a case might divert the president. Back in 1923, it had been Justice Sutherland who rejected Frankfurter's defense of the *Adkins* minimum wage, as well as his argument that Congress should force low wages up because they caused social problems. A year earlier, the high court had rejected his New York State minimum wage law in *Tipaldo.* Now, however, it had already agreed to hear a similar case, *West Coast Hotel Company v. Parrish,* a minimum-wage case brought by a woman in Washington State. Felix thought the progressive legislation might win this time.

But Roosevelt was not concentrating on individual cases. He was still thinking of action—action, even, for the fun of it. That spring Congress would pass the Fair Labor Standards Act, "putting a floor," as he would put it, under wages. And in a letter he wrote to Frankfurter on January 15, he let on he planned something else. "Very confidentially, I may give you an awful shock in about two weeks. Even if you do not agree, suspend final judgment and I will tell you the story."

Joe Robinson, the Senate majority leader, would help the president. He had already proven a valuable ally, and Roosevelt was talking about naming him to the Supreme Court.

The Four Horsemen were restless, but not ready to give up. Around this time, John Knox, the clerk, wrote in his diary of McReynolds: "The justice has been tipped off to something, but I don't know yet what it is. He is either fearing inflation or being forced to resign. He has had me go through his records back to 1903, he has been calling up his stock brokers, etc. A millionaire from Wall Street came down to advise him to ship part of his money to Canada and England." Later in the winter and in early spring McReynolds would cut off contact with many friends and acquaintances.

On January 26, Roosevelt spoke—not of courts but rather to announce he would give up on his plan to explore power pools between the government and the private sector. The statesmanlike concession that Willkie had made with such fanfare before the election had in the end served only to give Willkie's competitor, the TVA, extra time. With the old agreement between TVA and Commonwealth and Southern no longer in force, Commonwealth and Southern could no longer buy $800,000 worth of power from TVA; Willkie announced his company would need $10 million to construct steam plants to get its energy. In the same days Roosevelt's ally in Tennessee, Senator Kenneth McKellar, introduced legislation that would stop lower federal courts from blocking New Deal measures until the Supreme Court had ruled on a given measure's constitutionality. McKellar's specific intent was to halt the legal assault on the TVA. Senator Norris complained that "the power companies traveled from one court to another, I presume, to find a friendly court, and when they found one, they stayed there."

Roosevelt's confidence seemed only to grow. Again weather played a role. Some noted that, in that same January, an unusual thing came to pass at the White House: the lawn grew. So much that it had to be mowed—a first for January, as far as the records showed. The cause was the warmth from storms many hundreds of miles to the west—and now, as January moved to February, flooding. The Ohio-Mississippi flood, *Time* magazine noted, could only lend more credibility to Roosevelt's TVA and its water management projects. Ten years after Hoover's action, Roosevelt seemed the permanent flood rescuer, "the Great White Father," as *Time* wrote.

In February, the TVA offered Willkie a one-month extension on its contract to sell electricity to Alabama Power. But the contract also bound Willkie to negotiate with the TVA on very narrow terms, and Willkie could not see an advantage. Willkie fought back, saying that he would be more willing to stay at the negotiating table if Roosevelt would return to power pool discussions.

Roosevelt hosted the justices for a dinner at the White House. All came except Justice Brandeis, who made it a rule not to go out evenings, and Justice Stone, who had just returned after a convalescent voyage. Others in attendance were Donald Richberg—the man who had lost the *Schechter* case and, as it happened, penned a draft of the stinging inaugural address—and Sam Rosenman, who had also worked hard on the same speech. The president seemed to be enjoying his needling.

Three days after the dinner, Roosevelt finally acted. He announced that he would skip state ratification and simply send over to Congress legislation that would increase the number of justices from nine to a figure that could range as far as fifteen. For each justice who stayed past the age of seventy, a new one could be added. The concept was indeed similar to what McReynolds had suggested so many years ago, albeit for a lower court. The pretext for this action was the argument that the justices were too overloaded with cases. The *New York Times* reported that a congressman named Maury Maverick—the name suited the temperament—"ripped the mimeographed draft of the president's bill from the back of the message, pasted it on a house bill form, and threw it into the hopper."

The president's action was so direct that people used the same phrase they had used to describe Roosevelt's monetary forays in 1933: "a bombshell" had hit Congress, Turner Catledge of the *New York Times* reported.

Many were convinced that Roosevelt had indeed finally overstepped. His enemies now jumped to take advantage of the error. Senators William Borah and Charles McNary, prominent in the Republican Party, ominously signaled they were giving the proposal serious review. Hoover now was "eager to jump into the fray," Senator Vandenberg wrote in his diary. A number of Democrats spoke up as well. Senator William H. King of Utah summarized the skepticism when he said, "There is no necessity for it." Roosevelt's friends were chagrined, or worse. Marion Frankfurter, who had attended the press conference where Roosevelt made his original "horse and buggy" outburst, wrote to her husband: "I hate the whole bill so thoroughly, think it so cheap and dishonest, and I can't bear to have you accused of being in any way responsible for it."

The public reaction was strong. Those who had known Roosevelt from his days in the East Coast establishment felt especially betrayed, either by the Court action or the imperious tone the president was taking.

Around this time an old neighbor from the Hudson River and fellow yachtsman, Howland Spencer, was growing impatient with Roosevelt—he could not understand how Roosevelt would betray his social class in this way. His anger was exacerbated by the way Roosevelt called his Hyde Park home Krum Elbow. That name—"crooked elbow" in old Dutch—had historically been reserved for the west bank of the Hudson, where Spencer had his own grand estate. Yet when he met with Roosevelt, Roosevelt wouldn't listen to anything, Spencer later told reporters.

Walter Lippmann, one of the country's more esteemed opinion makers, said in his column that "Mr. Roosevelt's quarrel with the Supreme Court has no real relations with his power to avert another crash." Some old-fashioned liberals complained that Roosevelt's action was the height of illiberalism. Thousands of letters and telegrams arrived at the Court, nearly all opposing the legislation, author Marian McKenna reports. "Do not desert us. Please hold the Court—hold the fort. Even if you are old and tired, you can't quit now for three years."

Some of those around Roosevelt were surprised by the vehemence of the reaction. Harold Ickes wrote in his diary of the liberal critics, "I often think that the definition of a liberal is a man who wants what is unattainable or who wants to reach his objective by methods that are so impracticable as to be self-defeating. So many liberals want merely to be in opposition. They do not want to advance from objective to objective." That impractical liberalism Ickes had no time for.

Roosevelt for his part was also proceeding boldly. At a victory dinner for the Democratic Party—also in early March, the former season of presidential inaugurations—Roosevelt made the point that his court change was directly related to his plans for the TVA. "I defy anyone to read the opinions in the TVA case, the Duke Power case and the AAA case and tell us exactly what we can do as a National Government in this session of the Congress to control flood and drought and generate cheap power with any reasonable certainty that what we do will not be nullified as unconstitutional."

That month Hallie Flanagan of the Federal Theater Project lent Roosevelt credence when she trained her spotlight on Willkie. Her device was a new kind of theater: the "Living Newspaper"—a sort of dramatized documentary, in this instance about the power industry. In thirty-three scenes, *Power,* the play, portrayed the story of electricity as the story of exploitation. In some scenes citizens protested high utilities bills—a stretch, since utility prices were coming down at the time of the production. An Insull character was featured, too, bilking a consumer.

Some noted that Flanagan was herself overstepping, at least when it came to Willkie. The Willkie character in *Power* was an old man who doddered about in a white wig—in other words, something closer to the old caricature of Insull. This slip-up, *Time* noted, revealed that the producers could scarcely have been acquainted with the actual Willkie. Still, the play, a series of pageants reminiscent of Bertolt Brecht, was impressive and popular; sixty thousand people bought tickets in New York even before it opened. Afterward, Harry Hopkins, WPA director, went backstage to praise the play, saying, "People will say it's propaganda. Well, I say, what of it? It's propaganda to educate the consumer who is paying for power." The same issue of *Time* reported the forward march of Roosevelt on his power project: the president had recently let Congress know that his western equivalent to the TVA, the Bonneville Dam, would be ready for operation by the end of 1937. The *Oregonian*, the newspaper in the state where Bonneville was going up, praised *Power* effusively: "Even the private ownership boys and girls will have to admit that many flashes of genius are evident in the mechanics of '*Power*.'" The suggestion was clear: the Roosevelt utilities model was on its way to becoming national.

On March 9, Roosevelt delivered the first Fireside Chat of his new administration. The authors this time were Rosenman, Cohen, Richberg, and Tommy Corcoran. The president targeted the recalcitrant Horsemen again. The American form of government, the president said, was a "three-horse team, provided by the Constitution so that their field might be plowed." Now one horse—the Court—was not going along. The president himself was just another of the horses. It was the American people who were in the driver's seat. They could, and should, act to bring the horse of the Court in line. Then there would be some sort of national clarity, and all the other courts—the recalcitrant federal courts—would mount fewer obstacles to the New Deal. The country needed protection because "the dangers of 1929 are again becoming possible," the president said. He had mentioned the horses before, and was not going to abandon this line of argument. The only way it would be able to legislate protection from "those dark days" was to restructure. No matter what Lippmann argued.

The Court, Roosevelt said, was not ready to handle "our modern economic conditions." The four justices who had opposed his gold clause action could have "thrown all the affairs of this great nation back into hopeless chaos." They were reading into the Constitution "their own economic predilections." His plans were for the younger generation. In the chat, the president used the word "modern" a full five times, and the adjective "young," or "younger," four. The nation must "save the Constitution from the Court and the Court from itself." The Court needed "new blood."

Finally, Roosevelt pointed out that the number of justices on the court had been changed before in American history, several times. Additionally, the White House had found a fact that would be useful in its judicial battle: that old recommendation from McReynolds.

One of the sources for the chat was a piece by Stuart Chase on the necessity for stronger government. Chase sent the piece to the *New York Times,* also getting a copy to the president. The president shared Chase's work with Frankfurter: "Looking over the frothing rhetoric," Chase had written, "to the real land and the real people, you find that: Six million farmers were left in a legal vacuum. . . . Fifteen million industrial and clerical workers, more or less, were stripped of wage and hour protection." Chase concluded; "If we really cared about America, I think we should act."

Before Congress decided whether to act, the Old Men did. Privately, they made their disapproval known. "You can rest assured," Harlan Stone wrote, "that those who assert that age has affected the work of the Supreme Court, or that it does not do its work with the highest degree of efficiency of any Court in the world, cannot get to first base." The recent proposals were "about the limit."

Hughes now led an overall publicity drive showing the justices to advantage; the *New York Times* carried a picture of Brandeis with his wife, old and dignified, and a photo of Van Devanter, jaunty amid the brush on vacation. On March 22, Justice Hughes fired back at Roosevelt in a letter to the Senate Judiciary Committee. The Supreme Court was not behind or old; it was "fully abreast of its work." And: "The present number of justices is thought to be large enough so far as the prompt, adequate, and efficient work of the Court is concerned."

Hughes had outlawyered Roosevelt, as Ickes would write in his diary, going after the weakest argument in Roosevelt's case, that the justices were not competent, and documenting the evidence against that argument. Frankfurter was irritated, drafting a letter to Brandeis, "As for the chief—I have long written him down as a Jesuit—I deplored his letter and certainly its form." For the moment, the justices, and the opponents of Roosevelt too, seemed to have gained the advantage. Robert Jackson wrote that the letter "pretty much turned the tide" against FDR's plan.

The justices were not the only ones to stand strong. Mellon was feeling increasingly untouchable, as if behind pillars, already in another world. There were pillars, quite literally—sixty-two Ionic columns—at the new home of the Mellon Institute, which he would dedicate on May 6. Presiding that day, Mellon would recall his goals of the 1910s and the 1920s: the institute, and his French language study. Now he would remark, drolly, that while the mellon French "still is what it was, originally," the institute had progressed nicely. *Time* would note that the institute would continue to function as it had before, by hiring out scientists to do work for industrial companies. The magazine would also point out that the institute's "industrial fellowship" system was not only good for the economy, but also occasionally created a permanent job: "If a Mellon 'research' ends profitably, the worker is apt to get a good job with the manufacturer who paid the bills. If the worker is also clever he can get the University of Pittsburgh to award him a doctorate on the strength of the research he performed at the Mellon Institute to earn his living."

At the same time, however, there was also a shift in the air—a sense that battle was pointless, since Roosevelt might prevail, one way or another. Court watchers could see that the justices were also changing in ways that would make their opinions more acceptable to Roosevelt.

On March 25, Sutherland marked his seventy-fifth birthday; the newspapers were making note of all the justices' birthdays. The *New York Times* noted pointedly in the first paragraph of its birthday story that the justice planned to spend his day working. (Six justices, one such story reported, were over seventy, eligible to retire. Four, another noted, had already passed the three-quarter-century mark. Brandeis, past eighty, would turn eighty-one in November.) It may have been a bitter anniversary for Sutherland, for he already knew that shortly his own brethren would turn on him.

On March 29, Robert Jackson, Stanley Reed, and James W. Morris, all from the Justice Department, filed into the new Court building to hear what they knew would be a momentous case. Ten months before, the Court had struck down New York State's minimum wage law in *Tipaldo*. In the autumn, it had refused to rehear the case, but it had agreed to hear another, brought by Elsie Parrish, a hotel worker demanding back pay under Washington State's minimum wage law.

Now just what Felix Frankfurter and Drew Pearson had hoped for happened. Justice Roberts, at sixty-one the youngest, made a switch, joining four others in upholding Washington's minimum wage law. Where a New York laundress had had no rights, a Wenatchee, Washington, chambermaid now had them. The justices had been criticized with imposing their economic philosophy, but new opinion, when it came, was also an economic interpretation of the law. Oliver Wendell Holmes and Frankfurter were finally vindicated. As Hughes wrote in his opinion, "the economic conditions which have supervened" were important, and it was now "imperative that in deciding the present case the subject should receive fresh consideration."

The slap came with the nature of the opinion and the specific rejection of Sutherland's old *Adkins* position. "The Constitution does not speak of freedom of contract. It speaks of liberty," read the majority opinion. "Our conclusion," the Court said, "is that *Adkins v. Children's Hospital,* supra, should be, and it is, overruled." Sutherland, Van Devanter, Butler, and McReynolds dissented. But now the Four Horsemen were clearly alone on their charge, a minority. Sutherland offered a dignified defense. He would not want to impugn his colleagues' good faith, but, he said, "the meaning of the Constitution does not change with the ebb and flow of economic events." If Roosevelt wanted a change, he must lead an amendment to the Constitution.

Also that month, Raymond Moley testified against the president's legislation. Moley said that he disliked "the dead hand of the past" which was represented by the majority of this court. Still, there was arrogance in the concept that young or new was always better. "Our New Deal will be an Old Deal sometimes." And he opposed what was now commonly called the court-packing plan. Moley was after all a man of the law, a criminologist. For years now he had been arguing that Roosevelt's reforms were reforms of the rule of law, not arbitrary changes. Since 1935, Moley said, Roosevelt had been acting arbitrarily far too often. The court-packing legislation was the last straw. Washington had to find a new course: "Let us make democracy work by working through the instruments of democracy."

Yet the news of Wenatchee was good for Roosevelt, and everyone knew it. The Justice Department lawyers in the new courtroom were thrilled. Frankfurter's feelings were more complicated—and related to the fact he was hoping for his own seat on the Court at the time. In a short period he himself might have to demonstrate his own judicial independence. Instead of praising the justices for coming around to the view he believed more accurate, he criticized them for having changed out of political reasons. The Washington case, he wrote, "made me feel as though something very dear had died—my faith that the Court's processes had integrity." Roberts himself would later vigorously deny, in a memorandum to Frankfurter, that the change was political: "no action taken by the President in the interim had any causal relation to my action in the *Parrish* case." The vote in the Parrish case had been taken in December 1936, before both inauguration and the president's announcement of his plan.

April at first took everyone's mind off the Court—there was so much going on elsewhere. In Harlan County, Kentucky, the old question, "Which side are you on?" still had meaning: the drive to unionize coal miners was accelerating. On April 6, the White House hosted the Pine Mountain Settlement School of that county to pre-sent musical entertainment to assorted congressmen, senators, and other guests at the White House. That was what frustrated FDR's opponents—the way he pursued political goals through culture.

On April 7, another crowd gathered to cheer Lewis, this time at Michigan's state fairgrounds; he had reached a settlement with Chrysler. Ford now was the only remaining of the Big Three not to be unionized. William Green, the old head of the AFL, was locked in a struggle with Lewis, over both Lewis's tactics at organizing and his sit-down strikes against companies. Leading some of the more militant projects was John Brophy. It was clearer than ever that Lewis and his team were the face of new unionism, and Green the old.

The next week, the Court brought back attention to itself with its own union news: it would uphold the law that provided the legal framework for all these events, the National Labor Relations Act (NLFA). Looking at four cases, it upheld them by the same 5–4 margin; on a fifth, the group was unanimous. The finding also meant that Washington now had the authority to regulate manufacturing. And the Wagner Act really did give the unions the right to fight with the companies. That September the United Auto Workers membership would reach 375,000, more than ten times the 30,000 of September 1936. Lewis, again, rejoiced. Within a year he would have organized nine in ten workers in the coal industry. Florence Reece's picture of towns divided into union people and company thugs had been sanctioned by the highest court in the land.

Ogden Mills, the treasury secretary who had succeeded Mellon, wanted to point out the consequences of unions' higher wages for the economy. He did not agree with the advocates of spending—now coming to be called Keynesians—that consumer spending was always better than investing by the producer. After all, as Ford Motor Co. itself pointed out, if Henry Ford had brought his early employees together every week and shared out the profits, there would have been nothing left to spend on investment. Ford, the company, would not have grown. Ford executives were correct when they said that "the little shop would have stayed little."

Also that month, two other justices celebrated birthdays: Justice Van Devanter turned seventy-eight on April 17, and Hughes celebrated his seventy-fifth birthday. Van Devanter lived at 2101 Connecticut Avenue, in the same building as William Borah, one of the senators who opposed the Roosevelt plan. Borah urged Van Devanter to resign, hoping that this would make Roosevelt's packing plan look more unnecessary.

Now too Jackson had another go at Mellon. Having failed in its tax suit, the administration was still set on "getting" the industrialist. Jackson, who had been promoted to assistant attorney general in the Justice Department, led a suit against the Aluminum Company of America, the largest suit of its kind since 1911. Mellon, his son Paul, and other relatives—Sarah Mellon Scaife, Richard K. Mellon—were defendants. Jackson sought the dissolution of the company. Jackson would prosecute the case, but it was filed by another lawyer from Justice: Walter Lyman Rice, the same lawyer who had led the case against the Schechters. Rice had gone after the smallest possible foe; now he was trailing the largest.

Mellon was a minority shareholder. But Jackson and Rice would show reporters a chart dramatizing Mellon's influence at Aluminum Company of America—it looked stupendous. The attorneys argued that such a large company could not be good. Mellon's lawyers countered forcefully, noting that the chief symptom of a damaging monopoly was high prices for the product. But the price of aluminum had come down nearly continuously since the company's founding half a century ago.

Tax prosecutions were also moving ahead. Though there was still a sense of vengeance, there was also still the practical need for the revenue. Mellon's theory that higher rates sometimes narrowed revenue streams—that you could not charge over "what the traffic will bear"—was being borne out. The president told Congress that though Washington had raised its tax rates, the Treasury was still short $600 million. Roosevelt blamed not the arrangement but the wealthy themselves. Roosevelt, Morgenthau would tell Treasury officials, "wants to say flatly that our estimates and our methods of estimating are correct, but the citizens—that's the word he used—found a trick way of finding loopholes." Roosevelt insisted that these "loopholes be closed and that they be retroactive." If revenues were wanting, Roosevelt didn't mind investigating, prosecuting, or legislating his way to them. The concept of deficit budgets still sat easily with neither Morgenthau nor Roosevelt.

Panicked for cash, Morgenthau now had his Treasury set about trying to create dozens of Mellons. Roswell Magill of the department audited individual returns in New York and found, according to Morgenthau's diary, that citizens were using old tax breaks—legally, mostly. But Roosevelt was now set on erasing the old distinction between evasion and avoidance that the Treasury had danced around so long. Roosevelt also set out to prove that the intention of taxpayers who failed to complete complex returns correctly was malign: where there was ambiguity, taxpayers ought to be presumed guilty.

This was especially disingenuous of the president, for Roosevelt himself would submit an ambiguous tax return for the year 1937. His income from the presidency that year would be $75,000, about Willkie's level. But there were other issues that complicated the return. As he would write to Commissioner Guy Helvering, "I am wholly unable to figure out the amount of the tax for the following reason . . ." His own tax problem—one involving the timing of tax obligations—was something only experts could solve. "As this is a problem of higher mathematics," wrote FDR, "may I ask that the Bureau let me know the balance due? The payment of $15,000 doubtless represents a good deal more than half of what the eventual tax will be."

Morgenthau prepared memoranda with the lists, and Roosevelt was eager when they met on May 17: "Henry, it has come time to attack, and you have got more material than anybody in Washington to attack." Roosevelt was worried about the conservative

Democrats, and he needed Morgenthau: "Now it's up to you to fight." Morgenthau was excited to laughter at Roosevelt's energy.

"Why are you laughing?" the president asked. "Because you are such a wonderful showman," Morgenthau replied. "I don't know what's going to happen. I can't guess what I have got that is so useful to you." The men pulled together a list of businesses and individuals who had, as Roosevelt reportedly put it, "found means of avoiding their taxes both at home and abroad." Herman Oliphant of the department had written up a tax evader—it was apparently clear in this case—who had failed to declare both dividends and profits, as well as bribed relations. Roosevelt was irate, pounding the desk. "Why don't you call him a son of a bitch?" the president asked. He wanted the man behind bars, Morgenthau would later report in his diary.

"I want to name names," Roosevelt told Morgenthau. While he and Morgenthau decided for the moment to stop short of publicizing the list, Roosevelt read much of it to people around Washington, thereby ensuring news leaks. One influence on the president was Felix Frankfurter, who, along with his protégés, usually cited an old Oliver Wendell Holmes quote to justify tax increases: "I like taxes. They are the price I pay for civilization."

At 9:45 on May 19, while Roosevelt was still in bed, a messenger brought a letter to the White House: Willis Van Devanter was resigning. Taking advantage of the new law that allowed justices to retire on full salary, he would be leaving June 2, at the end of the term. As historian Marian McKenna reports, Van Devanter tried to calm them and defended his move, telling them of his neighbor on Connecticut Avenue that "Borah favors it."

Perhaps the news emboldened Roosevelt on the tax front. After an initial hesitation, he decided in the end that there was no substitute for giving out names. On June 24, 1937, tax commissioner Guy Helvering named sixty-seven "large wealthy tax payers, who by taking assets out of their personal boxes and transferring them to incorporated pocketbooks have avoided paying their full share of taxes." Thomas Lamont, John Raskob, Pierre Du Pont, William Randolph Hearst—and Andrew Mellon, again—were all listed. The tax decisions of these men were, the Treasury acknowledged, "perfectly legal," but still not conscientious. The publicity was followed by a law passed unanimously by both houses of Congress, limiting or eradicating tax-favored mechanisms, from breaks for trusts to breaks for personal holding companies and country estates.

Many observers were upset anew at this effort at shaming the men. The idea that one could ignore the law seemed grossly unfair. What Roosevelt and Morgenthau—two wealthy men themselves—were doing was worse than moving the goalposts; they were formally moving the goalposts in a game already played. One of those to register his shock was J. P. Morgan, who was returning home from a trip to Britain on the *Queen Mary.* While he was still on board, reporters asked Morgan what he thought of Roosevelt's tax plans. Apparently caught off guard, and perhaps not attuned to the mood of the second term, Morgan told reporters that there was nothing wrong with avoiding taxes. "Legal Tax Dodging Upheld by Morgan," blared a headline, "Would Mend 'Stupid' Law." Stepping onto dry land, Morgan had second thoughts. He issued a defensive statement: "My interview on shipboard with newspapermen last Monday took place before I had seen President Roosevelt's message. . . . What I feel strongly is that, when a taxpayer has complied with all the terms of the law he should not be held up to obloquy for not having paid more than he owed."

Far more confident in his reaction was Alexander Forbes, the president's cousin and a professor at Harvard Medical School. Forbes wrote to the *Boston Herald* that the "true patriot" would "claim every exemption the law allows and he may have more to spend on enduring contributions to the betterment of mankind." This repetition of Mellon's message, coming from a relative to boot, stung Roosevelt: "I do not hesitate to brand you one of the worst anarchists in the U.S.," he wrote to Forbes. "And, incidentally, I use 'anarchist' in the pure Greek sense" (from the root, which means "without government"). When it came to Morgan, Roosevelt was also scornful, asking mockingly of Charles Burlingham of the New York Bar what he made of "Morgan's exposition of Christianity when he landed the other day . . . ask yourself what Christ would say about the American Bench and Bar were he to return today."

The tax battle was not the only one going on in June. The unions were making advances—even in Willkie's world. Commonwealth and Southern announced that Willkie and Lewis's CIO had agreed to a raise for all operating employees at Consumers Power Company, a Commonwealth and Southern subsidiary. Consumers employed 6,600 workers, so it was an important challenge for Commonwealth and Southern. But one it could afford: operative revenues were up.

But two things were becoming clear now. The first was that Roosevelt was having trouble in his court-packing fight. The legislation, he was finding, simply couldn't be sold to the country, no matter what Stuart Chase wrote. And it was not popular in the Senate, either. Instead of welcoming the bill, the Senate Judiciary Committee rejected it. This despite Robinson's Herculean effort, and even though seven of the ten members of the committee, a firm majority, were Democrats. The language of their rejection was unforgettable: the senators said that Roosevelt's act "should be so emphatically rejected that its parallel will never again be presented to the free representatives of the free people of America."

Roosevelt rallied, hoping to jolly Congress along by inviting lawmakers to a weekend on Jefferson Island. The lawmakers enjoyed themselves, and some of the rage diminished. There was more debate, as the summer warmed. But the president's legislative leader in the fight, Senator Joe Robinson, succumbed one night to a heart attack, and soon it was clear that Roosevelt had lost this fight. "The Court Bill was dead, dead as a salt mackerel shining beneath the pale moonlight. As dead as the ashes of Moses, the world's first law giver," commented the *Jackson Daily News* of Mississippi. Hiram Johnson of California put it even more clearly, crying out in the Senate chamber: "Glory be to God."

The second reality was that while Roosevelt was losing his court-packing battle, he was still winning his war against the Court. The recovery seemed to have slowed, but there was the record of the achievement of the spring, when the market moved close to 200. Cordell Hull, secretary of state, had persisted, and his trade agreements lifted burdens from many economies. There had been economic gain from that too, even if Hull didn't have the Nobel. That year the *Economist* magazine would write, "In this tariff-ridden world the sight of any nation deliberately seeking to lower its tariffs is both rare and refreshing. . . . It is fully possible, for example, that Great Britain has already gained more from the concessions given by the United States in her treaties with other countries than could be obtained in a direct Anglo-American treaty." The phrases amounted, almost, to an offer of forgiveness for the administration's behavior at the 1933 London conference.

Frankfurter had been right. Time was doing for Roosevelt what he could not convince Congress to do. With Van Devanter going, Roosevelt could name his first man to the court—Hugo Lafayette Black of Alabama, chosen later that summer. Very shortly it emerged that Black had belonged to the Ku Klux Klan. What, Roosevelt's friends at Howard asked themselves, could the president mean by such an appointment? The Socialist Norman Thomas and the National Association for the Advancement of Colored People expressed concern. But Hugo Black got through. And now it was clear that within several years Roosevelt would be able to shift others, perhaps Stanley Reed and Robert Jackson, or Frankfurter, over to the Court as well. Roosevelt had assailed the justices for their tenure in office, their longevity. But he was prevailing over them because of his own longevity in office.

When observers thought about it, they realized that the outcome of Roosevelt's advance generally had been clear as far back as June. Then Roosevelt had finally acted on his old threat and put forward legislation to replicate the TVA. Senator Norris had introduced "the seven TVAs bill," as it was shortly known. From the Atlantic to the West Coast, there would be seven little TVAs, including a Southwestern Authority, to cover the Colorado River, Hoover's old territory. A nationwide chain of TVAs would be hard to undo once built. Willkie noted that the project would double the national debt, his biographer Joseph Barnes reports. Besides, steam power was more economical anyhow.

There had been another June signal of FDR's inexorability when 9,200 doctors attended an American Medical Association meeting in Atlantic City. The mood was a happy one: the physicians were looking at a new drug that seemed to conquer bacterial infections. It was called sulfanilamide. An expert from Pittsburgh, Ralph Robertson Mellon—no kin to Andrew—reported that this drug apparently could cure "certain types of pneumonia, typhoid, brain abscesses, scarlet fever and meningitis." Other doctors reported similar wonders.

But the doctors were distracted even from the miraculous sulfanilamide when a representative of the New York State Medical Society, Joseph Kopetzky, spoke. Kopetzky, an ear doctor, suggested a plan that would alter their very independence as professionals: the nationalization of health care. Under the plan, as reported by *Time*, "Every one of the 150,000 U.S. doctors must become an officer in the Federal Public Health Service." The federal government would pay for whatever service citizens could not pay. There likely would be a new secretary of social welfare. The doctors debated long into the night as to how the American Medical Association might reply. They paid such attention because they heard that Roosevelt had already seen Kopetzky. The doctors might not know everything, but by now they understood that once Roosevelt made a project his, he would not give up—unless someone stopped him.

12

The Man in the Brooks Brothers Shirt

January 1937
Unemployment (January): 15.1 percent
Dow Jones Industrial Average: 179

One morning in January 1937, about two weeks before Roosevelt was inaugurated, several reporters in New York showed up at an unusually large new office building overlooking New York's East River. They had come downtown to meet a new figure: Rex Tugwell, Wall Street man.

Tugwell had resigned his post as undersecretary of agriculture after the election, and this time Roosevelt had not blocked him. Tugwell guessed that the cost of having him around had simply become too high for the president. The greenbelts and projects like Casa

Grande were impossibly controversial. "Franklin did not doubt my loyalty any more than he ever had, I am sure, but he had been half persuaded—and Eleanor even more than he—that I had totalitarian leanings," Tugwell would later write.

Tugwell's Resettlement Administration, which had stood alone, was now slated to become a suboffice of the Agriculture Department. Living with that fact alone would have been hard to tolerate. Tugwell also had personal matters to deal with. He was thinking of leaving his wife and marrying his assistant from the RA, Grace Falke. He told himself that New York would be a better place to sort things out than Washington. He had an apartment at 460 Riverside Drive. It was time to go.

Still, the overnight transformation from New Deal bureaucrat to New York executive at 120 Wall was bold even for Tugwell. After all, the projects he had started were still functioning, and his staffers were still dressed in the khakis of fieldwork. In May Dorothea Lange would travel to the area around Casa Grande and photograph the long-limbed, destitute children of migrants—evidence of the need for resettlement projects like his model farm. Yet here was Tugwell, dressed in blue serge, standing before reporters at a nineteenth-floor office. The reporters learned that Tugwell would be a vice president at American Molasses Company, a sugar concern. The company belonged to New Deal friends: the Taussig family. Charlie Taussig was an old New Dealer, and Adolf Berle sat on the American Molasses board. The visitors saw that Tugwell had twenty roses on his new desk, a gift from the board of American Molasses.

Tugwell reminded the reporters that this was not his first experience in the private sector. He had worked for his father's canning business, the same one that he had fined as an agriculture department official three years earlier. He clearly feared he might be caught flat-footed on the facts about the company—"But don't ask me about cotton. I'm a molasses man," he had joked earlier with reporters. One reporter now asked him what he thought about getting a Social Security number. After all, the Social Security program payroll taxes were beginning and the numbers were a novelty for the country. Here Tugwell did blunder: "I'm out of that class," he replied, confused. Taussig corrected the slip—Tugwell would be a salaried employee and get a number. He would get a number and would pay into the new program like all the rest.

Taussig also explained Tugwell's move to the press: "In these changing times every business needs the service of men trained in economics, with a broad objective social viewpoint." The *Times* reporter gave the impression that Tugwell didn't mind leaving government; after all, it reported him as saying, "many of the things he had planned are being carried out." The headline the next day read, "Tugwell Bit Hazy about His New Job."

Tugwell was not the only New York intellectual feeling "hazy" and reexamining his old convictions. The news from abroad dominated the headlines: Stalin was executing one old hero of the Soviet Union after another in secret or semisecret proceedings. The whole idea of being on the Left was changing. Legitimately frightened by Hitler, some found themselves moving close to Soviet Russia despite themselves. A new divide was emerging among the intellectuals.

Some still supported Stalin. In March 1937 Corliss Lamont, the son of Thomas W., wrote a plea for liberal unity stating that "we believe that the Soviet Union needs the support of liberals at this moment, when the forces of fascism, led by Hitler, threaten to engulf Europe." Paul Douglas's ex-wife, Dorothy Douglas, signed it, as did the writer Lillian Hellman; Robert Lynd, the author of *Middletown;* and Louis Fischer, who had been in Moscow during the 1927 visit.

But others were anxious. Stuart Chase still wrote admiringly about the Soviet experiment from time to time. But he was shifting his attention to a new topic:

words and their meanings. "We have circled all around 'capital,' and 'capitalism,' " he wrote, "but made little progress in defining them." That June, Harvey Chase, his father, would write to Roosevelt about Stuart's forthcoming book, which would be titled *The Tyranny of Words.* Chase Sr. briefed the president: "No longer a socialist, communist, or collectivist, he has become a semanticist."

And some of the old Left were simply appalled. Suzanne La Follette, a member of the clan of liberal reformers, now pointed to Russia as the very opposite of her definition of liberalism. She penned an angry public letter to the *Nation,* which had not taken a clear stand against the trials: "I shall not be surprised if within ten years the *Nation*'s left-handed endorsement of Stalin's liquidations of the October Revolution is something that its editors would prefer to forget." Paul Douglas was likewise shocked. The next year Douglas would happen to be reading an item in the *New York Times* about a Trotskyite leader whom the Russian secret police had executed, and recognized the name with a start: Betty Glan. It was the Russian woman who had come up to him on his 1927 tour. Murdering one's corevolutionists seemed the very opposite of the liberalism the American Left saw as part of the spirit of revolution. What was the point of revolution, anywhere, if it led to this? Even if the New Deal had been proceeding perfectly, the Soviet Union would have caused them to question their old precepts.

And the New Deal was not proceeding perfectly. The national economy might have moved forward, but it still was not back to 1929 levels. "Everything was not happy in New York," either Tugwell would write of the city in that period. The bookish types did not care that it was New York (the Yankees) versus New York (the Giants) in the World Series. Now—perhaps because it was not a campaign year—the intellectuals in the city found themselves talking about the New Deal with the business community, whose doubts were stronger and older. In Washington, Roosevelt might seem invincible and 1937 might seem the year for permanent revolution. But that was Washington. In New York, for the intellectuals at least, 1937 would be a year of self-doubt.

The doubt began with a personal shock—the reminder that not everyone approved of the way the intellectuals and New Dealers had executed their ideas. Columbia had granted Tugwell several leaves to serve in Washington, but now the university was telling him it could not welcome him back. His former dean at the Wharton School, who now led the Economics Department at Columbia, gave him the news. "I meekly sent in my resignation and Columbia's hands were washed of me permanently," he would write later. This was "perhaps the hardest to bear."

Roosevelt had so often advised him not to mind the bad press, but now Tugwell could see that that press would have a permanent cost. As Harcourt Brace, a publisher of a textbook he had written, would report that year, sales of the book had gone down when a school superintendent in Gary, Indiana, William Wirt, attacked Tugwell as a leftist. These were consequences he now had to confront. He was after all an academic—a man who liked to experiment, to speak his mind. Uncertain, Tugwell boarded the *Scanpenn* in late January with Charlie Taussig. The ocean liner headed for the Caribbean. But the feeling could not be the same as it had been traveling the warm waters with Roosevelt on his *Potomac*.

Within a month, Tugwell had cause to reconnect with his old chief. While in Barbados, Tugwell learned that his twenty-year-old daughter, Tanis, had fallen ill with double pneumonia and entered Presbyterian Medical Center. It was hard to secure a flight north. Turning to the old familiar hand for help, Tugwell cabled Roosevelt. And the hand was there: Roosevelt asked Juan Trippe of Pan Am to hold a plane at Trinidad for Tugwell. Even this was not enough, for only a fishing smack was available to take Tugwell to Trinidad. In the end Pan Am helped out with a special flight from Barbados to Trinidad as well. Two Costa Ricans were bumped, but Tugwell made it to Tanis's bedside. The daughter recovered. It is hard not to read a bit of wistfulness in the act of the ex-undersecretary turning to his ex-boss. Tugwell's relationship with Roosevelt was still in the news, but often as a form of ridicule: in March the papers carried some material on Roosevelt's farming diversions in Georgia. The president had two mules, the story noted, "Hop" and "Tug," after Hopkins and Tugwell. In March the papers carried bad news: the first of Tugwell's projects, twelve rural settlements, were writing off 25 percent of their costs as unsustainable. The freestanding communities he had envisioned were not becoming reality. Tugwell's doubt in this instance was not about the reception of the New Deal; it was about the actual success of the programs themselves.

Tugwell did not speak publicly of regrets. Nor did he criticize the Roosevelt team—he was loyal. But Moley, who was now at *Newsweek*, a new competitor of *Time*, was definitively breaking with the president. Speaking at a meeting of advertisers, he talked about the lexicon of the New Deal. He praised the New Deal—up to 1935. Since then, however, there had arisen "new and fantastic counterparts" to the early New Deal. These later counterparts were too intrusive, and Moley could not approve of them.

In May the city itself provided reminders of the conflicts inherent in the New Deal. Roosevelt had created the Works Progress Administration to help labor; the same thought was behind his signature of the NIRA and the Wagner Act. But he had warned that the government would not always be able to afford to pay for the jobs. Now Roosevelt wanted to balance the budget, and some WPA jobs had to go. Instead of accepting the change, as perhaps Roosevelt expected them to do, the WPA workers were mimicking their private-sector brothers and striking. This seemed like ingratitude. How much, after all, could the government pay them? As if that was not sufficient, the WPA workers were striking not merely over labor but also over the political positions that newspapers were taking.

The *New York Daily News* editorialized against the WPA, arguing that it ought to be abolished in favor of an entity less subject to the sway of "politicians and communists." Local workers of the WPA promptly announced a "mighty mass protest" against the paper. Writers on staff at the Federal Writers' Project staged a sit-down strike across the street from the *News* at 235 East Forty-second Street. Even Hallie Flanagan of the theater group went along. In the same weeks she noted that the WPA workers were striking for "life, liberty and the pursuit of happiness."

Most New Deal supporters were more uncertain—in New York or elsewhere. They were taken aback at the brutality of unionization across the country. Some automakers had already unionized—General Motors, for example. But in Detroit, Henry Ford was still holding out. The basis of Ford's protest was that Ford could not afford the cost of unionization, logically enough. Companies like Ford were taking the position that it was illegal for strikers to strike within the factories—it was a trespass, a violation of company property rights.

Emboldened by the Wagner Act and its ambiguous language, the workers went farther—onto company property. Ford's company police retaliated by beating the protestors at its River Rouge plant, including a UAW organizer named Walter Reuther. *Time* reported that the workers were demanding a higher per-hour wage than offered by Ford or any other automaker. Reuther's battered face would become a national emblem of company brutality.

On Chicago's South Side, Douglas was rethinking things. "These sit-down strikes resembled the seizure of factories by the Italian unions in the fall of 1920," he would write. On Memorial Day weekend workers at Republic Steel paraded, pushing to get the company to unionize as U.S. Steel already had. Police turned on the crowd with revolvers and clubs, clubbing and killing even workers fleeing the scene. Douglas, still a professor, was asked by the editor of the *Chicago Times* to look into the Little Steel Massacre. He agreed, and moderated a protest meeting held at Insull's opera house.

Two problems stood out, observers noted. The first, again, involved the legality of the strikers' behavior. Both sides, Wall Street and unions, had worked so long and hard to see the Wagner Act passed; the idea all along had been to create a legal basis for protest. Now, instead of staying within the safe confines of the law, the protestors were pushing the envelope, seeing how far they could take the country. The strikes had the effect of escalating the battle: "They frightened most employers," concluded Douglas. To him and others the pattern—concession followed by escalation and radicalization—seemed far too much like not only Italy but Russia after Kerensky or Germany after Weimar for comfort.

The next problem was that the Communists were clearly active in the union movement. Lewis defended the practice, but many of the original labor supporters were apoplectic—Green at the American Federation of Labor, for instance. Douglas was also disturbed. He would complete the report on the Little Steel Massacre, Douglas recalled telling the Chicago paper in his memoir, but only if Communists were not assigned to work on the project with him. The boot steps of the fascists of Europe rang in Douglas's ears. He feared both the Communist Left and the Right. He had heard Benito Mussolini announce Italy's invasion of Ethiopia at the Piazza Venezia in Rome. Now he was becoming convinced that for the American Left to ignore all this was wrong, and that "isolationism was impossible and pacifism self-defeating against dictators." Frances Perkins, too, uttered her concern: "unwise and demoralizing," she concluded of the sit-downs.

Another problem, less articulated, was a simple economic one. To raise wages without increasing output hurt company profits and made goods more costly for consumers. Perhaps some of the companies could afford that, but it was no incentive to produce. Some consumers might be able to afford higher prices—but others, those working in nonunion industries, might not be able to. The idea that an increase in labor's wages would automatically restore the economy to its 1929 level was taking a long time to prove itself.

At the TVA, too, there was trouble. Roosevelt had fueled the tension between Lilienthal and Morgan by leading both men on. But this year, Lilienthal's biographer notes, the president was "studiously uninterested" in the fight at the TVA. Lilienthal found the going "tough"; both he and H. A. Morgan, his ally, hunted for evidence that A. E. was cooperating with Willkie. By September, A. E. Morgan would go on one of his retreats and leave his affairs to be managed by Harcourt Morgan. Wrote one TVA official in his diary of the dream: "Our own TVA is almost *bellum omnia contra omnes,* a war of all against all." Scarcely what the New Dealers had hoped for.

Every day there were more dark reports from Russia. Stalin had been holding his trials for some time. He had also announced an increase in his war budget of a full third. In June came news that he was trying eight of his most important generals, including Mikhail Tukhachevsky, a World War I hero well known to Americans. And the trial would be secret. Tukhachevsky had escaped the Germans four times in that war, but it did not seem likely he would escape his fellow Soviets. The intellectuals in New York had enormous trouble with this announcement, coming as it did on top of Trotsky's flight from Russia. They had always thought that in its way Leningrad was like New York and Paris. But the Soviets were proving that their cities were disastrously different.

Mary McCarthy typified the sudden indecision. McCarthy at first supported Stalin, as did her friends. And like her peers, she made a show of spurning the bourgeoisie; at the end of the summer of 1937, in fact, she was living an ostentatiously bohemian life with Philip Rahv, a Russian émigré, in a borrowed apartment on East End Avenue. In the summer, McCarthy handed in to the *Nation* a review that would be published just below one by Rahv. Her review attacked an American newspaperman who had published a book titled *The American Dream.* The author, Michael Foster, had written about "the quiet decent people . . . the silent ten percent whose names are not often on page one, because they are so busy, and who pay very little attention to the shooters and the grabbers." McCarthy dismissed the author's emphasis on this group as "absurd."

At the same time, however, McCarthy was having her doubts about her own positions, most of which involved events in the Soviet Union. She was considering committing herself to Trotskyism. "Tukhachevsky's murder could not make us happy—on the contrary," she would write later. "More than I, Philip grieved, I suspect; a boyish part of him was proudly invested in the Red Army." Trotsky had been thrown out of Soviet Russia. New York writers created a Commission of Inquiry on Trotsky, essentially to generate documentation of the travesty of Soviet justice in his ejection. John Dewey, now in his late seventies, that fall would lead the commission to Mexico to interview Trotsky. "It was the most interesting single experience of my life," Dewey would say of the trip. Already, in the summer, Trotsky's claims about Stalin were being vindicated anew each day, in the news. Even he was perturbed. In early July the papers reported that he was off on a fishing trip in Mexico. He told the reporters that "I want to get away from civilization and the press."

Like Tugwell, McCarthy was also beginning to try on new roles. Building in McCarthy's mind was a fiction story that she would eventually title "The Man in the Brooks Brothers Shirt." The story takes place in the mid-1930s, on a train ride. A young woman writer is heading west to get a divorce, just as McCarthy herself had done. The charm of the character's first husband, a Marxist theater man, has worn off. Now she does not know what she is looking for. But she is certain that she detests the bourgeois—the type of man who, a decade earlier, would have been called a Babbitt. Then a perfect example of that bourgeois enters the club car, a man in a Brooks Brothers shirt. He is a businessman, a man who has seen the world, and a man who makes companies grow. In short, a description that makes him clearly out of the question as a romantic partner for Mary's character.

But gradually, over whiskey and then trout, the young woman finds herself interested. "I've never known anyone like you. You're not the kind of businessman I write editorials against," she tells him. In the morning, she wakes up to find herself in his berth. The theme of the story is clear: the divorce is not merely that of one person from another, but also from the tedium of left-wing politics. "The Man in the Brooks Brothers Shirt" was published later and McCarthy's biographer reports a rumor that the traveler in the story was Wendell Willkie. This was never proven. But the story, in any case, described the summer of 1937 perfectly: a summer in which intellectuals were either becoming Babbitt, or at least getting to know him better.

And one of the Babbitts of real life *was* Willkie, who now popped up frequently at dinner parties in New York. In these years, Willkie was not merely battling Lilienthal; he was also getting around: among those he would meet were the authors Carl Sandburg, Rebecca West, Dorothy Thompson, James Thurber, the publisher Helen Reid, Henry Luce of *Time,* and the correspondent William L. Shirer. He socialized—and exchanged ideas—with everyone, in a fashion that would not have been possible for a utilities lobbyist in Washington. He often left Edith behind. Shirer later recalled that Billie, as she was known, spent afternoons at the Metropolitan Museum of Art, across from the apartment. "You couldn't help but admire her. She was probably terribly hurt. But she wasn't going to ruin his career."

The most important of Willkie's new acquaintances in 1937 proved to be Irita Van Doren, the literary editor of the *New York Herald Tribune*. A year older than Willkie, she had ended her marriage with Carl Van Doren, the historian.

Irita Van Doren's relationship with Wendell Willkie was a version of the relationship between McCarthy's intellectual and her Brooks Brothers man. In the McCarthy story, the love affair is short-lived—the girl turns back in revulsion. In McCarthy's real life that also proved the case, for she shortly remarried. Her new husband was to be another intellectual, in fact the leading intellectual of her era: Edmund Wilson. Wilson had just the year before published a book likening Stalin's Russia to Roosevelt's America: *Travels in Two Democracies*.

But Van Doren and Willkie were not a fiction, but rather a real story. Their romance shortly became something like a marriage, with weekends at her house in West Cornwall, Connecticut, and visits to her apartment on the West Side. Willkie, like Insull before him, thought he might please his sweetheart by wiring her house for electricity. Just as Insull had once strung cable out to Libertyville, Willkie now asked industry friends to wire Irita's house in West Cornwall—she had only kerosene lamps. The fact that Willkie was, to the core, always a utility man showed up too in the manner in which he expressed his friendship for Dorothy Thompson, Sinclair Lewis's wife and Irita's friend. A few years later, after a visit to Dorothy's country place in Vermont, he sent her a refrigerator.

Willkie was so proud of Irita, he could not stop himself from bragging. Harold Ickes would later write, "Willkie likes to play with a lot of women and is quite catholic in his tastes." At some point, Willkie would even tell Lilienthal about his relationship with Van Doren. Van Doren and Willkie were close, Lilienthal would write in his diary: "Wendell told me, rather explicitly, how close." Lilienthal was not leering; the relationship seemed to him "touching and beautiful."

But what mattered most was what Willkie and Van Doren learned from each other. He probably exposed her to new economic ideas. He joked that their friendship certainly would not please "your old friends on *The Nation*." Still it was Willkie who was the principal learner in the relationship. With Van Doren as a tutor, Willkie studied political and literary classics. He told her that he was interested in the South and that he had thought of writing a history of forgotten figures in American history.

He had always been a follower of Woodrow Wilson's; he believed in reasonable reform at home and democracy abroad. That was what he shared with Newt Baker, one of the Democrats whom Roosevelt had beaten out for the nomination in 1932. Now Willkie ranged wider. A couple of years into his conversations with Irita—in 1939—he would even publish a review in her paper of a book about one of the old UK Whigs, William Lamb, Lord Melbourne, "the evening star of the great day of the Whigs." Willkie's message in the book review for his contemporaries was that business, and perhaps his own utilities industry, had brought some of its troubles upon itself by forestalling reform. He quoted Lord Melbourne as noting: "Those who resist improvements as innovations will soon have to accept innovations that are not improvements."

Willkie's publishing and his time with Irita were about more than history. In discovering the old British Whigs, he discovered their liberalism—a liberalism that antedated Wilson and focused on the individual. It resembled the liberalism of Europe that he had heard about in childhood. Revisiting that old liberalism, he could see that while Roosevelt might call himself a liberal, the inexorable New Deal emphasis on the group over the individual was not liberal in the classic sense. Liberalism had historically included liberal economics, and Roosevelt had turned away from that. Willkie was finding the intellectual ammunition for his battles, and Irita was helping him do it.

At the same time, other New Dealers and Democrats were also redefining themselves through new personal relationships, all of which, in one way or another, affected their political lives. Ickes was now a widower—his wife, Anna, a Republican legislator at the Illinois statehouse, had died in 1935. In 1937 he was busy trying to convert a tempestuous affair with a Smith College girl, Jane Dahlman, into a marriage. In May he purchased a real country estate, the 230-acre Headwaters Farm, in Maryland, complete with servants' quarters. Ickes had lived well before, but the new spread somehow mattered more, especially in a man whose projects had, over the years, targeted wealth. The spring of 1937 had brought news of the revenues from the forced sale of the furniture in Insull's Chicago penthouse: $26,000, far below the $100,000 expected.

It was now already a few years since Paul Douglas had ended his marriage with Dorothy Wolff. After the marriage, Dorothy headed left. With his new wife, Emily, Douglas was moving to the center—and becoming especially suspicious of the Communists. Hence, at least in part, his caution over the report on the Chicago strikes. Later he would write in his memoir that he regretted Roosevelt's recognition of the Soviet Union. Few on the Soviet trip, he now realized, had perceived the threat that Stalin's regime represented.

Tugwell's administrative assistant at the RA, Grace Falke, had come to New York as well. She would now become director of the arts project of the National Youth Administration (Harry Hopkins's sister had also worked there, for a time). And Tugwell was not about to forget her. In August of the following year Tugwell's wife of twenty-five years, Florence Arnold Tugwell, would travel to Yerington, Nevada, to obtain a divorce. Tugwell would marry Grace in November 1938—Fiorello La Guardia performed the ceremony.

By July of 1937, the New York intellectuals were looking out at their country again—and again seeing more trouble on the horizon. Still thanks to the Wagner Act, the unions were growing astoundingly: the United Automobile Workers membership would by September reach 350,000 or so, a full ten times its size the preceding year. But that only seemed to make them more bellicose. In Michigan, the CIO was making matters hard even for a sympathetic governor, who had been unwilling to call out the police at the sit-down strikes.

Roosevelt was so irritated at both sides in the steel discussions that he quoted Shakespeare at them: "a plague o' both your houses." Much of the country felt the same—they might like the idea of organized labor, but they felt the CIO went too far. Nor did it, necessarily, represent the average worker. Odette Keun, a European journalist, came over to study the TVA and the American labor movement. After a long visit she made a conclusion that shocked her: "Labor in America is conservative. It is one of the most flabbergasting discoveries I have made." This conservatism, she wrote, was partly due to the retrograde American Federation of Labor. But it also was "due to the temper of the American workingman himself. In general his sense of solidarity was for a very long time nonexistent; it is not at all effective yet." Keun was ambivalent about this discovery but took pains to report honestly what she had found: "the workingman en bloc is still no revolutionist. He still has not the fanatical hatred of the capitalist. He still has no essential feeling that the system is essentially unjust, infamous, execrable, and must be wiped off the face of the earth."

A problem that Keun did not address was also becoming obvious: the Wagner Act was, at least in the short run, continuing to hurt profitability at companies. In August General Motors, which had suffered some of the worst strikes, reported that sales were up. Way up, both for the three-month and the six-month period. Earnings, by contrast, were down. The new wages and the costs of the strikes had made the companies less valuable. Roosevelt's groups were once again at odds: unionized workers versus shareholders.

A new respect for conservatism was also evident in the book business, where many of the intellectuals worked. The Russian books had done very well at the beginning of the decade. Political books still sold: Drew Pearson's *Nine Old Men* was on the best-seller list in Atlanta. Self-help books, history, religion, and escape also sold. A regular standout, at the top of the list, was *Gone with the Wind,* the novel by Margaret Mitchell, the wife of Willkie's employee at Georgia Power. Another best seller was *Orchids on Your Budget,* a relentlessly cheerful personal finance guide that ordered Americans to persevere even in the face of tight budgets. Yet another was *How to Win Friends and Influence People,* Dale Carnegie's self-help book—his movement was not so different from Bill W.'s. Even *American Dream,* the book that McCarthy panned, was on the list of top sellers in Los Angeles. There were yet more self-help books: *Mathematics for the Million,* for example. Odette Keun was right: America simply was not conforming to the Left's expectations. The country's old self-improvement impulse was prevailing, flowering even.

Another recent book shed some light on the reasoning of the average American: *Middletown in Transition,* a lengthy revisit by Robert and Helen Lynd, the pair of sociologists who had reported on Muncie in the 1920s. Muncie, the Lynds reported now, had gone for FDR in 1936. But it had not changed its values as much as the New Dealers had expected it to. The town did not share the view of Robert Lynd about Stalin, he noted with appealing objectivity. It did not even seem to share Roosevelt's inauguration view that the country had ceased admiring worldly success. "By and large Middletown believes," the authors had written, "in being successful"—the Brooks Brothers shirt. In general, the town also believed that radicals wanted to "wreck American civilization."

An editorial in a Muncie paper had spoken not of Moscow but of the revolution in Spain, equally large in the news. "There is one reason why a revolution of the kind now existent in Spain is improbable in the United States . . . Spain has no middle class." But the United States still did. The paper also quoted Abraham Lincoln: "There is no permanent class of hired laborers amongst us. Twenty-five years ago I was a hired laborer." An editorialist at a Muncie paper noted that Theodore Roosevelt had said, "To preach hatred of the rich man as such" was "to mislead and inflame to madness honest men." The point was clear: the current divisions were not ones that Middletown wanted to see made permanent.

What the interviewers observed especially was that Muncie's citizens were unhappy at receiving two opposing lessons from governments. The first might be labeled: "Saving—the Private Man's Only Safeguard." The second was "Spending—the Nation's Hope." The citizens had trouble squaring those two ideals, and the contradiction made them anxious. There was, what's more, the "growing feeling of the insecurity of future investments due to national government policies. Stocks and bonds are now very uncertain." But perhaps most telling of all the material that the Lynds had amassed was yet another editorial from the Muncie papers—one about "the forgotten man." "Who is the forgotten man in Muncie?" asked the paper. "I know him as intimately as my own undershirt. He is the fellow that is trying to get along without public relief. . . . In the meantime the taxpayers go on supporting many that would not work if they had jobs."

One of the people whom *Middletown in Transition* may have impressed was Irita, who shortly would serve on a committee of publishers in New York that selected two hundred books for the White House library. The books, as Van Doren would tell the press, would be chosen for their timeliness. A selection of the editors would be *Middletown in Transition.* Once again, the country was sending a general political signal, muffled but important, that it wanted change. And once again the people with their ears to the ground, ready to transmit the message—even one they did not like—were in New York.

Tugwell, for his part, was spending a restless summer. American Molasses wasn't right for him: "There was no real place for me and no real work to do," he would later write. Within a year he would move to a more appropriate job, creating low-income communities as chairman of the city planning commission. He was watching, with a bit of anxiety, the developments at all his RA communities. The social worker from Casa Grande Farms was getting ready to select families to live in its pastel-colored houses. Greenbelt, in Maryland, was due to open in September. But within a month, too, Henry Wallace would formally "liquidate"—that was the word the magazines used—the RA as an institution. With it would go a distressingly large share of the Tugwell legacy.

Tugwell may not have known it, but his life had now taken a turn like Herbert Hoover's. Later he would write that he was, "after all these years, still bitter about the disappearance of the Resettlement Administration," and that he still harbored "in spite of myself, a good deal of resentment." The next decades would include rewarding moments—Tugwell was still the honest idealist he had always been. Taking his new family, he would serve as the last governor of Puerto Rico. But the years would also be spent writing articles, giving speeches, producing memoirs, nearly all to the same purpose. Like Hoover, Tugwell was trying to limit the damage that history would do to his reputation.

13
Black Tuesday, Again

August 27, 1937
Unemployment: 13.5 percent
Dow Jones Industrial Average: 187

On an August evening at his daughter's at Southampton, Andrew Mellon died. The financier was eighty-two; his son-in-law David Bruce reported that a combination of bronchitis and other diseases had felled him. "My greatest impression of him was his innate modesty," said Hoover. S. Parker Gilbert, Mellon's former undersecretary of treasury, announced that Mellon's death represented a national loss.

Wall Street already knew that. What it was asking itself was whether more loss was yet to come. The same day that it reported Mellon's death, the *New York Times* carried a story on the consequences of the undistributed profits tax. Companies that had formerly sought to retain employees through downturns now no longer had the reserves to do so. They had likewise ceased to invest in new equipment, normally a traditional move in slow periods. The headline on the story was: "Levy on Profits Halts Expansion." What would happen to the meager recovery? Stocks had begun dropping in mid-August. Now they accelerated their decline. As if to underscore the news of Mellon's death came the news of the death of Ogden Mills, the former treasury secretary who had become so angry at Roosevelt. Mills died on Monday, October 11, in his home at 2 East Sixty-ninth Street, just off Fifth Avenue. At Saint Thomas Church farther down the avenue, Theodore Roosevelt Jr., the president's angry cousin, would serve as usher at the funeral. So would Herbert Hoover.

By the next Monday the worriers had their answer. Bond prices plummeted farther than they had on any single day in three years. Businesses and investors did not want to buy money anymore because they did not want to use it. Bonds can measure the market's expectations of growth. Stock-exchange seats are another measure of expectations—the stronger the economy and the more profits to be made, the higher the worth of a seat. The same mid-October day came the report that a seat had been sold for the lowest price since 1919, $61,000. As for stock shares, they were down between $2 and $15, the greatest drop in six years. In recent times, before this panic, the market had been "thin"—relatively few shares had traded, at least when compared with pre-Depression days. This panic, though, was so broad and trading so furious that the ticker closed seventeen minutes late. The traders were finally awakening, just as everyone had hoped they would. But they were awakening only to run.

At first the country hoped that the problem would be Wall Street's alone. After all, there was some good news in 1937: the cotton harvest was strong. Investors were sending cash to the United States. At Casa Grande, the Farm Security Administration had found what it believed to be the most crucial component to success, a competent-seeming manager. Robert Faul was himself a homesteader; "stout, raw-boned, farmer type," as one of the planners summed up. Another official had written: "Very practical type; used to giving orders and not asking too many questions." A letter in Faul's file read: "He handled crews of 200 to 300 men and got a lot of cooperation from them." Faul had also run the Arizona Turkey Growers' Association. Under Faul, Casa Grande might show the profits that improved productivity could yield. In the West Coast offices of the FSA, Faul became an emblem of agricultural comeback.

People busied trying to figure out the reasons for the downturn. Some of them were monetary. Inadvertently, Marriner Eccles's Fed and Henry Morgenthau's Treasury had replicated some of the conditions of the period of the crash. The new Fed law had created stricter reserve requirements for banks. Forced to keep more cash, banks cut back on loans. Then Fed officials had asked banks to increase reserves again. Gold had poured in for years, and Fed officials began to fear inflation. So again, they sterilized, offsetting gold inflows with restrictive actions. By the month of Mellon's death, more than a billion dollars had thus been extinguished.

Then there were the other factors. The payments into the new program, Social Security, were also taking money out of circulation. The year before there had been a soldiers' bonus, but that cash was not there this year. The Fed had increased requirements of member bank reserves, taking money out of the system. The easy feeling of 1936 was gone. Meanwhile, the Wagner Act was making business more expensive for employers. Benjamin Anderson, the Chase economist, had seen a chart at Treasury comparing U.S. wages, prices, and the cost of living. The changes in these three areas had moved along in a fairly consistent way from 1934 to

1936. But with the Wagner Act, wages had jumped far ahead of the rest. In the first six months of 1937 alone, wages rose 11 percent. In the steel industry the rate was higher, 33 percent from October to May. Taxes of course were a problem too. Private investment had been low all decade.

All the changes brought by the New Deal meant that the United States seemed a less reliable place. William Gladstone, one of the English liberals about whom Willkie was reading, had written once that "credit is suspicion asleep." Now, especially after the president's second inaugural, suspicion was more awake than it had been in peacetime that century. Knowing that the government wanted to enter an area of the economy was one thing. But knowing that it could always make vast changes was more disquieting. Roosevelt had had to deal with strikes of labor in the spring. Now he had to deal with the more serious strike of capital.

What data there was reflected this fear. The gold continued to flow in—mostly because the gold standard dollar was a relatively safe dollar. But foreign companies did not float new stocks in America the way they used to. In the 1920s such flotations—something like the modern Initial Public Offering—had run more than $1 billion a year. In the Depression, the average was below $50 million. The difference between short-term interest rates and those with longer maturities tells something about risk—high long-term rates mean that lenders require a lot to make them willing to expose themselves. As one economic historian, Gene Smiley of Marquette University, would later note, the striking thing about the second half of the 1930s was the spreading of that gap between long and short term.

When the data came in, they showed that August had seen the steepest drop in industrial production ever recorded. The Dow Jones Industrial Average dropped from its 190 level in August down to 114 on November 24, about the level it had stood at in the month of the Schechters' triumph. In the period from September 15 to December 15, the jobs started to disappear, with unemployment moving back to 1931 levels. The Wall Street shock was spreading to Main Street.

The old brain trusters in New York did not deny the problem. Adolf Berle noted in his diary that he had conferred with Tugwell's employer: "Charles Taussig and I agreed that the economic situation begins to look like a major recession and ought to be tackled on very broad lines," wrote Berle on October 15. "Yesterday the 1929 panic was really repeated with more to come today," he added four days later. "The Stock Market people are most bewildered and frightened." Berle concluded that it was "plain now that business is dropping as well as the market—in other words, we're in for a rather bad winter."

One did not have to be an economist to do the math: when wages moved ahead, profits narrowed and shareholders lost. The chart had compared the United States to other countries—Japan, France, Britain. But the only other country with a similar leap in wages was France, where Leon Blum had earlier in the 1930s put in his own New Deal. The French New Deal had caused a collapse in the economy. Anderson recalled later that the Treasury official had said to him of the American picture, "That looks too much like France."

A French trade expert, Jacques Stern, happened to be in the United States that autumn. He looked more closely at specific industries but came to similar general conclusions. Taxes were designed to punish risk but permitted little reward. The railroad industry had had to increase salaries because of the new labor laws but had not been able to raise its rates. As Stern pointed out, these industries consumed one-fifth of all the steel and wood produced in the United States, so the slowdown hurt others. Utilities, of course, were not hiring either. The next year Keynes would offer a similar conclusion in a letter to Roosevelt. Keynes saw no use, he wrote, "in chasing utilities around the lot every other week." Roosevelt should nationalize them if the time was right, but if it wasn't, he should "make peace on liberal terms." The recession wasn't merely monetary. "It is a mistake to think businessmen are more immoral than politicians," Keynes wrote.

Even on the farms there was serious discontent. All across the country, officials from the Agriculture Department were encoun-tering panicked families. The cotton news had not saved everyone. Earlier that year some 15,000 farm families, the department was discovering, had left drought areas in the Dakotas, western Kansas, and eastern Montana to look for new lives in the Pacific Northwest.

A lady who had placed her pension in utilities had written Mrs. Roosevelt earlier that year: "Personally, I had my savings so invested that I would have had a satisfactory provision for old age. Now thanks to his desire to 'get' the utilities I cannot be sure of anything, being a stockholder, as after business has survived his merciless attacks (if it does) insurance will probably be no good either. . . . I am not an 'economic royalist' just an ordinary white collar worker at $1600 per. Please show this to the president and ask him to remember the wishes of the forgotten man, that is, the one who dared to vote against him. We expect to be tramped on but we do wish the stepping would be a little less hard."

Companies were also marking new lows. Leonard Ayres, the executive at Cleveland Trust who had called on Aif Landon with Anderson in 1936, tried to get a grasp on the story by comparing the profitability of corporations in the current decade to that in the preceding one. He found that close to two of three had been profitable from the midteens through the 1920s. Since the Depression, however, that ratio had dropped below one in three, so that "for nearly a decade now the great majority of corporations have been losing money instead of making it," he would note. The editors at the *Economist* in London were also watching, trying to put what was happening to the United States in perspective. In 1930, the per capita national income of the United States had been one-third larger than that of Britain, the magazine wrote. At the end of the 1930s, it was about the same. The problem, the magazine would conclude several years later, was "institutional obstructions to a free flow of capital." The 1930s, all in all, the magazine would decide, were a strange decade; maybe, as it wrote, the United States really had forgotten how to grow.

Roosevelt's aides were perturbed, for they were seeing the president behave as he had around the time of the London monetary conference. He could not make up his mind which problem was the worst, or which must be addressed, and in what manner. And he could not see that it was important to be consistent. Regal, he was content to allow his men to joust. And the jousters duly performed, some new to the tournament of New Deal politics, some of them in the same roles as in the past, and some now taking different ones. But to the observer—although probably not to the king—the economic trouble now was too close to that of the early 1930s for the debate to amuse.

On the one hand there was Eccles, arguing now that there was insufficient money in circulation, and more must be spent. Joining Eccles in this view were other economists and New Dealers—Isador Lubin, Leon Henderson, and Lauchlin Currie. Currie, who had trained at the London School of Economics, was the intellectual leader of the spending group. (Currie, like Hiss, was also a Soviet spy whose arguments won a good reception from naïve colleagues.) What mattered, in any case, at the moment, was the policy. In November, Roosevelt held a meeting at the White House to talk over all these ideas. Eccles got the impression that Roosevelt was ready to come out for more deficit spending. And now there were institutions that stood ready to do that spending, from the Agriculture Department through the WPA to the Farm Security Administration, the National Youth Administration, and the Civilian Conservation Corps. Roosevelt might also loosen credit, and had backed legislation to increase spending on housing. Eccles involved himself deeply in the minutiae of plans to spend on housing to spur economic activity. Earlier and now, their solution was controversial. Still, while the early 1930s had been rough, Eccles believed they had taught them all something.

On the other hand there were the budget balancers. One of them, at least on some days, was Roosevelt himself. For as Anne O'Hare McCormick wrote in 1937 after another visit to Hyde Park, Roosevelt was still "the Dutch householder who carefully totes up his accounts every month and who is really annoyed, now that he is bent on balancing the budget, when Congress can't stop spending." Budget balancing also appealed to Morgenthau, now in his fourth year at Treasury.

Men become their offices, just as Coolidge had written a decade earlier. Morgenthau's response to this situation was to try to behave as he thought a treasury secretary should. Morgenthau had watched as the gold flowed into the United States when its economy or interest rates lured, and he had watched it flow out. He found himself offended by the Keynesians and their loose talk about the dollar. And his disapproval mattered—"Eccles was in the doghouse," blamed for the new downturn, Currie later remembered. Now Morgenthau was pushing hard for a balanced budget. At a lunch with Roosevelt on November 8, he tried desperately to convey his sense of fear to Roosevelt through a story about something that mattered to both men: their sons' generation. Henry III was at Princeton. So was Philip Willkie, Wendell's son—indeed, in 1940, Philip would be voted "most likely to succeed" in the graduating class. John L. Lewis's son, John L. Lewis Jr., would graduate in 1941. The New Dealers might have seen themselves as anti-elite, but they created an elite of their own. Morgenthau told Roosevelt that this young generation had to explain to itself, in its own terms, what the parents were doing. So, as Morgenthau told Roosevelt, he, the treasury secretary, had tried to explain the New Deal to Henry. And he had found himself struggling a bit. What exactly was the correct New Deal response to a floundering country? What had the New Deal achieved, actually?

Finally what the secretary came up with to tell the boy was that "the United States had come through this terrific turmoil and that the individual in this country still had the right to think, talk, and worship as he wished." Roosevelt, Morgenthau noted, merely replied by saying, "And add to that the right to work." Later, Morgenthau wrote in his diary that he had not understood what message Roosevelt was sending. Within days, Morgenthau was even more distressed, for Roosevelt continued to press suggestions for spending. "If you want to sound like Huey Long, I don't," Morgenthau said hotly to President Roosevelt as the pair drafted a speech designed to reassure Wall Street of the Roosevelt administration's intentions. In his memoirs Morgenthau would later take care to note that the speech had been "checked and double-checked—every word, every syllable, by the President."

On November 10, the day after the Dow closed at 126, Morgenthau finally gave his speech to the Academy of Political Science at the Hotel Astor in New York. It may have been a comfort that Mellon's own beloved aide and pallbearer, Parker Gilbert, introduced him. The war against the Depression, Morgenthau told the crowd, had required deficit spending. But the emergency was ending, and the "domestic problems which face us today are essentially different from those which faced us four years ago." The government would have to cut spending—a painful argument for Morgenthau to make, given that the project would include many New Deal cutbacks. Still, he had to do it, for "We want to see private business expand." And then he unfurled a conclusion that he imagined would please Wall Street: "We believe that one of the most important ways of achieving these ends at this time is to continue progress toward a balance of the federal budget."

That was when the laugh had come from the crowd—someone's laugh of contempt. Morgenthau and his adviser Herman Oliphant, caught up as they had been in convincing Roosevelt to stick with their plan, later recalled their own shock at the audience response. They had pushed the president this far, and Wall Street had not rewarded them. It even seemed to be siding with Eccles. "We sit here and lose the feel of what the typical leadership of American finance is," Oliphant said of life in Washington, "and it's very illuminating to realize the hopelessness of trying to work with them." The hurt was also compounded by fear that they may have misunderstood the president. Or been made to seem the fool.

Eccles for his part was also hesitating. Morgenthau's November 10 speech sounded to him like something from another century—not at all what he and Roosevelt had discussed. The Fed chief's confusion was doubtless not reduced when the president, early the following month—and after more bad news nearly daily—told reporters that the idea of a business recession was merely "an assumption." In the same press conference the president also threw a bone to the business crowd—the announcement that he had recently received a very interesting memo from Willkie on a utilities matter, and that he intended to study Willkie's thoughts very intently. Later, Eccles concluded sadly in regard to the recovery plan what others had concluded about the president on other matters: "It seems clear that the President assented to two contradictory policies because he was really uncertain where he wanted to move."

There was yet a second contest among Roosevelt's men. It was between those who sought the cooperation of larger businesses and those who wanted to attack them. In New York, Berle considered that a solution might be guaranteeing wages in certain industries, "beginning with housing." Such a plan, Berle reflected in his diary, "probably also means taking over the railroads," as well as, perhaps, government "taking over both housing and construction." He realized that would be an unparalleled infringement on private property. "But," he wrote with a diarist's sigh, "I do not see that it can be helped." Balancing the budget would likewise have to go out the window.

Tugwell too was all for planning among economic leaders—and perhaps still hoping for an inroad back to Washington. The president, casting about, had him to dinner that month, and Tugwell took the chance to go over some basics with FDR. ("Rex was trying to educate the President in general economics," noted Berle wryly.) In January, Tugwell and Berle would shepherd John L. Lewis, Thomas Lamont of J. P. Morgan, and others to the White House for a conference. The headline describing the meeting replicated headlines from the Hoover years: "Leaders with a Program for National Recovery to See Roosevelt Today: Cooperation Is Aim." Donald Richberg was also pushing for cooperative planning.

Just as back in 1932, Roosevelt seemed to be receptive to the planning idea. Was charity the answer? In the autumn he had called for a nationwide charity drive: the names on the benefit lists were reported to include Lillian Wald, Ida Tarbell, Arthur Hays Sulzberger, Mrs. Roosevelt, Mrs. Dwight Morrow, and "Wendel L. Willkie." At a press conference on the fourth day of January, the ambivalent president told reporters that he was thinking about renewing cooperation between government and business. Sensing ambivalence, Roosevelt scripted his way forward: "Don't write the story that I am advocating the immediate reenactment of the NRA. But the fact remains that in quite a number of the code industries under NRA it was perfectly legal for the heads of all the companies in a given industry to sit down around the table with the Government." Was allotting shares of business a good idea? asked a reporter, putting his finger on the old NRA problem. "Keep competition."

At the same time, however, Roosevelt was listening to and following the advice of another set of advisers: the anti–big business crowd. From afar a bitter Moley—who, unlike Tugwell, was no longer coming down for occasional visits—noted that Corcoran, Cohen, Ickes, Hopkins, and Robert Jackson were telling the president he must use the opportunity of the downturn to move in for the kill when it came to big business. And the president was heeding them. In public remarks, Roosevelt's men were speaking of "corporate tentacles" or "aristocratic anarchy." Hugh Johnson, now out and embittered, called the group around Roosevelt "the janissariat," a reference to Christian youths conscripted by the old sultans.

Corcoran and Cohen, whatever their ambitions, were mostly insiders, not ambassadors. Jackson, however, was someone whom Roosevelt increasingly viewed as a possible heir. Jackson, like Morgenthau or Eccles, had collected a set of specific instructions from Roo sevelt—in Jackson's case to define and prosecute antitrust violations, and, especially, to go after individuals. Sometimes—when he knew the targets involved, or liked them—Roosevelt suggested that Jackson soften. And always, Roosevelt took care not to harm those with special power to harm him. Learning from Jackson of a possible action against motion picture combines, Roosevelt said, "Do you really need to sue these men?" and asked that they be brought in for a talk. But other times he egged Jackson on.

In the fall of 1937 Roosevelt was thinking about whether Jackson might be a good man to run as candidate for his old job of New York governor. There was talk that Bob Wagner would run for that job, but Roosevelt wanted to keep him as an ally in the Senate. If Jackson, an attorney from Jamestown from a Democratic family, could capture the governor's seat, then Roosevelt could have a twofer. Roosevelt's feelings about Jackson made sense, for there were similarities between the two, as the columnists Joseph Alsop and Robert Kintner would note: "Both are upstate New Yorkers. Both are country squires turned political leftwingers." Another commonality: "H" in Jackson's name stood for Houghwout. He was descended from Dutchmen, like Roosevelt. At the end of November, Jackson accompanied the president on a fishing trip. Hopkins and Ickes—who at times had feuded bitterly—were also aboard the *Potomac,* sharing a cabin. The four prepared political strategy: specifically, an assault on the wealthy. Roosevelt caught a large mackerel early on, but it was Jackson who had the biggest catch of the trip, a barracuda of more than twenty-five pounds. If any of them considered the incongruity of planning a class war on a yacht, they did not mention it.

Just after Christmas, Harold Ickes gave a radio speech assailing America's wealthy, charging that sixty families who ran the nation were on strike against the rest of the country. He was correct about the strike part. The sixty-family part Ickes had taken from a sensationalist book recently published by Ferdinand Lundberg, a writer for the *Herald Tribune,* Irita's paper. Those families, Ickes said, were demanding that government give the country back the "suicidal license" it had had in 1929. Some listeners took the Ickes assault strike in good humor. One was Charles P. Taft, the son of the president. Taft told the *New York Times* that he felt about Ferdinand Lundberg's decision to include the Taft name in his list as Mark Twain had upon hearing the news of his own death: that the charge was "grossly exaggerated." But the message was clear: there would be more political attacks in 1938.

Next came Jackson, who now hoped to net the biggest fish of all for his boss: Willkie. The pair would go up against one another in a town hall meeting with a live audience. But it would also be a *Town Hall Meeting of the Air,* a show that the National Broadcasting Company aired across the country via its Blue Network. The public did not know it, but for Willkie the debate was a public reprise of those original meetings with David Lilienthal at the dark table in the Cosmos Club: "How Can Government and Business Work Together?" Both Jackson and Lilienthal were good fits for the town hall format, men who might actually have appeared at a genuine town hall meeting, the sort that Rockwell painted. Willkie, the son of Elwood's German American attorneys; Jackson, who had skipped college and had become a lawyer the old-fashioned way, training as a clerk—these were two old country lawyers who through merit had risen and now were debating the biggest issues in the land.

The event took place on the first Thursday in January. Jackson started out graciously. Probably no two men had ever looked at the relations of government to business through "more differently colored glasses" than he and Willkie. Yet Jackson admired Willkie's "consistent willingness to stand up man-fashion and submit his views to the test of dispassionate but frank discussion." Then Jackson moved on, blaming business for its strike and for creating insufficient work; and, "if industry will not provide it, the people are determined to provide for themselves." Small companies he would defend—indeed he as a young man had represented them himself once. Big companies, however, needed reining in. They were selfishly keeping their profits to themselves.

Next, Jackson moved to the topic of subsidy. Roosevelt's opponents claimed that the New Deal had created the idea of subsidy. That was a fallacy. The United States had always subsidized the private sector. In the nineteenth century, pioneers had enjoyed a form of early WPA; they got "a quarter section of public land just for occupying it." World War I had provided another sort of WPA, spending and jobs. The actual WPA was just a follow-on. Those who recalled Roosevelt's 1932 speech about the end of the frontier saw that Jackson was reprising an old theme.

Jackson also made a stab at explaining why there was conflict between government and business: "The man who is in government is brought in contact with the problems of all kinds and conditions of men. Everybody's business is his business." The private businessman, by contrast, "has been intensely preoccupied with a very narrow sector of the world." Again, Jackson assigned blame: the conflict was business's fault. Jackson concluded, somewhat irrationally for a trustbuster, with a plea for a "high volume, low price industrial economy." Those who opposed his notions were trying "to destroy this kind of American life."

Here Jackson had overshot. For Willkie this time was better prepared than he had been at the Cosmos Club, or against Lilienthal. What's more, he was aware that the stakes were higher than they had ever been. Just days before, on January 3, the Supreme Court had refused to let power companies use injunctions to stop federal lending to cities for power plants. "Mr. Ickes's delight over the court verdict was unconcealed," the *Times* had written. Sixty-one projects in twenty-three states that had been held up could now proceed, and even Justice Sutherland had gone along. With Ickes at work, and Lilienthal rising, the next year or two would probably mean the end of Commonwealth and Southern.

The executive started out gently, calling Jackson's bluff about the town hall. Was the very idea of business cooperation with government really so American? "I wonder," Willkie asked the audience, "if it seems strange to you tonight that we should be discussing the question of whether or not government should cooperate with American business . . . They might ask, with some surprise, if it was not the function of American government to encourage the development of private enterprise." Hoover's monstrous Department of Commerce was odd. And the New Deal might dress itself up as a community project. But conceptually it lay very far from the country's old community roots.

Willkie moved on, putting people he already knew out there, if only to show their relationship. Jackson was an antitrust man, and he ought to understand that taxation hurt the smaller business. For small businesses had no extra resources to handle tax work. The Roosevelt tax program was punishing the very same people whom Jackson was being so solicitous about. Especially problematic was the undistributed profits tax, which punished cautious business for failing to spend: "If there is any strike of capital it comes from those millions of small investors, not from the wealthy few." And punishing the rich at punitive rates encouraged them to escape. "As a matter of fact because of income tax laws which take up 83 percent of a rich man's investment in private enterprise most of the very rich have been investing more and more in the flood of tax-exempt government securities"—just as Mellon had said.

In their attacks on business, the New Dealers were forgetting their decency and their dignity. What, after all, asked Willkie, was the purpose of going after the rich? It seemed "a little ironical for government officials to be lecturing big business on the desirability of low price and large volume"—since, after all, the private sector had developed these concepts in the days of Ford. Phrases like "Bourbons" or "moneyed aristocrats" or "economic royalists"—the phrase Roosevelt had used in the campaign—were unworthy.

Then Willkie came to his own area—utilities. The government acted as though utilities were demons. In fact, of all industries, utilities, which were growing so fast, were central to recovery. Over the course of this bitter decade, the utilities had done their part. Prices overall since the period before World War I had risen 40 percent; yet the price of electric power was down by the same share. In Washington, where Jackson's own Justice Department pursued holding companies, electricity prices were low because the area was served by the Great Northern American Holding Company. The average rate was 3 cents per kilowatt-hour. But outside Washington, where a smaller company served the customer, the rate was 4 cents. The savings was one and a half million a year on the city electricity bill. Big private companies served Americans. Was it really economical for the country to destroy the holdings?

In the same weeks Roosevelt and Democratic leaders were successfully erasing the deficit of the Democratic Party by hosting a series of Jackson Day Dinners in honor of Andrew Jackson across the country. *Time* magazine noted that the president had already formed a concrete vision of his own role in history. Roosevelt very much enjoyed, the editors wrote, "projecting himself far into the future and viewing himself retrospectively in the grandeur he will have assumed 100 years hence." Just a week after the town hall meeting, Roosevelt would appear before the press to explicitly emphasize that he still wanted a "death sentence" for utility holding companies. As for holding companies, he would ask, "Why should there be any holding companies?" The very next day, following Roosevelt's press conference, Willkie would begin to do what he had always known he might have to: break up much of Commonwealth and Southern, selling 60 percent of the company to the Tennessee Valley Authority. It would be Willkie's last great offer, a "desperate" offer, as he would characterize it.

Willkie wanted to let the audience know that he was still pro-reform. He agreed that some of the ideals of the New Deal were all right. It was even all right to have some of the New Deal programs. They existed now and could be modified to be made even more useful. In the case of Social Security, for example, Willkie was anxious that the system be adjusted so that the money paid by workers went exclusively to fund pensioners, and not be diverted to other government projects.

Jackson was especially wrong to contrast big business with small business. It just didn't make sense—what mattered was the effort of enterprise, big or small, and everyone ought to be for that. The suggestion was clear: it was wrong to argue that a Jackson, or a Lilienthal, was inherently more virtuous than a Mellon, a Schechter, or even an Insull. It might be amusing to make the national battles seem to be those of the businessman versus the government man. But observers should not delude themselves; government men and businessmen had similar sins and virtues. "I find no halo on the head of either," said Willkie, in a vein similar to Keynes.

Finally, Willkie made a more general point. New Deal fervor was overshadowing the reality of what it was doing to business. The government tended to underestimate the terrifying affect its random targeting of businesses had on the general economy. That was the cause of the capital strike. "If there is a smallpox epidemic in a city," Willkie said, "you cannot convince a man he is in no danger because at the moment only 15 percent of the city is affected." So it was with business: if investors even suspected the government would take over a certain industry, they would withhold their investment—that "idle money." The New Deal was an inspiring phrase, but Willkie—like Morgenthau with his son—wondered what it actually described. Oliver Wendell Holmes was one of the Roosevelt crowd's favorite justices. Now Willkie quoted Holmes back at them: "A good catchword can obscure analysis for fifty years."

One listener asked whether the threat of the concentration of government wasn't stronger than the threat of concentrated business. Jackson, defensive, said that the United States would never have concentrated government in Washington if business had not become so powerful. The states alone were obviously not sufficient to regulate big utilities, Jackson argued: "How can a single state regulate Mr. Willkie?" Americans needed to regulate to stop the concentration of business.

No, said a voice from the audience. "That might be in Italy or Germany, but it isn't that in the United States of America." Applause. Another questioner in the audience drove the uncertainty problem home again in a question to Jackson. William Knudsen of General Motors had said people now had the money to buy cars but were still not buying because of the general uncertainty. What would Jackson do about that? Someone in the crowd asked why the federal government did not acknowledge that it wasn't including in its statements on the TVA the cost of money that taxpayers put into the authority. As a result, the questioner reminded, an "incomplete balance sheet is presented to the American people." The yardstick was fake. The TVA was questionable, just as the workings of private sector utilities were questionable. "What is the difference between this procedure and Mr. Insull's?"

The published transcript shows that Jackson confined his answer to replying that he wasn't an accountant, and he wouldn't get into details. But his recounting of the event a decade and a half later in his papers shows that he believed that the town hall debate had been a setup, and that that had helped Willkie. Each speaker had been given tickets, but Willkie had got extras—hundreds extra, Jackson believed—and packed the hall. What's more, Willkie had spoken seven minutes overtime, whereas he, Jackson, had punctiliously followed the rules. As Jackson recalled of the debate: "I had a talk with the President and told him about the program. He was annoyed at it and wasn't greatly surprised at Willkie. But he was surprised at the Town Hall outfit."

It is hard to imagine that Willkie's dominance in the debate can be entirely attributed to his overtime minutes, or even a hall packed in his favor. For the national response was strong—as strong as the response to some of Roosevelt's speeches. The next day the *New York Times* headlined the report "Industrial Leader Asks End of Government Catchword," noting especially Willkie's defense against "Slurs on Business." General Johnson howled that Willkie had made a "perfect monkey" of Jackson. Moley wrote that Jackson hadn't done badly, but "Willkie so utterly outclassed him that the Jackson build-up dissolved into the elements from which it came."

Willkie's friends in Henry Luce's empire were pleased. The Town Hall debate was exciting because it revealed that, given the right context, the argument against interventions could resonate. The problem in 1937 and 1938 was not that the New Deal was mismanaging or helping or punishing one sector of the economy over the other. It was, just as even Democrats now knew, that it was competing with the private sector, and frightening it. The solution to the depression within the Depression was not anything either of the two squabbling sides in the administration was contemplating. If Roosevelt wanted the economy to thrive in peacetime, he had to call off the competition.

Finally, the spluttering Republicans and disillusioned Democrats had found a voice. Willkie was different from the xenophobes and the isolationists. A year later, in 1939, the editors at the *Saturday Evening Post* would capture Willkie with the headline "The Man Who Talked Back." A messenger had been found. Now all that remained was to determine the rest.

14
"Brace Up, America"

January 1938
Unemployment: 17.4 percent
Dow Jones Industrial Average: 121

The country was now at an odd moment. There was a new sense of permanence about the Depression. Being poor was no longer a passing event—it was beginning to seem like a way of life. Roo-sevelt's prophesies about America seemed to be coming true—the country might be like old Europe, frontierless, something out of Dickens. The story of William Troeller, the boy who killed himself to spare the family food, recalled Oliver Twist asking for "more" at the orphanage.

Some suicides were committed by people who had played and played for time but now were giving up. In

East Orange, New Jersey, on Harrison Street, Emmet Faison, the owner of the failing Orange Engraving Company, concocted a desperate plan to save his company. He might die, but the life insurance payment would rescue his company from bankruptcy. After he poisoned himself, his family discovered that a clause in the contract precluded payment after suicide. The first crash had seemed like a nightmare; this crash felt like a life in the dark.

But two new factors were also at work. Americans were becoming experienced at finding light in the dark—even creating their own light. There was a new sense of concern about what was going on abroad. Very few Americans wanted to hear about trouble overseas. In 1935, Congress had passed a Neutrality Act, and Roosevelt had signed it. Republicans were still leading the isolationist charge. But the news from Europe and Asia was so awful that both parties were beginning to look like bad watchmen. The darkness in Europe might in the end prolong the night of the United States.

As 1938 unfolded, it was the first matter, finding one's own way at home, that dominated. In New York Bill Wilson was struggling. He and his wife, Lois, had no cash, and were moving about—they would move dozens of times in these years. But Wilson was not drinking. He thought he really could make a go of his new system, in which alcoholics helped one another. He wrote notes to himself, steps that each boozer would have to get through if he was going to achieve his own recovery. He was also dictating and writing out longhand on legal pads something larger—what came to be known as *The Big Book,* a primer for his movement. In *The Big Book,* he sought to solve a problem for his growing group, with members ranging from the pious to the atheist to the agnostic. He settled for a kind of spiritualism, writing of "God as we understand Him," or a "power greater than ourselves." That past winter, hearing word of Bill Wilson's new ideas about helping drinkers, a Rockefeller Foundation executive named Willard Richardson met with Wilson. Richardson, who was also ordained as a minister, sat with him at lunch, along with Dr. Bob, who had come in from Akron, and William D. Silkworth, a doctor who had helped Wilson dry out in New York. John D. Rockefeller Jr. himself wrote a check—for only $5,000, but it was enough for Bill and the others to start a foundation, the Alcoholic Foundation, to solidify the new project.

Jobs were becoming scarcer: by April unemployment, the scholars would later estimate, was again hitting a full two in ten. There were others who were further along than Bill Wilson in developing new communities, communities that operated outside political life, or tried to change politics from outside. One was Dale Carnegie, whose *How to Make Friends and Influence People* had held a place on the best-seller list for so many weeks. Aimee Semple McPherson preached virtue nationally from her base at the Angelus Temple in Los Angeles.

The most idiosyncratic community leader, though, was Father Divine. He took money from his constituents and spent it as he liked; several of them had left the cult and were constantly after him in court. His properties, often shabby, ill-kept places, burned down from time to time. And the papers still ridiculed Father Divine's use of the phrase "Heaven" to describe his empire, carefully placing quotation marks around each mention.

Nonetheless, Father Divine's "Heaven" was increasingly a fact: he was continuing to acquire property. Even as he tangled with the law, he bought houses and storefronts by the dozen, many of them beyond Harlem, in the country up the Hudson River. By 1937, he had twenty-five properties operating in New York's Ulster County, mostly as farms. One, at Elting's Corners, had burned in April of that year, but that was a small setback. He was making his economic community, something like the agricultural communities that Upton Sinclair had led in California. "Father is going to make Ulster County into a model community that will be an example for the United States government," his spokesman, John Lamb, told the papers. The *New York Times* reported that some in the area, nearly entirely white, were not happy. "News that the county was going to be used as a lab for a negro collectivization experiment in camp meeting tempo was received with wrath by the Ulster County farmers and businessmen yesterday," the paper wrote. But Divine was serious: he had announced plans for canneries, farms, and eventually even automobile manufacturing. His followers numbered in the thousands, perhaps the tens of thousands, but he envisioned far more.

Father Divine had failed in 1936 to affect the presidential election. But he was still trying to provoke when he could. One of his great precepts was schooling; he insisted his followers improve themselves through training and education. In June of 1937, several had competed in a spelling bee at a PWA adult education program.

More important to him was antilynching legislation, still, shamefully, blocked in Congress. In 1937, Father Divine had watched as Senator Hugo Black, a former Klansman, won confirmation and replaced Van Devanter. And now, yet again, southern lawmakers were filibustering rather than allow the legislation to pass. Walter White of the National Association for the

Advancement of Colored People was waging the fight of his life with Congress, but Roosevelt was not backing him sufficiently to break the filibuster. The delay was infuriating, for all through the decade, blacks had been lynched in the South. When would the country be ready to stop the violence? Father Divine's movement was a peace movement, which made sense for many blacks at the time. Foreign wars kept civil rights legislation from making it to the top of American priorities. In New York, several years later, a black doctor would post a sign on his car after a black man was attacked in the South: "Is There a Difference? Japs Brutally Beat American Reporters. Germans Brutally Beat Several Jews. American Crackers Beat Roland Hayes and Negro Soldiers." Father Divine was coming to believe that the filibuster itself must be altered if it continued to block legislation to stop lynching—that it was, as he would later put it, "not an expression of freedom of speech, but it is an act of abuse." Father Divine wanted to pressure politicians on it—all politicians. But Roosevelt was the president, and therefore he would press Roosevelt hardest of all.

Meanwhile, foreign stories were intruding. A seventy-five-year-old-treaty permitted U.S. boats to patrol Chinese rivers. Yet at the end of the last year the Japanese had sunk a U.S. gunboat, the *Panay,* on the

Yangtze River. Roosevelt accepted an apology but took the event as a sign of perhaps more trouble to come. In February, the papers reported that a car had carried the desperate Austrian prime minister, Kurt von Schuschnigg, over the Bavarian border to Hitler's mountain headquarters at Berchtesgaden. Hitler forced Schuschnigg to agree to free all the prisoners in Austria, a move that could only strengthen the Nazis within Austria and open the door for Hitler's entry there. The isolationist *Time* magazine frantically tried to put a good face on it, praising Schuschnigg for "yielding much without yielding Austria's territorial integrity." But it was getting harder for this position not to sound apologist. Within a month, *Time* looked the fool: German troops marched into Austria.

Time's flat-footedness reflected a similar awkwardness in the Republican Party. Its isolationism looked worse by the day. The Republicans had a sense now that fighting for economic liberalization at home, the old-fashioned kind, might be worthwhile, that it might help the party in midterm elections that year. But to fight for liberalization at home while ignoring the illiberal spirit of new governments in Europe was inconsistent.

Roosevelt planned to spend time that summer at Hyde Park. If there was no European war before July 16, he wrote Felix Frankfurter, then he hoped to sail around in July and see Frankfurter and Marion in August on the Hudson. Now Father Divine dropped his own well-timed bombshell. The papers reported that he had bought an estate on the Hudson—directly across the river from Hyde Park. The seller was Howland Spencer, Roosevelt's old acquaintance. Spencer's property name dispute with the president had not ended: Roosevelt was still calling his property Krum Elbow, the name that Spencer believed belonged only to his bank of the Hudson. Spencer told a reporter he was especially furious because federal officials had come "and placed brass markers on my property, naming it 'Spencer Point.' " With the announcement of his purchase of the 500-acre Spencer property, Father Divine let it be known that he would retain the property name of Krum Elbow.

Spencer was an uneven, difficult man. It was not even clear he owned the estate, for several of his relatives had taken out mortgages on it. Spite mixed with politics among his motives. "The president is heading for a Russian state," he told reporters who visited him at the announcement of his sale. "Father Divine on the other hand will not accept as his follower a man who is on relief until he has paid the government what is owed." For Spencer, it was a joke, and a nasty one, in which Father Divine was also a target: "We have a 'messiah' in Washington and now we have a 'god' at Krum Elbow."

Father Divine's motives were subtler. He told his followers after the sale that he wanted to use the old mansion to create a "divine, modern, mystic standard of living." There, men of all creeds and color might work and "be free and never become public charges." What, a reporter would ask him, did he think of the spat between Roosevelt and Howland? Divine said that he was "not interested in these things." But Father Divine was interested in moving into Roosevelt's field of vision. He was working all the time to remind Congress that its tardiness on lynching was the shame of the land. Now, from his property, Roosevelt would have that reminder—he would see Father Divine at work. The newspaper cartoonists jumped on the story, one publishing a map with the "Great White Father," Roosevelt, on one side of a river, and Father Divine on the other. They relished the contest between FDR and FD.

Shortly after the purchase, in early August, some 2,500 of Father Divine's followers noisily converged on the Spencer estate, creating, as the *Times* put it, "two days of jubilation such as this Rip van Winkle country never before saw." The Roosevelts showed themselves unperturbed: Mrs. Roosevelt wrote in her column that it must "be pleasant to feel that in future this place will be 'heaven' to some people, even if it cannot be to its former owner." Roosevelt said he was confident that the people at Heaven in Ulster County "will be good neighbors."

The same day that Roosevelt gave his press conference, however, he was likely preoccupied with other news. Not only Mellon, but also Sam Insull was gone—Insull had died suddenly in the Paris *métro* in July. Roosevelt, though, was still fuming over the failure of Congress to back him on the court-packing plan. In the spring he had sent Ickes, Hopkins, Corcoran, and Cohen to help left-leaning or progressive candidates in several races in the hope that they might unseat conservative Democrats. One of his targets was "Cotton Ed" Smith of South Carolina; Senator Patrick McCarran in Nevada and Senator Millard Tydings in Maryland were two others. Among the group were opponents of anti-lynching legislation. "Roosevelt Followers Who Rate Less than 100% Under Fire," wrote the *Times,* quoting senators as describing a "purge." But the lawmakers were getting renominated.

Germany was sending threatening communiqués about Czechoslovakia, and now Britain was warning that attacking the Czechs would mean another world war. In September, Prime Minister Neville Chamberlain concluded an agreement with Hitler at Munich. But most people, including Roosevelt, were aware that Munich might not stop war. In October, German troops occupied the Sudetenland in Czechoslovakia. A single Czech unit, a lieutenant and his ten men, barricaded themselves in to fight the Germans and die. In London, the pro-German ambassador Joseph Kennedy urged at the annual Trafalgar Day dinner that democracies and dictatorships get along. "I wonder if Joe Kennedy understands the implications," wrote Frankfurter to Roosevelt.

That autumn also came the midterm elections, followed not only in the United States but through the rest of the world. The Republicans had lost seats in both houses in all four preceding elections—1930, 1932, 1934, and 1936. This time, however, they gained mightily. "Eighty in House, Eight in Senate, Eleven Governors," announced the *Times*—not the final count, but still one that told all. The count still favored the Democrats decisively, but made a comeback seem possible to the Grand Old Party. Henry Wallace told the press that "the outstanding conclusion is that people do not like the business depression." The big story crowded a single-column foreign one: "Berlin Raids Reply to Death of Envoy." In the days of the U.S. election, Hitler had raided the synagogues and shops of Germany and beaten and killed Jews, in what would later be called "Kristallnacht."

The extent of the political patronage of 1936 had finally sunk in. So had Roosevelt's method of operation.

Roosevelt might quote Thomas Jefferson, but Jefferson had deplored the creation of unnecessary government offices. With their new strength, lawmakers prepared a law that would pass in 1939. The Hatch Act would limit political activities by government employees of the sort that had been so effective in the presidential election. The nation was beginning to know Roosevelt's pattern. Writer Turner Catledge laid that pattern out in detail: "First there is the early 'idea' period, when either the President or some group of his associates hatches the rather rough form of what is to be attempted. Then there is the selling stage, in which the person or the group who thinks up the idea has to 'sell' it to the other. There follows in third place the 'method' stage when the modus operandi is evolved. Then there comes the final 'publicity' stage when the program is announced and the argument is submitted both to Congress and the public in behalf of its adoption."

The next year, 1939, was a turning point for many old Roosevelt hands. On January 4, Roosevelt phoned Frankfurter to tell him he was nominating him to the Supreme Court—despite an earlier visit by leading Jews who had warned him that Frankfurter on the Court would provoke anti-Semitism. Touched by Roosevelt's move—and by his disregard for the cowardly Jewish group—Frankfurter was speechless. Later, Frankfurter would pen Roosevelt a letter on Supreme Court stationery: that the gift of the nomination was one he "would rather have had at your hands than at those of any other President barring Lincoln."

The same month, the Associated Press carried a lengthy story on Casa Grande. Robert Faul, the manager for whom the government had such hopes, had left in a rage. "Quits FSA, Likening Project to Soviet's," read the headline. Faul hadn't gotten along with his government managers, but there were other problems. The collective setup was not overriding other disadvantages, including some it had created. In the first year, Casa Grande had lost $3,069. A water shortage had plagued the settlement; the project planners had anticipated the problem and dug a deep well, but now a nearby Indian reservation was claiming that the water was theirs, and Casa Grande had not been able to use it.

What's more, the farmers found themselves railing at being treated like shift workers and had fought back against Faul. The farmstead feel of farming, the farmer and his own land, was missing. Factions formed: Okies challenged farmers who had arrived earlier. In nearby towns like Florence and Coolidge, public opinion began turning against Casa Grande. The more the farmers in Pinal County thought about the Casa Grande concept, the more it did not make sense to them. As in the case of the Schechters, poultry was a source of tension. Poultry, along with dairy, had lost money the first year. A visiting economist asked a struggling farmer near Casa Grande what he thought of the cooperative poultry coops. "It's all right, I guess," he would say. "But the thing I can't figure out is how a man tells his own chickens apart, runnin' them all together like they do there."

The people of Coolidge and Florence "kidded the settlers about being 'reds' when they met them in the gasoline stations, over the counters and in the barber shops," wrote a social scientist who documented the period later. The settlers, many of whom were not politically oriented, felt demeaned. "You know how it feels when first one person then another asks you if you're sick and tells you you look pretty bad? After a while you begin to think you're sick as hell and maybe going to die. Well, that's what happened to Casa Grande."

Even as Casa Grande was faltering, however, Bill Wilson's community was finding its feet. Now there were meetings, both in Akron and New York; even when Bill was not present, the principles—alcoholism was a disease, alcoholics could form a voluntary community to help one another—seemed to be working. In January Wilson sent four hundred copies of *The Big Book* out to interested parties, truly wanting the movement book to be a collaborative effort. He was getting the feeling the book was powerful; at least two people had found a way to recovery after reading the unpublished draft. On June 25, the *New York Times* published a review of his book, now officially titled *Alcoholics Anonymous.* It was a rave: "Lest the title should arouse the risibles in any reader," wrote the reviewer, "let me state that the general thesis of *Alcoholics Anonymous* is more soundly based psychologically than any other treatment of the subject I have ever come upon."

But Europe again was intruding. Hitler threatened the Jewish population in a speech at the Reichstag, signaling that there would be no letup after Kristallnacht. The pollster George Gallup noticed that public opinion was shifting: "A majority of Americans are now in favor of doing exactly what the Neutrality Act forbids"—supplying Europe with arms or food. In a March 1939 poll, 76 percent of those polled responded yes to the question: "In case war breaks out, should we sell Britain and France food supplies?" In April that share became 82 percent. Gallup was also doing work on relief payments for the poor, and here, the attitude was shifting as well. Two in three Americans favored returning relief administration to the states, and taking it away from Washington. Eighty-four percent believed that politics colored the administration of relief payments.

Meanwhile, Father Divine pressed closer to Roosevelt. As the 700-acre Vanderbilt estate just to the north of the Roosevelts was among the many large properties in the United States for sale, Father Divine would propose to buy the mansion, Corinthian columns and all, and its grounds. The real estate agent, a firm called Previews, publicly put Father Divine off, saying, "On the face this rumor is absurd. Properties marketed through Previews are always offered with full consideration for neighborhood standard and welfare." A singing group like Father Divine's—nearly all black—was not welcome in Dutchess County. "The mere submission of an offer by Father Divine is no indication that it will be considered," the firm said. But it was the White House, not the realtor, whom Father Divine was addressing—and in correspondence as formal as Andrew Mellon's missive. A letter would go to Mrs. Roosevelt, a telegram to the president:

Hon. Franklin D. Roosevelt,
Hyde Park,
New York

With respect and appreciation of your many humanitarian efforts and your very democratic administration in Washington, I wire you as a matter of courtesy to ascertain your views on a matter which intimately concerns your Hyde Park home. I wrote Mrs. Roosevelt at Washington yesterday, but waited to communicate with you until I had record of your whereabouts.

My followers wish to purchase the Vanderbilt estate at Hyde Park as a private residence for me and my staff and a place where I can receive distinguished guests. As this is very near your estate, I have withheld my approval of the plan until I could consult you. I would not for a moment wish to embarrass you or your friends in the least. Would you be so kind as to let me know whether or not it would be pleasing to you for this property to be used for such a purpose. I should appreciate a frank statement immediately if convenient.

Peace.
Rev. M.J. DIVINE
(Father Divine)

It was Mrs. Roosevelt who was the first to respond, on August 12. "My dear Father Divine," she wrote. "I have talked with the President in regard to your letter and your telegram to him, and he is writing you, telling you that there can be no reason against any citizen of our country buying such property as he wishes to acquire." But the president, she said, was also writing, in part to let Father Divine know that the Vanderbilt estate had a special feature, its arboretum, remarkable in the rarity of its trees. "For some time," therefore, she noted, the president had been "trying to interest some public or quasi public body in the acquisition" of the estate. Steve Early, the president's press spokesman, sent a similar, but more detailed, letter.

By this time, however, the Roosevelts were deeply distracted. Just as blacks feared, war again was postponing civil rights action. And within days of the Divine letter came the biggest news of the year, that Stalin had signed a nonaggression agreement with Hitler. Germany had immediately attacked Poland. The old English and American Left felt shock yet again; yet again, there was a reevaluation of the 1930s. W. H. Auden would capture the disillusionment in verse:

The clever hopes expire
Of a low dishonest decade

Roger Baldwin was on the beach at Chilmark on Martha's Vineyard when he learned of the news. "I think it was the biggest shock of my life. I never was shaken up by anything as I was by that pact—by the fact those two powers had got together at the expense of the democracies." He reflected—perhaps recalling what Emma Goldman had written him upon reading *Liberty under the Soviets* in the late 1920s—"I frankly admit that people as naïve as you are hopeless. They see the world and the struggle through romantic rosy eyes as the young innocent girl sees the first man she loves." Baldwin would now determine to change the ACLU and clear its board of people who supported undemocratic regimes—or were affiliated with an entity that was not democratic, ranging from the Ku Klux Klan to the Communist Party. In an intense search of his soul, Baldwin was realizing that his institution must alter its premises to function.

Willkie was also changing. Just a week before the Molotov-Ribbentrop Pact, he and David Lilienthal had finally signed off on the transfer of the Tennessee Electric Power Company to the TVA. "Tennessee, sixth floor," the elevator boys called out to the crowd arriving for the transaction at the First National Bank of the City of New York. Lilienthal, in a pin-striped suit and checked red tie, was all seriousness. Willkie, always the good sport, put a good face on the handover check of $44,728,300 at the ceremony. In his remarks he made it clear that Americans should be wary about this deal. Stockholders might do all right—that was what

Lilienthal was emphasizing—but what did the trend mean for the utilities customer? That was less obvious. The Dow's utilities index stood in the lower 20s, lower than during *Ashwander.* "We sell these properties with regret," he told the papers. And he issued a statement—in turn provoking Lilienthal—reminding the public that the New Dealers and the TVA had forced the sale on Commonwealth and Southern. Later, he would debate Felix Frankfurter, at the Harvard Club. He headed up to Irita's in West Cornwall afterward to show off the check.

But Willkie also felt relieved about the sale, because it gave him a chance to move on to broader projects. He thought about the articles he was now writing, which ranged far beyond the power issue. Over the weekend he and Edith headed to Saugatuck Harbor to visit Russell Davenport, an editor at *Fortune*, and his wife. Smoking furiously, Willkie talked about everything under the sun with Davenport. Davenport was also concerned with the future of liberalism—in fact, at Yale he had organized a Liberal Club. Though they had come together over the utilities question, now much of the talk was about Europe. Maybe it was time to start moving the Republican Party away from the isolationists. The pair also discussed the possibility of Willkie, a Democratic businessman, running for the presidency.

It seemed unlikely—unless, conceivably, the convention delegates were divided.

Whatever was coming next, Willkie was confident about it. Like Bill W., he was groping for a new format, a way to rally countrymen so that they could find courage. In June, he had published an article in the *Atlantic Monthly.* "Brace Up, America," Willkie exhorted. Maybe he could build a campaign around that. If Americans could revive their old sense of economic liberty, not much could stop them. Joblessness was drifting downward, back toward the levels of the election of 1936. American business was waiting for an excuse to recover; even the bitter peace that Willkie and Lilienthal had concluded seemed to provide such a one. That autumn, Willkie finally registered as a Republican, telling Edith, "Well, I've done it."

Father Divine also would not be put off. His followers would proceed, he wrote to Roosevelt, in the purchase of the Vanderbilt estate. And he would advise them "to use the property as described by you," allowing the ground floor as a public museum. Nonetheless, Mrs. James Laurens van Alen, the seller, blocked the transaction, announcing she had no intention of selling to Father Divine. The next news in the papers of the story would come in February 1940, when the president disclosed that the federal government had plans to acquire the Vanderbilt property, allowing the public to enjoy both the trees and the architecture of the mansion. Mrs. van Alen would give the property to the government.

Father Divine would not stop his real estate dreaming—he bought a fifty-room mansion in New York, and inquired about a property in Newport. And Bill Wilson persisted. Still, Father Divine's peace movement looked increasingly out of place; his followers would in coming years be arrested for failing to report to the draft board. Bill Wilson too found his attention altered by the war. He had served as a soldier at the end of World War I. Within two years, he would be trying to reenlist—though he was too old for combat, he would go to the trouble of collecting a recommendation from a colonel to serve at the army's quartermaster depot in Philadelphia.

Though these events were in the future, the change was already clear. In March 1940 the columnist Arthur Krock would pen an essay about the eighth anniversary of the New Deal. But the article was not about the alphabet agencies, or the rages of Jimmy Warburg, or the Supreme Court: "Foreign Problem Uppermost," it read.

15
Willkie's Wager

January 1940
Unemployment (year): 14.6 percent
Dow Jones Industrial Average: 151

One winter night in early 1940 a twenty-eight-year-old named Oren Root went to hear Herbert Hoover speak to the Young Republican Club of New York. Root was the grand-nephew of Elihu Root, who had served as William McKinley's secretary of war and Theodore Roosevelt's secretary of war and state. He lived with his parents and worked on Wall Street as a junior lawyer. Like his ancestor, Root was preoccupied with international events. Germany had invaded Poland a few months before, and now reports of civilian murder, torture, and flogging in the former republic were coming almost routinely. The issue of the hour was what the United States could do to stop a European war.

Hoover's facial features were now assembled in a permanent configuration of chagrin, but he had not given up on the game. Lately he'd busied himself collecting statements from Republican delegates who might be friendly toward a Hoover candidacy at the GOP's convention a few months hence.

Yet, Root discovered, Hoover was saying little that Republicans had not said before. Even worse—for, especially in an election year, presentation mattered a lot—the ex-president would take questions only in writing. After introductory remarks, Root recalled, Hoover "ran through" the questions written on the papers "as one would shuffle a deck of cards." Eventually Hoover came to a query about the policy of the United States in the event that German arms jeopardized the future of France and Britain. Since Germany had already invaded Poland and Czechoslovakia, and France was still thinking in the context of its Maginot Line, the question seemed reasonable. Hoover dismissed it, Root later recalled, "with the comment that it was too impossible an event to warrant comment."

The weakness of the performance shocked Root. After all, 1940 was the year when, finally, Republicans had a real chance at the presidency. They'd made those gains in 1938. The concept of a third term for one individual in the presidency seemed improbable, even if the figure was Franklin Roosevelt. To find Hoover once again hogging the Republican stage was unacceptable. Root thought about the other Republican possibilities—Thomas Dewey of his own New York, Robert Taft of Ohio, and Arthur Vandenberg of Michigan in the Senate. He wondered, he would later write, "whether they offered the answer to our problems." Dewey was a New York prosecutor—indeed, some of his more publicized cases had involved the same business as the Schechters', the live poultry market. He was zealous and personally brave: he had faced down mobster Dutch Schultz, who at one point had put out a contract on Dewey's life. He was a New Yorker, and New York was electoral king. But Dewey was hard to like. There was something simultaneously cold and juvenile about him—later, when he ran for president, Ickes would joke that he "threw his diaper into the ring." And, like many litigators, he was weak on policy itself. He took too few clear positions.

Root believed several things. The first was that the struggle for Europe was related to the struggle to get beyond the New Deal at home. The second was that though the country was not ready in January or February to talk about war, the presidency might end up going to the man who understood that the United States must involve itself in Europe and that foreign policy had to do with growth at home. Indeed, Root was willing to bet on it. And right now one person on his horizon fit that description: Willkie. Willkie was an old Wilsonian. Willkie understood that democracy was at stake. Root was braver than the leaders of his petrified party. He decided it did not matter that Willkie had become a Republican only the year before. Willkie was, at least, "positive and constructive." Root decided to float Willkie as a candidate. What did he have to lose?

Several months after the Hoover meeting, Willkie provided Root with a format for doing so. With the aid of Russell Davenport, he published his own political manifesto in *Fortune.* The title was "We the People." The manifesto spoke to Roosevelt directly. "In the decade beginning 1930 you have told us that our day is finished, that we can grow no more, and that the future cannot be equal to the past. But we, the people, do not believe this, and we say to you: give up this vested interest that you have in depression, open your eyes to the future and help us to build a New World." Root, feeling the adrenaline rise, wrote his own pro-Willkie petition, basing it on *Fortune* language: "Because Wendell Willkie does not believe in this philosophy of defeat we welcome him."

The petition that Root created said that Willkie would "be the defender of our power"—the power of the country as a whole—"and not of the power of any institution or favored group." There were fifteen places for signatures, and instructions that completed pages be returned to Root's residence at 455 East Fifty-seventh Street, New York City. Root mailed off his petitions to two groups in his world—the alumni of Yale's class of 1925 and Princeton's class of 1924. The reaction, he later recalled, was "immediate and overwhelming"; those who did not receive copies of the petition printed out more. The phone switchboard at Root's law office was also "swamped, to the exclusion of the firm's proper business." The partners at Root's firm were not pleased.

Willkie protested showily—he had Thomas Lamont, his friend from the New York Economic Club and a partner at J. P. Morgan, ring up the young Root to scare him off. But when the scare tactic failed to intimidate Root, Willkie went along. Other fans printed up tens of thousands of Willkie buttons. Through the energies of Root and others, Willkie Clubs were starting across the nation. Root even traveled to Oscaloosa, Iowa, on the train to help a Willkie Club get started.

Willkie liked Root's wager. He made it his own. He launched his campaign from Irita's West Side apartment—though the papers did not mention the venue.

Irita saw what the chance meant for Wendell. Edith also went along, graciously appearing as the spouse in public. The Katharine Hepburn film *State of the Union* later fictionalized such a threesome.

At the start, there were mainly scoffers. Root and Willkie might have been thinking about Europe, but many Americans still wanted to tell themselves that staying out might keep the European conflict smaller. The American Left was in a state of shock after the Nazi-Soviet pact. Willkie was seen as a setup, a puppet of a party in disarray. Felix Frankfurter called Willkie "Wonder Boy," the same phrase that Coolidge had contemptuously used for Hoover. The best putdown came from Alice Roosevelt Longworth, the daughter of TR and an establishment Republican. People said Willkie had come up from the grass roots, but she quipped that those grass roots were "the grass roots of ten thousand country clubs."

This argument, however, weakened when it became clear that, despite his Wall Street allies, Willkie was garnering at least a following across the nation. Though the economy had at times recovered, it was still, international observers noted, nowhere near as strong relative to other nations as it had been. The United States was not the power it had been. The reputation of the New Deal was continuing to drop.

In part this was because people were taking in the longer-term consequences of all the experiments. At Casa Grande, the settlers were still having trouble putting down roots. They had come to the farm to be homesteaders, and now they were more like tenant labor. That year, 1940, the farm would have a new regional director, Laurence Hewes. From James Waldron, the farm supervisor, Hewes heard what was coming to be a familiar story: "There is a definite split in the membership and very strongly opinionated, in fact almost bitter, groups have developed." The division: "one group wishes to get everything possible from the government in the way of wages, benefits subsidies, etc, and also to control association on a political basis. The other group, in our opinion, has a more fundamental outlook. They look to the future of the organization."

Hewes that year decided to find a "high-type educator"—what Hewes thought of as a $6,000-a-year man—to fix the social problems at Casa Grande. But a senator from Arizona who had an interest in the project insisted that the person to fill the job had to reside in Arizona. Hewes gave up the plan, certain he couldn't find such a person in this state. Morale worsened. Later in the year, Hewes on a visit, would discover the ultimate expression of the settlers' opinion of Tugwell's project. They had trashed the community house.

But such domestic minutiae were hard to concentrate on—even for Tugwell. All spring, Hitler seemed on the edge of invading the countries of Western Europe. The *New York Times* that winter was reporting the possibility of a record famine for the occupied areas. The paper noted that the Belgian Relief Unit foresaw "the worst suffering in the history of the Western World." Suddenly the war was becoming an issue, just as Root and Willkie had thought it might.

And Root's campaign had generated 200,000 signatures in advance of the June convention. Willkie's New York socializing had paid off. Henry Luce put his press empire behind Willkie, and *Life* magazine fronted him, printing a picture of Willkie's already large head that was bigger than life. Hubert Kay wrote in the magazine, "In the opinion of most of the nation's political pundits Wendell Lewis Willkie is by far the ablest man the Republicans could nominate for president at Philadelphia next month." Even Roosevelt was impressed, understanding that Willkie, unlike Dewey, Landon, and certainly Hoover, matched him when it came to charm. Over and again, people found that meeting with the utilities executive changed the course of their lives. Willkie would make such an impression on the poetess Muriel Rukeyser that she would, decades later, publish a 330-page epic poem about him. Especially inspiring to many Americans was Willkie's good humor. "With malice toward none"—the theme of a 1939 article—was a great change from the sour rage of the Liberty League. It was also a change from Roosevelt, whose lists of names were hard to forget.

As the Nazis rolled forward in Europe, Willkie gave a speech at home in Indianapolis. He charged that Roosevelt practiced a "technique of defeatism" and was militarily unprepared for war. The country was more sophisticated than it had been in the past—certainly more sophisticated than in the days of World War I, when Germans were still called Huns. Instead of counting against him, Willkie's German background was actually an advantage. As refugees from Prussian Germany, Willkie's family knew all too well about European tyranny. Willkie himself had served in World War I. When Willkie argued from Indianapolis that the Europeans' cause was the American cause, the Germans, the Poles, the Czechs of the Midwest all understood.

Roosevelt shot back in a Fireside Chat in late May. The unpreparedness argument was wrong—the United States had spent a billion on the navy under Roosevelt. The U.S. Navy was "far stronger today than at any peace-time period in the whole long history of the nation." The United States must pay attention and help "the destitute civilian millions"—if only through the

American Red Cross. Europe was not "none of our business." The United States could not retire "within our continental boundaries"—a defense policy that invited future attack. Still, these revisions did not stop Willkie's momentum. Dorothy Thompson, the journalist, was in Paris before the Nazis marched in that June. She sent home a dispatch arguing for a joint nomination by Democrats and Republicans of a Roosevelt-Willkie team. (Roosevelt would have none of it, writing to a friend who knew Thompson: "Do try to get that silly business of Wendell Willkie out of her head.")

Politics is not exclusively about absolute numbers; it is also about relative change. Because Willkie was such a dramatic dark horse, his new popularity raised enormous hopes for him at the Republican convention in Philadelphia. To be sure, Dewey was still a favorite. And Hoover was there, spoiling the party again, railing against U.S. involvement in the war: "Every whale that spouts is not a submarine," he intoned. "The 3,000 miles of ocean" was "still protection," a buffer between the United States and contentious Europe. Dewey spoke of the New Deal's "temperamental inability to follow a straight road toward a national goal." Robert Taft was a possibility, the name Taft being the one many Republicans believed most likely to beat the name Roosevelt. The delegates split over Dewey and

Taft, just as Davenport and Willkie had predicted on Long Island the summer before.

Still, what the delegates talked about was that by now Willkie-for-president clubs across the nation had swelled to almost five hundred in number. The novelty factor that had benefited the Roosevelt team for so long now served the Republicans. In 1940, a year when delegates could still change their allegiance at the convention, they did. Dewey won the first ballot, but only by a plurality. A majority was necessary. He led the second, and the third. Willkie led the fourth, but with insufficient votes. From the galleries and the floor, delegates and guests shouted "We Want Willkie!" The final ballot, and the only one with enough votes for a successful nomination, went to Willkie. The very exhilaration of the Willkie nomination—and the fact that he had beaten such long odds—now made the man seem invincible. From the convention on, every day, it seemed clearer that Willkie would fare better than Landon had.

Other observers at the time and especially later emphasized the differences between Willkie and Landon, or Willkie and other Republicans. Willkie's candidacy now was about the war; though he, like Roosevelt, shifted position from time to time, Willkie was emphatically not an isolationist. Unlike the Liberty League types, Willkie was able to show that free market ideas were innately American common sense. As a vice presidential candidate he had accepted the nation's highest-ranking Republican, Senate minority leader Charles McNary.

The Willkie-McNary campaign produced short film clips for movie theaters. One, featuring Willkie and McNary amid the corn in a field, sought to demonstrate his understanding of the farmer (and presumably the farmer subsidy). Another publicized the economic costs of higher taxes and the importance of freedom (this one featured the ringing of the cracked Liberty Bell). The theme in many of his speeches was the protection of freedom and growth in the United States. With more than one in ten Americans still out of work, the argument was compelling.

Besides, Willkie did not merely criticize the New Deal. He tried to show where it served well, and where it diverged from that common sense. In a speech before the American Newspaper Publishers Association, Willkie argued it was important that governments "leave men free." He also argued that the United States had to think about freedom in Europe, that the United States had "a vital interest in the continuance of the English and French way of life." At the end of the Republican convention, the situation on the domestic front seemed clear. Willkie was for limiting the New Deal; FDR was for expanding it.

At his own party convention in Chicago, Roosevelt tried to shift the terms of the debate. He said that he stood for the businessman but also stated, via the party platform: "We have attacked and will continue to attack unbridled concentration of economic power and the exploitation of the consumer and the investor."

It was a difficult moment for him and his old brain trusters. Earlier that year the *Town Hall Meeting of the Air*—the same in which Willkie appeared—had aired a show about Steinbeck's theme, the rural migrant. The title was "What Should America Do for the Joads?" Tugwell had been one of the debaters, but when it came to present solutions, he fell silent. The next day the *New York Times* commented on his vagueness: "Mr. Tugwell was the only one of the speakers who did not have a concrete suggestion for alleviating measures." Chase for his part was also outside, consulting at the Temporary National Economic Committee, an office created by Roosevelt during the 1938 downturn to study monopolies. (Chase was now advocating the establishment of a permanent PWA.)

But now people knew the president. After reelection, he might turn back to his old planning friends, and he might do something else—the unpredictability was the only thing you could be sure of. Again, Roosevelt had a response to this: to anchor extant constituencies. The president at the last minute traded John Nance Garner, his conservative vice president, for farming's powerful friend, Henry Wallace of the Agriculture Department. The move showed he understood the threat of Willkie's wager. In July he told fellow Democratic strategist James Farley, "You know, if the war should be over before the election and I am running against Willkie, he would be elected." The country seemed wild to know everything about Willkie—right down to the fact that his given name was really Lewis Wendell Willkie, and not Wendell Lewis. On August 3, George Gallup, the pollster, reported that Willkie would have the edge over Roosevelt if the election were held that day. Willkie had already decided to go back home to deliver his acceptance speech in Elwood.

He took his time writing it. It was the most final, and strongest, rebuttal to the progressives that had yet been offered. Before a crowd estimated at 200,000, and with the weather 102 degrees in the shade, Willkie asked the public to think about what it meant to be an American liberal. Was a liberal merely a left progressive? Or was a liberal someone who believed in liberalism in the classic sense, in the primacy of the individual and his freedom? Willkie railed against Roosevelt's "philosophy of distributed scarcity." And he argued, speaking of both the United States and Europe, that it was "from weakness that people reach for dictators and concentrated government power . . .

"American liberalism does not consist merely in reforming things. It consists also in making things. The ability to grow, the ability to make things." Redistribution was a loser's game: "I am a liberal because I believe that in our industrial age there is no limit to the productive capacity of any man." Growth, not government action, would lift the United States out of its troubles: "I say that we must substitute for the philosophy of distributed scarcity the philosophy of unlimited productivity. I stand for the restoration of full production and reemployment by private enterprise in America."

Listeners yelled their approval. Anne O'Hare McCormick, the columnist who had accompanied the labor delegation on its visit with Stalin in Moscow and had profiled Roosevelt at Hyde Park after his nomination eight years before, now produced a giant feature that the *New York Times* titled "Man of the Middle West." A new group, Democrats for Willkie, hailed him as "this leader of true liberalism." And in a way he was that leader—the liberal of individual freedoms rather than the leader of group rights. From Yale, Irving Fisher, ebullient as ever, now wrote to offer his services.

In an essay arguing against a third term for Roosevelt, Fisher recalled that Theodore Roosevelt had promised not to run in 1908, and had kept that promise. When it came to his old friend FDR, Fisher argued that the election of 1940 was dominated by "two sinister facts. One is that he has built up a political machine. The other is that he has put millions of voters under obligation to him." Fisher had managed to get a meeting with Willkie in July and wrote in his diary, "A red letter day, not because it's the Fourth, but because I saw Willkie." Willkie, unlike Roosevelt, was "pressed for time," Fisher noted—"His desk was a terrible mess." Still, Fisher felt he had connected with him.

Another supporter of Willkie's turned out to be Judge Joseph L. Dailey of New Mexico. Dailey had been the head of the western division of Tugwell's Rural Rehabilitation Service. Now he was lunching with Willkie. Gallup's math put Willkie ahead of Landon's record in Maine, a fact that some observers took to mean that Maine would, through Willkie, win back its old role as a signal state. Willkie's campaigners were ordering up buttons that read "Learn to Say: President Willkie."

Elwood proved to be the high point of the Willkie campaign. This was partly because, like most independents, Willkie was less equipped for a general election than a campaign primary. The marriage between him and the Republican Party as it existed in 1940 was an awkward one, and it made Willkie a weaker candidate. Willkie, after all, was a hawk when it came to Europe, and liked free trade; the party platform took the opposite positions. Willkie liked the private sector, and McNary liked it less so. Observers began to underscore Willkie and McNary's differences: McNary was from Oregon, a big government power state—indeed, a large dam would be named after him only a few years later. McNary had fought against steps to build the arsenal of the Allies. But policy was not the end of it, for there were also differences even in speaking style. Willkie talked everywhere, whereas McNary had a record as the silent senator in his quarter century in the Senate. Though Willkie justified his differences with McNary as part of politics—only a broad coalition could win in the United States—the disparity between them was so great as to make the ticket illogical.

The Republican film clip depicting Willkie and McNary as men of the soil—"Willkie and McNary Know Their Farming"—was a bit ridiculous, when people thought about it. Willkie had his farms in Rushville, Indiana, but he was a self-confessed "conversational" farmer—he managed his properties, mostly, by telephone. As for McNary, he was more a man of the farming business than a farmer; he had, for example, worked on increasing the commercial prospects for the Imperial prune. Two weeks after Willkie's Elwood speech, Henry Wallace gave his own acceptance speech in Des Moines, Iowa. The new vice presidential candidate was showing the country, in effect, that the Democrats had a real farmer on offer as well.

Willkie compounded his problems by softening his positions in other areas. Suddenly he was talking about supporting organized labor, a position that seemed at odds with his arguments against "vested interests" earlier in the year. Very late in the campaign, in a moment of pure political angling, John L. Lewis endorsed Willkie over the radio, railing against "Caesar"—Roosevelt. Herbert Hoover wired his congratulations to Lewis the same night.

Lewis's move seemed brilliant tactically but it made no sense when it came to policy, and therefore helped neither Willkie nor Lewis. Nor could the fact that Willkie used the opportunity of a speech before a labor audience in Mellon's Pittsburgh to strike out at Frances Perkins. Promising to name a labor secretary from among organized labor itself, Willkie added, "And it will not be a woman, either." Perkins later reported that Roosevelt consoled her, saying: "That was a boner Willkie pulled." Why, Roosevelt asked, reasonably enough, "did he have to insult every woman in the United States?"

There was a logic to Willkie's inconsistency that went beyond the blind desire to win. It was the logic of his, and Root's, 1940 wager. The German liberal or Social Democrat in him—and the American civil rights advocate—continued to watch Europe. That war needed stopping. With the determined ambition of a candidate, Willkie decided that he would subordinate domestic concerns, shift his positions, all in the name of winning control of U.S. foreign policy.

Here Roosevelt still worked from greater advantage. To start with, he had his interest groups lined up. The new Hatch Act notwithstanding, he was still spending on jobs across the country. Forty-two million American workers were now enrolled in Social Security—more than it took to win an election by far—and they looked forward to getting their Social Security payments. Farmers still believed—cheap Willkie-McNary advertisements notwithstanding—that Roosevelt would protect them. Roosevelt's choice of Henry Wallace as running mate was paying off. Nor did many in labor forget that it was Roosevelt whose law had given them the closed shop. "They will vote to continue the New Deal," said Jacob S. Potofsky of the Amalgamated Clothing Workers at the end of October. "Labor will not scrap its newly won rights because of one man's personal grudge." Roosevelt had created the modern farmer-labor coalition, and now it was there for him.

But the real reason Roosevelt started to gain was the coming war. For one thing, it promised yet more spending. The Lend-Lease law would be passed only after the election, but both events and Willkie were already forcing Roosevelt in the campaign period to make clear that he would spend to defend the United States and to help its allies. The downturn was ending. Gross national product was finally approaching the level of 1929, though a comparison was misleading, for now the population was millions greater. The cotton crop for the year looked to be good again. And business activity picked up tremendously in preparation for that spending. Even in World War I, government spending had had a tremendous influence. Business knew that if government was already bigger by so much than it had been in the 1910s, then a coming war would only increase its scale more.

There was another, less discussed factor. Roosevelt truly was doing what Willkie had asked back at the *Town Hall Meeting of the Air* debate in January 1938. He was toning down his rhetoric against business, again, and asking for a truce. In the war, Roosevelt needed a picture of a self-reliant United States, not a weak one. If that meant changing the New Deal, well, of course he would change it. This switch was already evident within the administration. About a year before, for example, Roy Stryker and his photo office had been visited by a German diplomat who wanted to send photos of America's weakness home to Hitler. Stryker felt a sudden resolve: "He was a very pleasant little Nazi. I had no intention of allowing the record of America's internal problems to fall into his hands. I had the file clerks show him a wonderful range of things—mountains and rivers and lush fields."

Now, in 1940, Stryker sent one photographer out at Halloween with the assignment of documenting the opposite of what the team had portrayed in the preceding decade: "Emphasize the idea of abundance—the 'horn of plenty'—and pour maple syrup over it you know, mix well with white clouds, and put on a sky-blue platter." The domestic political goal of highlighting trouble now was subordinate to international politics, and everyone, including Stryker, knew it. In 1942, Stryker would be even more direct in his orders to his photographers. They could still photograph the needy but should also take "pictures of men, women, and children who appear as if they really believed in the U.S. Get people with a little spirit. Too many in our file now paint the U.S. as an old person's home and that just about everyone is too old to work and too malnourished to care much what happens."

The old figures whom businesspeople had feared were now ignoring them, or asking them for help. Ickes was focusing on foreign matters; he would shortly begin to manage energy for the president in the war period, teaming up with oil companies rather than attacking them. Hopkins too was hard at work on foreign policy. Reports came out that Hopkins even lived at the White House—he had been there since the spring. Frankfurter's stay had been about the courts and legislation; Hopkins's visit focused on the crisis of defense. *Time* readers learned that Hopkins had been aboard when the president had sailed down the Potomac with the navy secretary, touring military facilities and talking about conscription.

Roosevelt knew that he needed more than an economy that looked good. He needed an economy that actually was strong. A war on business and a war against Europe could not happen at the same time. In World War II, as in any war, bigger businesses tended to do well, for they were the ones who became government partners. The smaller ones sometimes suffered—and sometimes didn't. The nimbler among them found a way to survive or even thrive while serving the war cause. Alfred Loomis, for example, was a great anglophile; he believed that the destruction of Britain was the destruction of civilization. Now he could put to work all the research he had been doing on radar, in the service of beating the Germans. From an antagonist of Roosevelt's, he turned into a servant.

Another example of this new dynamic in operation—albeit on a very small scale—showed up at Casa Grande. The farmers continued to squabble at the collective farm. It was clear that they might not stay together in the long run. But some of the edge was off. For the fighting was no longer about losses—it was about gains. Nineteen forty was turning out to be a good year. Water flowed from the Coolidge Reservoir, and prices for crops were rising. The management's hypothesis that livestock was a good idea was proving true. Though the farmers could not know it yet, 1941 would be even better, showing a profit of $15,791, nearly treble the expected rate, for the cattle. There would be profits in cotton and poultry. After their long wait, the settlers would eventually get a raise, to $65 a month. The concept of settlement still felt wrong, but the edge was off.

Washington too was giving up its fantasy. Within a few years, by 1943, lawmakers would make their impatience with Tugwell's experiment explicit. They would bar the Farm Security Administration from using any appropriated funds for resettlement projects—unless that cash was used to speed liquidation of such farms. Myer Cohen, assistant regional director at the FSA, moved on. He would take up work as administrator of the United Nations Relief and Rehabilitation Administration, helping war casualties in Europe. Though the farm would, in the good harvest year of 1942, offer $4 a day to workers who used to get less than $2, many of those migrants would not take work even at that price; as *Time* magazine would report, they could get $1.12 an hour building internment camps for Japanese Americans. Casa Grande was becoming history.

Over the fall, the Republicans began to realize their error. Willkie had wagered correctly indeed—it was true that the candidate who seemed to know best about the war would win. But Roosevelt had found the biggest flaw in his wager: it was easy for Roosevelt to supplant Willkie as that candidate. Roosevelt had more credentials. Willkie may have served as an officer, but Roosevelt had personally managed the navy as assistant secretary for seven years. He had led the country through the domestic war of the Depression. What the Depression had been to the Roosevelt candidacy in 1932, the war was to the Roosevelt candidacy of 1940: the single best argument to reelect Roosevelt and give him special powers. Even *Time* of the Luce empire, the very empire that had advanced Willkie, understood. That year *Time*'s editors had written, "Whether Mr. Roosevelt is Moses or Lucifer, he is a leader."

All these facts Roosevelt, a more experienced campaigner than Willkie, understood. But there was one additional, and very powerful, bonus for Roosevelt. Willkie was basing his campaign on the ten million unemployed whom he would cite all year as evidence of Roosevelt's failures. Though unemployment was heading down now, it was still over one in ten. A war, however, would hand to Roosevelt the thing he had always lacked—a chance, quite literally, to provide jobs to the remaining unemployed. On the junket down the Potomac, for example, he could count 6,000 men at work at Langley Field; 12,000 men at Portsmouth Navy Yard, where there had been 7,600; and new employment in the military or the prospects of it, for Americans elsewhere. Roosevelt hadn't known what to do with the extra people in 1938, but now he did: he could make them soldiers.

GOP leaders fought back. But as leaders and oppositions since have discovered, war trumps everything—economics as well as politics. The TVA, for example, could not be the target it had found itself in the 1930s, for now it was generating power for the war effort. By July 9, a subcommittee in the House of Representatives had already approved an extra $25 million infusion to the TVA budget so that the TVA might partner with—of all concerns—Mellon's own old firm, Aluminum Company of America. Past enmities were to be forgotten in the name of the production of aluminum sheeting, vital for such things as airplanes. By the end of the month Roosevelt was signing $68 million in cash for the TVA, "essential to the national defense." This was well over the original appropriation for the whole TVA project and double the amount in Roosevelt's original plan for construction of the Norris Dam at Cove Creek. Lilienthal held a cheery press conference in Washington—as it happened, two doors down from the Washington for Willkie headquarters—affirming forcefully that "no TVA director" or employee would participate in the election. But then, none would now need to.

All this made matters harder for the necessarily partisan Willkie. General Hugh Johnson, Roosevelt's old NRA head, was especially unhappy at the thought of the United States heading to Europe, an event that would certainly lead Congress to giving Roosevelt new powers. Johnson later told a radio audience that Roosevelt would give up the emergency powers "as willingly as a hungry tiger gives up red meat." As election night drew near, the nation was tenser than it had been in many preceding elections. The *New York Times* announced plans to flash results from its tower in Times Square—when the beam swept north, it meant Roosevelt was leading. A beam sweeping south would signal that the lead was Willkie's. A steady beam to the south made Willkie the winner. A steady beam to the north was a Roosevelt victory.

At the end, the beam shone steadily north. The Republicans were bitter, for they concluded, accurately enough, that the outcome would sideline not only their party but their record of accuracy when it came to the economy. They had been right so often in the 1930s and they would not get credit for it. The great error of their isolationism was what stood out. And their bitterness made them look small.

But Wilkie had polled 22 million votes, more than any Republican in history, even more than Hoover in 1928. Even discounting for the population increase, it was an impressive showing. Willkie recognized that the political news was that he had come as far as he had. "I accept the result of the election with complete good will," he told the press at the Hotel Commodore's Parlor A. The New Deal had clearly changed the country forever. Now the government would always be there on the national stage. But the election of 1940 showed that the less-governed America of Coolidge and Mellon—or Father Divine, Joseph Schechter, and Bill W.—was still strong.

Back in Elwood, Willkie had reminded the country of the original forgotten man, so obscured in recent years. That forgotten man—Summer's man, the individual so beloved of the old liberals—was important too, and he had no political party. Then Willkie had posed a rheotrical question. Was the government with its support now the central thing about the country, the thing that "the forgotten man wanted us to remember"? It wasn't. A government might help, when necessary, but a government was secondary, "not enough."

"What that man wanted us to remember," Willkie said, "was his chance—his right—to take part in our great American adventure." The country now seemed to remember again what it always knew: that the adventurer was the force who pushed the country forward. It was the adventurer's America too that the soldiers would shortly be defending. And no one wanted to serve more than the Forgotten Man.

Coda

Roger Baldwin, the ACLU cofounder, worked hard for the peace movement in the 1930s. His change of heart at the news of the Soviet-Nazi Pact changed the course of both the ACLU and American history. Baldwin now brought strong anti-Communists onto his board. In 1940 the ACLU expelled a board member, Elizabeth Gurley Flynn, who was a Communist; Baldwin concluded that "an organization devoted to civil liberties should be directed only by consistent supporters of civil liberty." At the end of 1959, Baldwin told scholar Lewis Feuer, "We went wrong, we were starry-eyed. We didn't see the potentiality of totalitarianism." Some have called Baldwin's anti-Communist shift early McCarthyism, but it gave the ACLU a legitimacy that would enable it to play an important role in civil rights battles after World War II.

Stuart Chase went on to write in a number of other areas. His best-known book was titled *The Tyranny of Words,* and his work on semantics produced admiration from people of all political backgrounds. One observer wrote admiringly, "Mr. Chase's logic wobbles, but his sentences march." From the 1950s, he served on the planning commission of his town, Redding, Connecticut. In the 1960s he strongly supported President Lyndon Johnson's Great Society programs. In 1961 Chase traveled to the Soviet Union with the singer Marian Anderson, the publisher of the *Encyclopaedia Britannica,* William Benton, the anthropologist Margaret Mead, and other intellectual and cultural figures to meet with Soviet citizens in the hopes of improving understanding between Moscow and Washington. Chase died in 1985, at age ninety-seven.

Despite his age—he was turning nearly fifty—**Paul Douglas** enlisted in the Marine Corps as a private during World War II. After the war, he was elected U.S. senator and became one of the first on the Hill to insist on racial integration of his staff. Douglas championed civil rights, and led the successful drive to protect the Indiana Dunes for recreation and the environment. He would later write that he had doubts about the U.S. recognition of the Soviet Union:

As one who must bear perhaps an infinitesimal share of responsibility for this decision, I have often wondered whether it was wise. Certainly recognition helped pave the way for Russia's combining with Great Britain and the United States to defeat monolithic Nazism. But Hitler's invasion of Russia in 1941 might have brought about the same result without any recognition.

But whether or not this decision was wise, I no longer believe in [Henry] Clay's doctrine of always recognizing existent government . . . My disillusioning experience with Russian recognition was one of the factors that led me to oppose the recognition of Communist China and its admission into the United Nations.

(**John Brophy** also had doubts, writing in his autobiography of his time in Russia, "I had to reassess some impressions in the light of later events.")

Father Divine faded after World War II. In 1959 Krum Elbow was sold to a real estate developer.

Samuel Insull never made it back to wealth despite his acquittal. He honored his debts as he could, selling or signing over to banks nearly all his assets, right down to the swans from the lake at his country house and a Wedgwood dinner service of 205 pieces. He died of a heart attack in the Place de la Concorde stop of the Paris *métro* in the summer of 1938, less than a year after Mellon.

The War Relocation Authority hired **Dorothea Lange** to photograph the internment of Japanese Americans during World War II. All told, Lange and WRA colleagues produced some 13,000 photographs of the camps created under Roosevelt's Executive Order 9066. She and her husband, Paul Taylor, were horrified at what they saw and drafted letters from Roosevelt that sought to explain to the internees the justification for the action. Lange's photos of the Japanese Americans were impounded and rarely seen, until recently.

David Lilienthal became head of the TVA and was eventually nominated to head the Atomic Energy Commission. The TVA sold the town of Norris to a private investor for $2,107,500 in June 1948. Angry senators tried to establish that he had harbored Communists at the TVA during his confirmation hearings. Lilienthal offered a rebuttal that silenced them and helped bring about his confirmation. He published his own book about his concept of American civics, *This I Do Believe.* Eventually, Lilienthal went into the private sector, joining Lazard Freres on Wall Street and consulting for foreign governments. He and his wife, Helen, retired to Princeton.

Paul Mellon would become as great a collector and donor as his father. At Yale, he created the Yale Center for British Art. He attended the dedication of the National Gallery in 1941. At the dedication, President Roosevelt said that the "giver of this building has matched the richness of its gift with the modesty of his spirit, stipulating that the gallery shall be known not by his name but by the nation's." The president noted that Abraham Lincoln, even in the middle of the Civil War and a national financial crisis, had pushed for an opulent Capitol dome. "Certain critics, for there were critics," noted Roosevelt, "found much to criticize. There were new marble pillars. . . . But the president answered, 'If the people see the Capitol is going on, it is a sign that we intend this Union shall go on.' "

Mellon wrote later that the president's remarks seemed to him "a little wry, considering the behavior of his tax bloodhounds," but "perhaps he had come to realize that Father really had been a public-spirited man." Andrew Mellon's gift inspired others to join him, including the Kress family and Joseph Widener. During the war, the gallery remained open on Sunday evenings for "benefit of men in the armed forces and war workers in the city."

Raymond Moley worked in the public sector, seeking to help such groups as the Tax Foundation and the American Enterprise Institute. Moley published numerous books, including an enlightening portrait of politicians he had known, *Twenty-seven Masters of Politics*. He advised Herbert Hoover and supported the candidacy of several Republicans for president—Willkie, Barry Goldwater, and Richard Nixon. Moley died in 1975.

Walter Lyman Rice became an attorney for Reynolds Metals Co., a company in the industry he had prosecuted during the New Deal. He also served as ambassador to Australia.

The **Schechters** went back into business after their Supreme Court victory, according to their descendant Glen Asner. In a note to the author, Asner wrote: "Their major political concern in the 1930s was anti-Semitism. They believed that if Roosevelt had not solved the problems of the Depression, the U.S. could have gone the way of Nazi Germany. Their overarching political concern was the condition of the Jews. That said, the main topics of discussion around the house were horses, stocks, and business." Asner thinks it likely that the Schechters voted for Roosevelt all four times.

John Steinbeck published his novel about Okies and the migrant camps in 1939. The *New York Times* commented that *The Grapes of Wrath* might read to some like "a disquisition by Stuart Chase." The book became an American classic.

Roy Stryker and the FSA offices produced 130,000 photographs. Upon leaving government, Stryker supervised a massive photo documentation project for Standard Oil of New Jersey.

William Graham Sumner died in 1910.

After **Rex Tugwell** left government, many of his projects were completed, including Casa Grande, with mixed results. The Casa Grande settlement was sold off in the early 1940s. Later a student of Tugwell's, Edward C. Banfield, published a monograph about

Casa Grande's failure, *Government Project.* Banfield went on to be a leading scholar in the urban studies field. Tugwell contributed a foreword to Banfield's book, criticizing the settlers for their susceptibility to the anti–New Deal press. Of the settlement's history, he concluded, "It is not a nice story." Thoughtful as always, he also took some of the blame himself: "As I look back now after almost two decades it seems to me that we were doomed to failure from the start." Of Casa Grande generally, he wrote: "We can see in it many lessons if we will."

The story of a less ambitious planned community is told by Cathy Knepper in *Greenbelt, Maryland: A Living Legacy of the New Deal.* Greenbelt was in a different class from Casa Grande, for it never aimed for economic independence; rather, it was a cooperative suburb whose inhabitants traveled elsewhere to work. Still, both in the 1930s and 1940s, Greenbelt attracted the hostility of members of Congress; in the 1940s a select committee investigated the Farm Security Administration, which oversaw Greenbelt at the time, finding the agency "communistic." In the period of McCarthyism a Greenbelt citizen, Abe Chasanow, became a target.

Tugwell became governor of Puerto Rico and, later, a professor at the University of Chicago. He and his second wife lived for a time in Greenbelt. They named one of their sons Franklin.

Wendell Willkie joined a New York law firm, today still known as Willkie, Farr and Gallagher. He became Roosevelt's emissary and traveled the world for the president during wartime. Roosevelt, in arranging a meeting concerning Lend Lease with the British prime minister at Chequers, wrote to Churchill that Willkie was "truly helping to keep politics out over here." Willkie was as susceptible to Stalin's charm as Roosevelt, or the 1927 travelers. With the aid of Irita Van Doren, he published a book about a vision of the globe without war, *One World.* The Council on Books in Wartime called his book an "imperative" read. *One World* sold 1.6 million copies within a few months, making it one of the best-selling books in history. Twentieth Century Fox, of which Willkie was chairman of the board, bought the movie rights. Admiring the success, David Lilienthal wrote to Willkie seeking advice about publishing his own books.

Willkie also helped found Freedom House, a New York–based think tank, to fight for the advancement of democracy. He befriended the boxer Joe Louis, who had endorsed his candidacy in 1940. He represented an American Communist, William Schneiderman, before the Supreme Court, defending the man's right to be a citizen. Willkie won the case for Schneiderman and assailed federal prosecutors for going after the Communist; the action, he said, set "an illiberal precedent." He ran for president in 1944 but found little support.

Willkie died in 1944 of a heart attack, while his son Philip, a lieutenant in the navy, was still at sea. Sixty thousand people filed outside Fifth Avenue Presbyterian Church in New York on the day of his funeral.

Bill Wilson's Alcoholics Anonymous spread across the world. By 1975, there were 22,000 local groups in the United States and more abroad. In June 2005, the twenty-five millionth copy of *The Big Book* was published.

Acknowledgments

My first debt is to Sarah Chalfant and Andrew Wylie, the two agents who have stood by this book for over half a decade. I also owe Tim Duggan of HarperCollins, who understood the initial idea better than anyone and waited in good faith and encouraged even when all there was of the book were a few mismatched chapters. Tim's ear improved immensely the quality of the product.

I owe much to the people who have employed me as a columnist while I worked on the book. At the *Financial Times*, John Gapper, Chrystia Freeland, James Montgomery, John Lloyd, and the late Peter Martin were especially supportive. So was Robert Thomson, now of the *Times* of London. Elizabeth Tucker of Marketplace Radio was always an ally. At Bloomberg, my column's new home, Matt Winkler, Tom Keene, Jim Greiff, and Matt Goldenberg overlooked the gaps in column output. John Wasik of Bloomberg has been an especial friend. Douglas Holtz-Eakin and Walter Russell Mead, colleagues at the Council on Foreign Relations, cheered me on. Ira Stoll of the *New York Sun* did as well. My thanks to Steve Chapman, Bruce Dold, and Marcia Lythcott of the *Chicago Tribune*.

Five years of Great Depression work required enormous research. I am lucky to have had funding for that work. Thank you to Ingrid Gregg, President of the Earhart Foundation, who, in the spirit of Amelia herself, took this risk—and thanks to David Kennedy before her as well. The Annette and Ian Cumming Foundation provided me with funds for numerous costly tomes. I appreciate too the support of the American Enterprise Institute, which made me a visiting fellow and part of its National Research Initiative in 2002 and 2003. Kim Dennis was there from the beginning and Henry Olsen at the end. Any author is lucky to have the counsel of AEI's Chris DeMuth, who both advised and gave the chance to present the Schechter theme early on in a Bradley lecture, as well as to deliver, later, yet a second Bradley lecture on the Forgotten Man himself. Additional and crucial support has come from the Alice and Thomas Tisch Foundation. The author appreciates the fact that the foundation was enthusiastic early, when this project seemed improbable. The Manhattan Institute has been supportive: thank you, Roger Hertog, Mabel Weil, Larry Mone, and David Desrosiers for your friendship and patience.

At the American Academy in Berlin, Gary Smith hosted me as J. P. Morgan fellow in economic and financial journalism; this provided needed international perspective.

Ron Chernow was a wise adviser. The work of state think tanks, especially the Mackinac Center and its scholars, Larry Reed and Burt Folsom, inspired this project. Diana Furchtgott-Roth of the Hudson Institute has always lovingly goaded the project forward: "what is the pub date?"

Several researchers, principally Menachem Butler and Johanna Conterio, mined documents and checked, as did Seth Johnson and Charles Siegel. Anna Williams found the photos and taught me all about illustrations in the process. Andrew Relkin provided Web support. Ellen Phillips checked chicken facts and advised on Jewish law. The Lilly Library at Indiana University graciously hunted documents. Thank you to Eileen Giuliani at the Stepping Stones Foundation, who hosted me and my daughter for a visit. Virginia Lewick at the Franklin D. Roosevelt Library and Museum helped out a number of times; the library generally was friendly, a tribute to Roosevelt's spirit.

I owe an enormous debt to Gene Smiley, professor emeritus at Marquette University, and Jeffrey A. Miron of Harvard, both of whom read this book in its entirety. Smiley's *Rethinking the Great Depression* is the ideal thinking man's text. Arthur Levitt also took the time to read this—and told me that as a child he himself went to visit Father Divine. Thank you, Arthur. I owe a debt too to Paul Johnson, who has advised me on all my books, including this one, and another to George Will, who has served as a mentor for many years now. To read Harry Evans's wonderful *They Made America* was to get a lesson in how to profile an innovator; his *American Century* likewise inspired me. I could therefore not believe my good luck when Harry also proved willing to look over the manuscript—thank you. Mark Helprin gave up the most valuable commodity of all, time, to review this project—and to offer some much-needed pointers on style.

People who also read all or parts of the text include Marian McKenna, Wendell Willkie II, Chris DeMuth, Anne Applebaum, and Robert Asahina. Bob Asahina especially understands this story, having studied the Forgotten Man in his own time in his illuminating book, *Just Americans. Just Americans* shows how both

Hoover's refugee centers and the New Deal tradition of resettlement made it easier for the nation to stumble into the tragic error of interning Japanese Americans. Joe Thorndike, the tax historian at Tax Notes, helped me out at every stage (Thorndike's virtual tax museum, at www.taxhistory.org makes for fascinating browsing.) Dan Yergin boosted morale, as did Steve Forbes, who shared his library though he disagrees with the gold thesis. Steve's colleague and my friend Tim Ferguson was always in the background. Glen Asner of the Schechter family, a historian himself, gave me feedback about the family culture. Thank you to Estelle Freilich, a Schechter daughter, for permission to reprint the poem by Joe Schechter's wife. Marian McKenna in Calgary also helped me out with the Schechters. Her book *Franklin Roosevelt and the Great Constitutional War* is the gold standard work on the court packing. From Capitol Hill, Ike Brannon and Donald Marron contributed ideas and encouragement. Tunku Varadarajan, Amy Finnerty, John Lipsky, David Malpass, and Adele Malpass were always helpful and interested. Fred and Susan Shriver heard more about this project than they ever hoped to. The late Robert Bartley walked me through a number of these chapters by e-mail and telephone. Allan Meltzer, author of the authoritative *History of the Federal Reserve*, talked over some of the book. Alan Baker of the *Ellsworth American* of Maine advised as well.

All errors are mine.

Every book has a friend. This book's was Elizabeth Bailey, who understands the importance of long projects.

Books are hard on everyone, but hardest of all on the author's family. My father and mother, closer to the period than I, encouraged me. So did Marylea and Rolf Meyersohn, and my cousins, Elizabeth Meyersohn, John Meyersohn, Julie Hartenstein, and Maggy Siegel. My grandmother Ruth Brill Shlaes Reis sent along her summary, which accurately represented many Americans': "Everything was all right in those years, but only if you had a job." Beatrice Barran deserves special thanks. My siblings Noah Shlaes, Jane Shlaes Dowd, and Bruce DeGrazia let me know that they would accept no other outcome than a finished book. But the biggest thank-yous go to our children, Eli, Theo, Flora, and Helen, and my husband Seth—the man never forgotten.

New York, 2006

Bibliographic Notes

Introduction

*Across the East River . . . a utilities executive named Wendell Willkie . . .*Willkie has numerous biographers. For details of his business side, Joseph Barnes's *Willkie* (New York: Simon & Schuster, 1952) is indispensable. Other sources are Ellsworth Barnard's *Wendell Willkie: Fighter for Freedom* (Marquette, Mich.: Northern Michigan University Press, 1966), and *Wendell Willkie* by Mary Earhart Dillon (New York: Lippincott, 1952). Oren Root's *Persons and Persuasions* (New York: W. W. Norton, 1974) gives a feel for the 1940 campaign. *Dark Horse* (*Dark Horse: A Biography of Wendell Willkie:* Garden City, N.Y.: Doubleday, 1984) by Steve Neal is excellent. So is Charles Peters's

Five Days in Philadelphia: The Amazing "We Want Willkie!" Convention of 1940 and How It Freed FDR to Save the Western World (New York: PublicAffairs, 2005). *This Is Wendell Willkie,* edited and with an introduction by Stanley Walker (New York: Dodd, Mead, 1940) contains many primary Willkie documents.

1
The Beneficent Hand

Average unemployment: 3.3 percent During the 1920s and the Depression, Washington did not keep the sort of systematic unemployment data that it collects today. Europeans found this fact disconcerting. "No trustworthy statistics of unemployment exist and, strangely enough, little effort is being made to collect them," the editors of the London *Economist* grumbled. But the reason came out of two realities of American life from the period: the country was still heavily agricultural, and farmwork was harder to quantify than industrial work. Until the mid-1930s, state governments, taken together with cities, were still a larger presence than Washington, and were still the most logical ones to do the counting. There was another consideration, reasonable in hindsight: creating a statistic gives people something to politicize.

Still, there were some national numbers to talk about. The Labor Department collected figures, as did the Census Bureau, the Commerce Department, state governments, Paul Douglas, and William Green at the American Federation of Labor—as well as several newspapers, including the *New York Times* and the *New York Sun.* Usually unemployment was reckoned in numbers rather than percentages.

By the 1940s the Bureau of Labor Statistics was constructing firmer data, much of the work being led by scholar Stanley Lebergott. Later Lebergott became the leading authority in the field, and it is his updated figures that are used at the headings of the early chapters, as well as in the final chapter. They come from table A-3 on page 512 of his *Manpower in Economic Growth* (New York: McGraw-Hill, 1964). Lebergott documented that unemployment more than doubled from 1920 to 1921, reaching 11.7 percent on average in 1921, above the level for 1930. For the 1930s, I have gone with month-by-month figures calculated by Richard K. Vedder and Lowell E. Gallaway, in *Out of Work: Unemployment and Government in Twentieth Century America* (San Francisco: Independent Institute; New York: Holmes & Meier, 1993). These authors use Lebergott and government numbers as their basis. Economists on the right such as Michael

Darby, Harry Scherman Fellow at the National Bureau of Economic Research, argued later that Lebergott and the BLS both overestimated the number of unemployed by counting as unemployed people who actually had full- or part-time work in make-work programs such as the WPA. But I have gone with the traditional numbers. The material on black employment is especially important. Looking back over pre-Depression Censuses, Vedder and Gallaway found little difference between blacks' unemployment rates and whites'. Even discounting for the very real possibility that the black unemployed were undercounted, the gap between the two groups was nowhere near as dramatic as it would be in the modern period, after World War II.

Dow Jones Industrial Average: 155 Data on the stock market in this book come from Dow Jones indexes. The numbers used are the beginning of the month or year in question. Readers can search for the DJIA themselves at djindexes.com. John Prestbo of that company has also authored a useful book on the history of the Dow, *The Market's Measure: An Illustrated History of America Told through the Dow Jones Industrial Average* (New York: Dow Jones, 1999). Prestbo's insight about the volatility of the Dow in the 1930s helps to explain the economic costs of political uncertainty.

Floods change the course of history Much of the flood material is covered in John Barry's outstanding *Rising Tide: The Great Mississippi Flood of 1927 and How It Changed America* (New York: Touchstone/Simon & Schuster, 1997), which gives a useful, albeit hostile, portrait of Hoover. *Time* magazine and the *New York Times* also covered the flood action extensively. The more general biographical material comes from Hoover's own abundant work, other sources, and the authoritative multivolume biography of Herbert Hoover by George Nash. Hoover's own memoirs (New York: Macmillan, 1952) tell us about his fishing habits.

The idea of philosophical continuity from Coolidge to Hoover On Coolidge, the best resource is Robert Sobel's outstanding biography of the thirtieth president, *Coolidge: An American Enigma* (Chicago: Regnery, 1998). Much of the detail on Coolidge is drawn from this book, including Coolidge's nickname for Hoover, "Wonder Boy." Coolidge also wrote an autobiography that sheds light on his views about presidential restraint. When it comes to Mellon, his own *Taxation: The People's Business* (New York: Macmillan, 1924) is clear on Mellon economics, as are his son Paul's memoirs, *Reflections in a Silver Spoon* (New York: William Morrow, 1992). *Time* magazine covered Mellon thoroughly in the 1920s.

By 1908, Hoover had outgrown Bewick Moreing Hoover's attitude toward Russia comes out clearly in his memoirs and in George Nash; Lewis Feuer also details them in his article "American Travelers to the Soviet Union, 1917–1932," *American Quarterly* 14, no. 2 (Summer 1962): 119–49. Robert Thomsen's biography *Bill W.* (Center City, Minn.: Hazelden Press, 1975), provides detail on this stage of Wilson's life.

The paramount symbol of such immigrant independence Details on the Bank of United States, including its status, can be found in Allan Meltzer's *History of the Federal Reserve,* vol. 1 (Chicago: University of Chicago, 2002).

Blacks too were part of this story Father Divine's story is delightfully and meticulously told by Robert Weisbrot in *Father Divine: The Utopian Evangelist of the Depression Era Who Became an American Legend* (Boston: Beacon Books, 1983).

2
The Junket

One late July day The passenger list of the *President Roosevelt* was published in the *New York Times,* as was the news of the *Duilio.* Material on the founding of the Consumers Union can be found in Richard

Vangermeersch's short biography of Chase, *The Life and Writings of Stuart Chase* (New York: Elsevier, 2005).

The story of this trip to Russia was told by a number of the travelers, both upon their return, to newspapers, and later in articles and books. Two reports were published by those on the trip—a short report by the American Trade Union Delegation, titled *Russia After Ten Years,* and a longer report by the technical advisers, the group that included Paul Douglas and Rex Tugwell. This second book, by Melinda Alexander et al., *Soviet Russia in the Second Decade* (New York: John Day, 1928), contains Tugwell's long essay on agriculture and four essays by Paul Douglas. Douglas, Brophy, Maurer, Tugwell, and others would revisit the trip in their memoirs.

So the travelers carefully gave themselves a label In regard to Communist involvement in the planning of the trip, Sylvia R. Margulies reports in her *Pilgrimage to Russia* (Madison: University of Wisconsin, 1968) that the Comintern was involved from the beginning, asking the American Communist Party to create a delegation of labor leaders. Even at the time, the American Federation of Labor was suspicious; its executive council wrote that it seriously doubted "the good-faith of such a self-constituted commission." There is no evidence that Paul Douglas, Rex Tugwell, or many of the other travelers were aware of the extent of Communist involvement. On the American fascination with Soviet Russia generally, David C. Engerman's *Modernization from the Other Shore* (Cambridge, Mass.: Harvard University Press, 2003) is enlightening. In regard to trips and intellectual pilgrims, three books stand out: Peter G. Filene's *Americans and the Soviet Experiment* (Cambridge, Mass.: Harvard University Press, 1967); Paul Hollander's *Political Pilgrims: Travels of Western Intellectuals to the Soviet Union, China and Cuba* (New York: HarperCollins, 1983); and Margulies, *Pilgrimage to Russia.* Two articles proved crucial in the decision to place emphasis on the summer 1927 junket: Feuer, "American Travelers," and Silas B. Axtell, "Russia and Her Foreign Relations," *Annals of the American Academy of Political and Social Science* 138 (July 1928): 85–92. Feuer is especially good on Stuart Chase and on W. E. B. DuBois's trip. Axtell became disillusioned during the 1927 trip and suspicious of the objectivity of some of his fellow junketeers. Of the descriptions of the Soviet Union, none speaks so much to us as Emma Goldman's *My Disillusionment in Russia* (1925; repr., Gloucester, Mass.: Peter Smith, 1974).

Roosevelt had taken History 10-B The material on Roosevelt's undergraduate curriculum comes from his collected letters as edited by Elliott Roosevelt. On the evolving American policy toward utilities, Preston J. Hubbard's *Origins of the TVA* (New York: Norton Library, 1961) is especially useful.

his feeling for Harvard Law School Frankfurter's statement about the quasi-religious feel of Harvard comes from Harlan B. Phillips, *Felix Frankfurter Reminisces* (New York: Reynal, 1960).

There was another element to Frankfurter's personality Feuer, "American Travelers," analyzes this stage in Frankfurter's work.

Frankfurter had a virtual monopoly The detail about the clerk who did not come from Frankfurter comes out of Dennis J. Hutchison and David J. Garrow, eds., *The Forgotten Memoir of John Knox* (Chicago: University of Chicago Press, 2002).

"You learn no law in Public U" Joseph Lash cites the poem in *From the Diaries of Felix Frankfurter.*

"victims of 'American capitalism'" Both Peggy Lamson and Robert C. Cotrell, *Roger Nash Baldwin and the American Civil Liberties Union* (New York: Columbia University Press, 2000) provide material on Roger Baldwin's Soviet stay. The story of the posting of the news of Sacco and Vanzetti in Russia comes from Lamson, *Roger Baldwin* (Boston: Houghton Mifflin, 1976).

"the 'Three Musketeers' in Moscow in 1927" The letter is at Hyde Park in the presidential archives.

"there is a bit more to eat" Tugwell's analysis of Soviet agriculture can be found in Alexander et al., *Soviet Russia in the Second Decade.*

"We had been good the day before" Tugwell recalls this and other details in *To the Lesser Heights of Morningside: A Memoir* (Philadelphia: University of Pennsylvania Press, 1982).

3
The Accident

"Closing Rally Vigorous" The literature on the 1929 crash is vast. John Kenneth Galbraith's clear and short *The Great Crash: 1929* (Boston: Houghton Mifflin, 1988) argues that the market crashed because of a speculative bubble and was rescued by government action. Frederick Lewis Allen, *Only Yesterday* (New York: Harper, 1931), gives a good feel for the mood at the time. Economists have long differed on the causes of the crash and downturn. Milton Friedman and Anna Schwartz argued famously in the early 1960s that deflation caused the Great Depression; Allan Meltzer has more recently, in his *History of the Federal Reserve,* supported that thesis with newer sources. A less technical version of the deflation argument can be found in Lester V. Chandler's underappreciated *America's Greatest Depression, 1929–1941* (New York: Harper and Row, 1970).

Chandler's book includes some astounding charts on unemployment, including the fact that by January 1934, six in ten of all unemployed Massachusetts males had been unemployed for more than a year. In 1940, 15 percent of unemployed males had been out of a job two years. All these authors are to some extent building on Irving Fisher, who helped develop modern monetarism.

Gene Smiley's *Rethinking the Great Depression* (Chicago: Ivan R. Dee, 2002) is also highly useful. Barry Eichengreen's *Golden Fetters* (New York: Oxford University Press, 1995) assigns blame to the gold standard. Thomas E. Hall and J. David Ferguson's *The Great Depression* (Ann Arbor: University of Michigan Press, 1998) is admirably clear. In the camp arguing that inflation was the problem, the best-known book is Murray Rothbard, *America's Great Depression* (Auburn, Ala.: Mises Institute, 1962). Benjamin Anderson's *Economics and the Public Welfare* (New York: D. Van Nostrand, 1949) also stands out, especially for its analysis of the tax problem. As early as 1949, the year this book was first published, Anderson provocatively titled chapter 31 "The New Deal in 1929–1930." Anderson's blow-by-blow account of the 1937–38 recession is especially important. (The difference between the downturn of 1921 and 1929 is noted in Don Lescohier's article in David Shannon's anthology, *The Great Depression.*)

Economists and historians likewise differ on the costs of the Smoot-Hawley tariff. Some argue that the Fordney-McCumber tariff of the early 1920s did not weigh down the U.S. economy and believe others exaggerate the damage of Smoot-Hawley. I agree with economist Charles Kindleberger on this, who argues that the Smoot-Hawley Act was damaging per se but also because it made clear that, when it came to the world economy, "no one was in charge." Overall the best economic guide to the crash and early Depression years is Michael A. Bordo et al., *The Defining Moment* (Chicago: University of Chicago Press, 1998), a volume of papers presented at a National Bureau of Economic Research conference in 1996. On the effect of public spending after the crash and later, Herbert Stein's *Fiscal Revolution in America* (Washington, D.C.: AEI Press, 1990) and his *Presidential Economics* (New York: Simon & Schuster, 1984) are the classics.

Alfred Loomis More on Loomis's philosophy and movements in the market can be found in Jennet Conant's insightful biography, *Tuxedo Park.*

Irving Fisher . . . argued that stock prices were too low Fisher's relationship with Sumner was a strong one. As Fisher's son Irving Norton later noted in *My Father, Irving Fisher* (New York: Comet Press, 1956), Fisher dedicated his book *The Nature of Capital and Income* to Sumner, writing, "To William Graham Sumner, who first inspired me with a love for Economic Science." Nobel Prize winner Edward C. Prescott outlined the increasingly accepted argument that stocks were not outrageously high in 1929 in a 2003 paper coauthored with Ellen McGrattan, "The 1929 Stock Market: Irving Fisher was Right," Research Department Staff Report 294, Federal Reserve Bank of Minneapolis.

the actual money supply available dropped The figures are from Hall and Ferguson, *Great Depression.*

"Mr. Hoover's Economic Policy" Tugwell's article is discussed in Bernard Sternsher, *Rexford Tugwell and the New Deal* (New Brunswick, N.J.: Rutgers University Press, 1964).

"much-needed construction work" The Roosevelt letter is quoted in *The Hoover Administration: A Documented Narrative.* The construction data are from "Public Construction Highest in Five Years," *New York Times,* April 21, 1930.

Meanwhile, the horizon darkened further The unemployment estimate comes from Vedder and Gallaway, *Out of Work.* As *Time* magazine reported, states were crucial in unemployment data collection in those days, but only ten states compiled numbers.

Smoot-Hawley raised the average tariff Douglas Irwin points out that Smoot-Hawley returned tariffs to turn-of-the-century levels in National Bureau of Economic Working Paper 5895, "From Smoot-Hawley to

Reciprocal Trade Agreements," 1997. The 40 percent drop in trade is mentioned there.

Bankers Athletic League "Bank of U.S. Five Wins League Title." *New York Times,* February 25, 1930, reports the basketball match. On December 12, 1930, the paper reported the story in "Bank of U.S. Closes Doors." More of this material can be found in Milton Friedman and Anna Schwartz, *A Monetary History of the United States, 1867–1960* (Princeton, N.J.: Princeton University Press, 1971), and Meltzer's *History of the Federal Reserve.*

"Let it fail" The *New York Times* of Dec. 21, 1930, describes the demonstration at Freeman Street. The remark is recorded in Meltzer, *History of the Federal Reserve.* See "Throngs at Bank," *New York Times,* December 23, 1930.

4
The Hour of the Vallar

One late summer day Marriner Eccles's own memoirs, *Beckoning Frontiers* (New York: Alfred A. Knopf, 1951), and the descriptions of his biographer, Sidney Hyman, helped here. Descriptions of the "money drought" come out of contemporary papers; a good summary of the scrip network appeared in *Time* magazine's January 9, 1933, issue, in an article titled "For Money." Also useful is Wayne Parrish and Wayne Weishaar, *Men without Money: The Challenge of Barter and Scrip* (New York: G. P. Putnam's Sons, 1933). Paul Douglas mentions the Hyde Park Co-op's founding in his memoirs, but the story is also detailed on the Co-op's website, *http://www.coopmarkets.com/history.htm.*

"households, farmers, unincorporated businesses, and small corporations" Ben Bernanke, especially, studied the damage when banks could or did not lend; the quote is from his *Essays on the Great Depression* (Princeton, N.J.: Princeton University Press, 2004).

"There will be this situation" U.S. Senate hearings before a subcommittee of the Senate Banking and Currency Committee on S. 2959, 72nd Cong., 1st sess., January 1932. Much of this material can also be found in Chandler, *America's Greatest Depression.*

Other components of the downturn Douglas Irwin in *The Defining Moment.*

National unemployment . . . was something like 16 percent and rising The data on unemployment are from the U.S. Department of Commerce annual figures, cited in Chandler, *America's Greatest Depression.*

interest in the Soviet effort was mainstream Peter Filene does a valuable job of evaluating and quantifying U.S. interest in Soviet Russia in his *Americans and the Soviet Experiment.*

The energized professors wanted to fashion for Roosevelt a dramatic message In regard to Franklin Roosevelt's brain trust, there is a rich body of material to draw from. Tugwell wrote multiple memoirs, including one of his rural childhood, *The Light of Other Days* (New York: Doubleday, 1962). Adolf Berle's role comes clear in *Navigating the Rapids* (New York: Harcourt Brace Jovano-vich, 1973). Scholars Jordan Schwarz and Bernard Sternsher beautifully capture this period in their books. Michael Namorato and Sternsher cover Tugwell. Of all the members of the brain trust, Ray Moley feels the most modern. Much of his story comes out in *After Seven Years* (New York: Harper and Brothers, 1939), the book he wrote upon breaking with Roosevelt. Stuart Chase's *New York Times* obituary gives him the credit for the phrase "New Deal," but it was in the air at the time.

"I turned the tables" "Roosevelt Confers on Russia Policy," *New York Times,* July 26, 1932.

"backbone of the nation" Some of the text of the address was reprinted in "Mills' Address Closing Republican Campaign," *New York Times,* September 11, 1932; more of the text elsewhere.

puff piece in the New York Times "Capper and Wallace Disagree on Effect of Hoover's Speech," *New York Times,* October 6, 1932.

Over lunch, Eccles asked Eccles describes the lunch meeting in *Beckoning Frontiers.*

news that Stalin was moving farther forward with the collectivization of agriculture "50,000 Soviet Reds Will Direct Drive to Socialize Farms," *New York Times,* January 30, 1933.

5
The Experimenter

They met in his bedroom This scene comes from Henry Morgenthau's diaries, which were pulled together by the historian John Morton Blum as *From the Morgenthau Diaries* (Boston: Houghton Mifflin, 1959–67). I first read the story of President Roosevelt's gold experiment in John Brooks's outstanding *Once in Golconda: A True Drama of Wall Street* (New York: Harper and Row, 1969). Ickes, Tugwell, and others also commented on the experiment in their writings about the period.

Some of the projects were mere extensions of Hoover's efforts Moley describes the cooperation in *Twenty-seven Masters of Politics* (New York: Ticknor and Fields, 1991).

signing a bill that included deposit insurance Paul Mellon describes the deposit insurance discussion. As Irving Norton Fisher notes in *My Father, Irving Fisher,* 1956), Roosevelt went back and forth on the insurance more than once. "I think it is true that FDR is ready to change his mind easily. I believe in this for myself, but is very worrisome to have a President do it. He seems now to be wobbly about Bank Deposit Guarantee, because New York bankers via [Treasury Secretary] Woodin have questioned it again."

Columbia's president . . . had been flatteringly reluctant Tugwell describes the interaction with Columbia in *The Diary of Rexford G. Tugwell, 1932–1944* (New York: Greenwood Press, 1992).

"no president ever took over a nastier . . . mess" Warburg's reaction is described in his own *Hell Bent for Election* (Garden City, N.Y.: Doubleday Doran, 1935) and also in Brooks, *Once in Golconda.* Fisher's notes about his meeting with Roosevelt are in Robert Loring Allen, *Irving Fisher: A Biography* (Cambridge, England: Blackwell, 1993) and his son's memoir, *My Father, Irving Fisher.*

"culmination of my life work" Fisher's son reprints Fisher's letter to his wife in *My Father, Irving Fisher.*

"Congratulate me" John Brooks relays this incident, but so does Warburg; for an economic description of events, Peter Temin's *Lessons from the Great Depression* (Cambridge: Massachusetts Institute of Technology, 1989) is especially clear.

The news about homeowners continued to be bad Ben Bernanke, "Nonmonetary Effects of the Financial

Crisis in the Propagation of the Great Depression," *American Economic Review* 73 (1983).

6
A River Utopia

The Tennessee Valley Authority On the early TVA, Roy Talbert Jr., *FDR's Utopian: Arthur Morgan of the TVA* (Jackson: University Press of Mississippi, 1987), proved useful.

TVA's first rate schedule "Muscle Shoals Electric Rates," *New York Times,* Sept 15 1933.

"I recall that we were two exceedingly cagey fellows" To get a feel for Lilienthal, the reader need only dip into *The Journals of David Lilienthal* (New York: Harper and Row, 1964). The description of his early meeting with Willkie is contained in an appendix to the first volume.

more than seven dams as large as Dnieprostroi These facts are drawn from David Lilienthal's *Democracy on the March* (New York: Harper and Brothers, 1944).

7
A Year of Prosecutions

Sam Insull and Andrew Mellon The key work on Samuel Insull is Forrest McDonald's *Insull* (Chicago: University of Chicago Press, 1962). More recently author

John Wasik has usefully explored Insull's contribution to technological innovation in *Merchant of Power* (New York: Palgrave Macmillan, 2006). An unpublished memoir by Burton Berry, the State Department official who traveled home with Insull, gives insight into Insull's state of mind. Insull's papers are at Loyola University. The material on Mellon's prosecution is drawn mostly from the newspapers of the period; on the justice of the prosecution, Hoover had some comments to make as well in his memoirs. *The Diary of Rexford G. Tugwell*, edited by Michael Namorato, describes the period in which he was closest with Roosevelt.

Pecora had made a stunning start The Pecora hearings were extensively covered by the press, as in "Pecora Appointed for Stock Inquiry," *New York Times*, January 25, 1933, or "Senators Question Him," *New York Times*, May 24, 1933.

The same day that Mellon went free, Insull gave himself up "Insull Put in Jail," and "Mellon is Cleared of 1931 Tax Evasion," *New York Times*, May 9; "Insull Released," *New York Times*, May 12.

Keynes, the British economist, visited Perkins describes Keynes's visit in *The Roosevelt I Knew* (New York: Viking Press, 1946).

a Democratic sweep "Democrats Hold 69 Seats in Senate," *New York Times*, November 8, 1934.

"a sweeping victory of immense importance" "First Felony Case Is Won under NRA," *New York Times,* November 2, 1934.

"ever tasted worse champagne" This material comes from the published diaries of Tugwell and Ickes.

"Previous to the last national election" "M.S. Eccles Heads Federal Reserve," *New York Times,* November 11, 1934.

8

The Chicken Versus the Eagle

the Brooklyn Schechters The quotes from the Schechters are drawn from testimony of their trials in the lower courts. Because the Supreme Court heard the Schechter case, that material is readily available; I found it in New York University's library. In *Quarrels That Have Shaped the Constitution* (New York: Harper Perennial, 1988) John Garraty treats the Schechter case. G. Neil Reddekopp's "The Schechter Case and the Constitutionality of the NIRA," an unpublished paper (Department of History, University of Calgary, 1977), proved useful. Drew Pearson sketches a hostile portrait in *Nine Old Men* (New York: Doubleday, 1936). The *New York Times* covered the case from the local and national angles.

Early Willkie material can be found in several biographies, including Joseph Barnes, *Willkie* (New York:

Simon & Schuster, 1952), and Steve Neal, *Dark Horse* (Garden City, N.Y.: Doubleday, 1984). On Dorothea Lange and Roy Stryker, Milton Meltzer's *Dorothea Lange: A Photographer's Life* (Syracuse, N.Y.: Syracuse University Press, 2000) provides details.

the firing of Jerome Frank In recent decades the declassification and analysis of the Venona code papers has shown that Hiss was a Soviet spy. Jerome Frank, however, was more like Tugwell, a romantic. See John Earl Haynes and Harvey Klehr, *Venona: Decoding Soviet Espionage in America* (New Haven, Conn.: Yale University Press, 1999).

Her cheerful mood accorded with that of the country. The 1935–36 rally was the longest of the decade, and the most significant in terms of point increases, going from 110 or so to 190 in 1937. Rallies and declines in this instance are measured in the classic way: from bottom point to top, or top to bottom. The "FDR Rally" that came at the time of FDR's first election was greater in percentage terms, but did not last as long, and did not come close to recovery levels. For more, see John Prestbo, *The Market's Measure*. Prestbo also notes that the Dow of the 1930s was the most volatile of decades on record for the century. This clearly has to do with monetary policy and international changes, but also with an unpredictable White House.

9
Roosevelt's Wager

Felix Frankfurter moved in Max Freedman, ed., *Roosevelt and Frankfurter: Their Correspondence* (Boston: Atlantic Monthly Press, 1967) proves very useful in understanding the relationship between the law professor and the president. So does Elliott Roosevelt's compilations of his father's letters. Harold Ickes's *Secret Diary* gives a good feeling for the 1936 campaign. Perkins's *The Roosevelt I Knew* provides many details.

"we should consider the truth" McReynolds's written dissent in *Ashwander* can be found in Court records; his statement about the power of government to compete with utilities was reported in "Deliberateness of Chief Justice Keeps Court Room Throng in Long Suspense," *New York Times*, February 18, 1936.

Now Tugwell had an idea Meltzer, *Dorothea Lange*, is useful for details on the photographers and their work.

By November, the new CIO had opened an office Brophy's work at the K Street office is detailed in Melvyn Dubofsky and Warren van Tine, *John L. Lewis: A Biography* (Champaign: University of Illinois, 1986).

The TVA paid no taxes, he noted Some of this material comes from Barnes, *Willkie.*

In September, Roosevelt spoke at Harvard See Freedman, *Roosevelt and Frankfurter.*

10
Mellon's Gift

Mellon's paintings must be spared David Finley's *A Standard of Excellence* (Washington, D.C.: Smithsonian Institution, 1973) sheds light on Mellon's collection habits, as does Paul Mellon's memoir. Two biographies of Duveen are also helpful, S. N. Behrman's *Duveen* (New York: Little Bookroom, 2002) and Meryle Secrest, *Duveen: A Life in Art* (Chicago: University of Chicago Press, 2005).

"I built a temple" Paul Mellon published the poem in his memoirs, *Reflections in a Silver Spoon.*

Mellon, David Cannadine's outstanding biography of Mellon, came out as I was just finishing this manuscript; it contains all details of the tax and National Gallery stories.

11
Roosevelt's Revolution

Having learned the importance The construction of the speech is detailed in Samuel Rosenman's *Working with Roosevelt* (New York: Harper Brothers, 1952).

legislation that would increase the number of justices McKenna's *Franklin Roosevelt* is extremely useful. For insight into the justices, I liked Hutchinson and Garrow, *Forgotten Memoir of John Knox*. Frankfurter's correspondence with his wife, Marion, on the court-packing plan can be found in H. N. Hirsch, *The Enigma of Felix Frankfurter* (New York: Basic Books, 1981).

Power, *the play* Rosenman, *Working with Roosevelt*, p. 144. *http://newdeal.feri.org* has some material on *Power*.

One of the sources for the economic part of the chat was a piece by Stuart Chase The February 15, 1937, submission by Chase to the *Times* is published in Freedman's edition of Roosevelt and Frankfurter's correspondence.

"the Court's processes had integrity" Frankfurter's note to Roosevelt is in Hirsch, *Enigma of Felix Frankfurter*.

Ogden Mills, the treasury secretary These thoughts come from Ogden Mills, *The Seventeen Million* (New York: Macmillan, 1937).

"trick way of finding loopholes" Roosevelt's exchanges with Morgenthau on taxes are in Blum, *Morgenthau Diaries*.

"I am wholly unable to figure out" Roosevelt's tax return and his letter to Commissioner Helvering are publicly available at taxhistory .org. Joseph Thorndike,

the creator of the history project, has also posted the returns of several other presidents.

12
The Man in the Brooks Brothers Shirt

"But don't ask me about cotton" Rex Tugwell's start at American Molasses was covered in "Tugwell Bit Hazy about His New Job," *New York Times,* January 5, 1927. Some of the detail on Tugwell in this period comes from Michael Namorato, *Rexford G. Tugwell: A Biography* (New York: Praeger, 1988). More, including the story of Tugwell's separation from Columbia, can be found in *Diary of Rexford G. Tugwell.*

"we believe that the Soviet Union" The source for this is Peter Filene, *American Views of Soviet Russia* (Homewood, Ill.: Dorsey Press, 1968); the petition and signatories are reprinted on p. 117.

Stuart Chase still wrote Richard Vangermeersch, *Life and Writings of Stuart Chase* (New York: Elsevier, 2005) contains the material about Harvey Chase's correspondence with Roosevelt. Chase's *The Tyranny of Words* (New York: Harcourt Brace, 1938) marked the beginning of a new stage for the author.

Betty Glan Douglas writes about his discovery of Betty Glan's death in *In the Fullness of Time: The*

Memoirs of Paul Douglas (New York: Harcourt Brace Jovanovich, 1971).

Mary McCarthy typified Mary McCarthy's own writings, especially her *Intellectual Memoirs* (New York: Harvest, 1993), give a feel for the period. Steve Neal, *Dark Horse*, covers this period in Willkie's life especially well.

"Wendell told me, rather explicitly" The Lilienthal remark is from his diary, also quoted in Neal, *Dark Horse.*

A new respect for conservatism *New York Times*, "Bestsellers of the Week, Here and Elsewhere," July 5, 1937; "Sellers of the Week, Here and Elsewhere," July 12, 1937.

"There was no real place for me" This material is from *Diary of Rexford G. Tugwell.*

13
Black Tuesday, Again

On an August evening at his daughter's "Andrew Mellon Dies at Age of 82," *New York Times*, August 27, 1937.

Wall Street already knew that The concept of regime uncertainty developed by the scholar Robert Higgs helped me enormously in understanding the period 1937–38. Both Morgenthau's and Adolf Berle's diaries and papers are useful for these years, as are *Beckoning Frontiers*, Eccles, and Anderson, *Economics and the Public Welfare.* Anderson's insights are astonishing.

At Casa Grande Crucial to the story of Casa Grande is Edward C. Banfield's book *Government Project* (Chicago: Free Press, 1961). Banfield, a young scholar, detailed every stage of the Casa Grande project, and its sorry outcome; Tugwell wrote an ambivalent introduction. Herb Stein's explanation of the monetary environment is also important. Some of the details from Roosevelt's fishing trip appear in Robert Jackson, *That Man* (New York: Oxford University Press, 2003).

Currie was also a Soviet spy Currie was one of the few Roosevelt Administration figures who actually served as a genuine spy for Moscow. The federal government's Venona Project, made public only in the 1990s, showed that Currie had reported to the KGB. But for the moment this mattered less than the more immediate problem of the U.S. economy.

"Eccles was in the doghouse" The source for this is a 1981 interview with Lauchlin Currie on London Weekend Television, reprinted in *Journal of Economic Studies* 31, no. 3 (2004).

"more differently colored glasses" The transcript of the debate between Jackson and Willkie appears in *This*

Is Wendell Willkie (New York: Dodd, Mead, 1940). Ray Moley's *After Seven Years* is also important when it comes to understanding the state of mind of business.

the Supreme Court had ruled See, for example, "Decision of the Supreme Court on PWA," *New York Times*, January 4, 1938, or "Utilities' Grief," *Time*, January 10, 1938.

14
"Brace Up, America"

Bill Wilson was struggling Francis Hartigan, *Bill W.* (New York: St. Martin's Press, 2000), sheds light on this stage in Wilson's life.

"Father is going to make Ulster County" Much of the material on Father Divine comes out of the papers of the period; the *New York Times* covered him extensively. See also Weisbrot.

"people do not like the business depression" See, for example, "Ickes Says Roosevelt Won in Vote," *New York Times*, Nov 11, 1938.

Casa Grande Banfield, *Government Project*, conveys the Casa Grande detail.

The Roosevelts . . . were deeply distracted Roosevelt's preoccupation with war comes through in both his letters generally and his correspondence with Frankfurter.

Roger Baldwin was on the beach at Chilmark Roger Baldwin's biographer, Robert Cottrell, details both his Soviet trip and his disillusionment with Soviet Russia in the late 1930s, as well as the purge at the ACLU, in *Roger Nash Baldwin.* Other details on Roger Baldwin come from Lamson, *Roger Baldwin.*

"Well, I've done it" Neal, *Dark Horse,* provides the detail about Willkie's decision to register as a Republican in 1939. Some biographical material about Davenport comes from there as well.

15
Willkie's Wager

Root decided to . . . float Willkie as a candidate Oren Root's memoirs, *Persons and Persuasions* (New York: W. W. Norton, 1974), supply the story of his enthusiasm for Willkie. Material on the campaign comes also from the Willkie Papers at Indiana University.

"Wonder Boy" Felix Frankfurter's attitude toward Willkie comes from Freedman, *Roosevelt and Frankfurter.*

"Learn to Say: President Willkie" "Willkie Serves Notice on Democrats," *New York Times,* July 17, 1940.

"They will vote to continue the New Deal" Potofsky's position is reported in Dubofsky and Van Tine, *John L. Lewis.*

"essential to the national defense" See "For TVA Speed Up as Defense Move," *New York Times*, July 10, 1940; and, for the details of Lilienthal's press conference, "Roosevelt Signs TVA . . . ," *New York Times*, August 1, 1940.

Coda

Roy Stryker The Carnegie Library of Pittsburgh's online exhibit of urban photography provides details of Stryker's bio.

Sixty thousand people filed outside "Willkie Tribute Paid by 60,000 Here," *New York Times*, October 10, 1944.

Selected Bibliography

In researching this book I found period papers and magazines most useful. Newspapers and magazines used as sources include the *Washington Post,* the *New York Times,* the *New York American,* the *Richmond Times-Dispatch,* the *Nation,* the *New Republic, Newsweek,* and *Time.*

Archived Papers

Roosevelt's papers are at the Franklin D. Roosevelt Presidential Library at Hyde Park, New York. The Lilly Library at Indiana University at Bloomington, Indiana, houses Wendell Willkie's papers. Some of the Insull material comes from Loyola University Archives; especially interesting is an unpublished manuscript by Burton Y. Berry, the State

Department executive who escorted Insull home from Greece ("Mr. Samuel Insull," Berry, Burton Y. Loyola University Archives, Insull Collection, Folder 17-2).

Alzada Comstock's papers are at Mount Holyoke College Archives and Special Collections, 8 Dwight Hall, 50 College Street, South Hadley, Massachusetts.

Rexford Tugwell's papers are at the Franklin D. Roosevelt Presidential Library at Hyde Park, New York, with the president's and Mrs. Roosevelt's.

Irita van Doren's papers, in the Library of Congress, contain some material from the period of her relationship with Willkie.

Much of the Schechter material comes from Supreme Court documentation from the October term, 1934, found in *Case No. 854, A.L.A. Schechter Poultry Corporation, Schechter Live Poultry Market, Joseph Schechter, Alex Schechter, and Aaron Schechter, Petitioners v. United States of America.*

Online Resources

Dow Jones Indexes, djindexes.com; Dow Jones Industrial Average and utilities index.

"Bridging the Urban Landscape," online exhibit, Carnegie Library of Pittsburgh, http://www.clpgh.org/exhibit.

The Franklin and Eleanor Roosevelt Institute's online library, the New Deal Network, http://newdeal.feri.org.

Books and Articles

Adler, Stephen J., and Lisa Grunwald. *Letters of the Century.* New York: Dial Press/Random House, 1999.

Alexander, Melinda, et al. *Soviet Russia in the Second Decade: A Joint Survey by the Technical Staff of the First American Trade Union Delegation.* New York: John Day, 1928.

Allen, Frederick Lewis. *Only Yesterday.* New York: Harper, 1931.

Allen, Frederick Lewis, and Agnes Rogers. *I Remember Distinctly: A Family Album of the American People in the Years of Peace, 1918 to Pearl Harbor.* New York: Harper, 1947.

Allen, Robert Loring. *Irving Fisher: A Biography.* Cambridge, Mass.: Blackwell, 1993.

Alter, Jonathan A. *The Defining Moment: FDR's Hundred Days and the Triumph of Hope.* New York: Simon & Schuster, 2006.

Anderson, Benjamin. *Economics and the Public Welfare: Financial History of the United States, 1914–1946.* New York: D. Van Nostrand, 1949.

Antwerp, William C. Van. *The Stock Exchange from Within.* Garden City, N.Y.: Doubleday Page, 1914.

Arkes, Hadley. *The Return of George Sutherland.* Princeton, N.J.: Princeton University Press, 1994.

Axtell, Silas B. "Russia and Her Foreign Relations." *Annals of the American Academy of Political and Social Science* 138 (July 1928): 85–92.

Badger, Anthony J. *The New Deal.* Chicago: Ivan R. Dee, 1989.

Banfield, Edward C. *Government Project.* Introduction by Rexford Tugwell. Chicago: Free Press, 1951.

Barnard, Ellsworth. *Wendell Willkie, Fighter for Freedom.* Marquette: Northern Michigan University Press, 1966.

Barnes, Joseph. *Willkie: The Events He Was Part of, the Ideas He Fought For.* New York: Simon & Schuster, 1952.

Barone, Michael. *Our Country: The Shaping of America, from Roosevelt to Reagan.* New York: Free Press, 1990.

Barry, John M. *Rising Tide: The Great Mississippi Flood and How It Changed America.* New York: Touchstone/Simon & Schuster, 1997.

Bartlett, Bruce. *The New Deal and Tax Policy, 1913 to Present.* Washington, D.C.: Heritage Foundation, 1980.

Baruch, Bernard. *Baruch: My Own Story.* New York: Henry Holt, 1957.

Barzini, Luigi. *The Italians: A Full Length Portrait Featuring Their Manners and Morals.* New York: Touchstone, 1964.

Behrman, S. N. *Duveen: The Story of the Most Spectacular Art Dealer of All Time.* Introduction by Glenn Lowry. New York: Little Bookroom, 2002.

Beito, David T. *From Mutual Aid to the Welfare State: Fraternal Societies and Social Services, 1890–1967.* Chapel Hill: University of North Carolina Press, 2000.

Bellush, Bernard. *The Failure of the NRA.* New York: W. W. Norton, 1975.

Benton, William. *The Annals of America.* Vol. 15, *The Great Depression.* Chicago: Encyclopaedia Britannica, 1968.

Berle, Adolf A. *Navigating the Rapids, 1918–1971.* Edited by Beatrice Bishop Berle and Travis Beal Jacobs. New York: Harcourt Brace Jovanovich, 1973.

Berle, Beatrice Bishop. *A Life in Two Worlds: The Autobiography of Beatrice Bishop Berle.* New York: Walker, 1983.

Berlin, Isaiah. *Two Concepts of Liberty: An Inaugural Lecture Delivered before the University of Oxford on 31 Oct., 1958.* Oxford, England: Clarendon Press, 1958.

Bernanke, Ben. *Essays on the Great Depression.* Princeton, N.J.: Princeton University Press, 2004.

———. "Nonmonetary Effects of the Financial Crisis in the Propagation of the Great Depression." *American Economic Review* 73 (1983).

Bernhard, Virginia, David Burner and Elizabeth Fox-Genovese. *Firsthand America: A History of the United States,* Volume II. St. James, N.Y.: Brandywine Press, 1992.

Bernstein, Irving. *A Caring Society: The New Deal, the Worker and the Great Depression.* Boston: Houghton Mifflin, 1985.

———. *The Lean Years.* New York: Penguin, 1966.

Best, Gary Dean. *The Critical Press and the New Deal: The Press versus Presidential Power, 1933–1938.* New York: Praeger, 1993.

Biles, Roger. *Crusading Liberal: Paul H. Douglas of Illinois.* DeKalb: Northern Illinois University Press, 2002.

Billington, Monroe Lee. *Thomas P. Gore: The Blind Senator from Oklahoma.* Lawrence: University of Kansas Press, 1967.

Bird, Caroline. *The Invisible Scar: The Great Depression and What It Did to American Life, From Then to Now.* New York: David McKay, 1964.

Blakey, Roy, and Gladys Blakey. *The Federal Income Tax.* New York: Longmans, 1940.

Blum, Walter J., and Harry Kalven Jr. *The Uneasy Case for Progressive Taxation.* Chicago: University of Chicago Law School, 1953.

Bordo, Michael A., et al. *The Defining Moment.* Papers presented at a National Bureau of Economic Research conference, 1996. Chicago: University of Chicago Press, 1998.

Brande, Dorothy. *Wake Up and Live.* New York: Simon & Schuster, 1936.

Brinkley, Alan. *Voices of Protest: Huey Long, Father Coughlin, and the Great Depression.* New York: Vintage/Random House, 1982.

Brinton, J. W. *Wheat and Politics.* Minneapolis: Rand Tower, 1931.

Brooks, John. *Once in Golconda: A True Drama of Wall Street.* New York: Harper and Row, 1969.

Brophy, John. *A Miner's Life.* Edited by John O. Hall. Madison: University of Wisconsin Press, 1964.

Brown, Josephine Chapin. *Public Relief, 1929–1939.* New York: Octagon, 1971.

Brownlee, W. Elliott. *Federal Taxation in America: A Short History.* Princeton, N.J.: Woodrow Wilson Center, 1996.

———, ed. *Funding the Modern State, 1941–1995.* Princeton, N.J.: Woodrow Wilson Center, 1996.

Burg, David F. *The Great Depression: An Eyewitness History.* New York: Facts on File, 1996.

Busch, Francis X. *Guilty or Not Guilty: An Account of the Trials of Leo Frank, DC Stephenson, Samuel Insull, and Alger Hiss.* New York: Bobbs Merrill, 1952.

Cassels, Alan. *Mussolini's Early Diplomacy.* Princeton, N.J.: Princeton University Press, 1970.

Catchings, Waddill, and William Trufant Foster. *Profits.* Boston: Houghton Mifflin, 1925.

———. *Road to Plenty.* Boston: Houghton Mifflin, 1928.

Chandler, Lester V. *America's Greatest Depression, 1929–1941.* New York: Harper and Row, 1970.

Chandler, William U. *The Myth of TVA: Conservation and Development, 1993 to 1980.* New York: Ballinger, 1984.

Chase, Stuart. *A New Deal.* New York: Macmillan, 1932.

———. *The Tyranny of Words.* New York: Harcourt Brace, 1938.

Cheever, Susan. *My Name Is Bill—His Life and the Creation of AA.* New York: Simon & Schuster, 2004.

Chenoweth, Lawrence, *The American Dream of Success: The Search for the Self in the Twentieth Century.* Boston: Duxbury Press, 1974.

Chernow, Ron. *House of Morgan.* New York: Simon & Schuster/Touchstone, 1990.

Christman, Henry M., ed. *Kingfish to America: Share Our Wealth; Selected Senatorial Papers of Huey P. Long.* New York: Schocken, 1985.

Collins, Robert M. *The Business Response to Keynes, 1929–1964.* New York: Columbia University Press, 1981.

Conklin, Groff, ed. *The New Republic Anthology, 1915–1935.* Introduction by Bruce Bliven. New York: Dodge, 1936.

Conrat, Maisie, and Richard Conrat. *Executive Order 9066.* San Francisco: California Historical Society, 1972.

Cotrell, Robert C. *Roger Nash Baldwin and the American Civil Liberties Union.* New York: Columbia University Press, 2000.

Cook, Blanche Wiesen. *Eleanor Roosevelt,* Volume 1. New York: Penguin, 1992.

Cortissoz, Royal. *An Introduction to the Mellon Collection.* New York: Merrymount Press, 1937.

Couch, Jim F., and William Shughart. *The Political Economy of the New Deal.* Cheltenham, England: Edward Elgar Publishing, 1998.

Crossman, Richard H., ed. *The God That Failed.* New Foreword by David Engerman. New York: Columbia University Press, 2001.

Currie, Lauchlin, interview on London Weekend Television. Reprint, *Journal of Economic Studies* 31, no. 3 (2004): 264.

Cushman, Barry. *Rethinking the New Deal Court.* Oxford, England: Oxford University Press, 1998.

Dickson, Paul, and Thomas B. Allen. *The Bonus Army.* New York: Walker and Company, 2004.

Diggins, John P. *Mussolini and Fascism: The View from America.* Princeton, N.J.: Princeton University Press, 1972.

Dillon, Mary Earhart. *Wendell Willkie*. Philadelphia: J.B. Lippincott, 1952.

Doris, Lillian, ed. *The American Way in Taxation: Internal Revenue, 1862–1963.* Buffalo, N.Y.: William S. Hein, 1994.

Douglas, Paul H. *In the Fullness of Time: The Memoirs of Paul Douglas.* New York: Harcourt Brace Jovanovich, 1971.

Dubofsky, Melvyn, and Warren Van Tine. *John L. Lewis: A Biography.* Champaign: University of Illinois Press, 1986.

Eccles, Marriner. *Beckoning Frontiers.* Edited by Sidney Hyman. New York: Alfred A. Knopf, 1951.

Economist editors. *The New Deal.* New York: Alfred A. Knopf, 1937.

Edsforth, Ronald. *The New Deal: America's Response to the Great Depression.* London: Blackwell, 2000.

Eichengreen, Barry. *Golden Fetters: The Gold Standard and the Great Depression, 1919–1939.* New York: Oxford University Press, 1995.

Engerman, David C. *Modernization from the Other Shore.* Cambridge, Mass.: Harvard University Press, 2003.

Ellis, Edward Robb. *The Epic of New York City: A Narrative History.* New York: Kodansha, 1997.

Emerson, Thomas I. *Young Lawyer for the New Deal.* Savage, Md.: Rowman and Littlefield, 1991.

Evans, Harold. *The American Century.* New York: Alfred A. Knopf, 1998.

———. *They Made America: From the Steam Engine to the Search Engine, Two Centuries of Innovators.* New York: Little, Brown, 2004.

Ezzell, Patricia Bernard. *TVA Photography: Thirty Years of Life in the Tennessee Valley.* Jackson: University Press of Mississippi, 2003.

Fauset, Arthur Huff. *Black Gods of the Metropolis: Negro Religious Cults of the Urban North.* Philadelphia: University of Pennsylvania Press, 2002.

Feuer, Lewis S. "American Travelers to the Soviet Union, 1917–1932: The Formation of a New Deal Ideology." *American Quarterly* 14, no. 2 (Summer 1962): 119–49.

Filene, Peter G. *Americans and the Soviet Experiment.* Cambridge, Mass.: Harvard University Press, 1967.

———, ed. *American Views of Soviet Russia.* Homewood, Ill.: Dorsey Press, 1968.

Filler, Louis. *The Anxious Years.* New York: G. P. Putnam's Sons, 1963.

Finley, David. *A Standard of Excellence: Andrew W. Mellon Founds the National Gallery of Art at Washington.* Washington, D.C.: Smithsonian Institution, 1973.

Fischer, Louis. *Men and Politics: An Autobiography.* New York: Duell, Sloan and Pierce, 1941.

Fisher, George W. "Irving Fisher: An Economist." *American Journal of Economics and Sociology* 63 (Dec. 31, 2004).

Fisher, Irving Norton. *My Father, Irving Fisher.* New York: Comet Press, 1956.

Flanagan, Hallie. *Arena: The Story of the Federal Theater.* Foreword by John Houseman. New York: Limelight, 1985.

———. *Dynamo: An Adventure in College Theater.* New York: Duell Sloan and Pearce, 1943.

Fleming, Thomas. *The New Dealers' War: Franklin D. Roosevelt and the War Within World War II.* New York: Basic Books, 2001.

Flynn, John T. *The Roosevelt Myth: Fiftieth Anniversary Edition.* Introduction by Ralph Raico. San Francisco: Fox and Wilkes, 1998.

Freedman, Max, ed. *Roosevelt and Frankfurter: Their Correspondence.* Boston: Atlantic Monthly Press, 1967.

Friedman, Milton, and Anna Schwartz. *A Monetary History of the United States, 1867–1960.* Princeton, N.J.: Princeton University Press, 1971.

Friedman, Milton, and Rose D. Friedman. *Two Lucky People: Memoirs.* Chicago: University of Chicago Press, 1998.

Galbraith, John Kenneth. *The Great Crash: 1929.* Boston: Houghton Mifflin, 1988.

Garraty, John A. *Quarrels That Have Shaped the Constitution.* New York: Harper Perennial, 1988.

Garrett, Garet (Bruce Ramsay, ed.). *Salvos Against the New Deal: Selections from the Saturday Evening Post, 1933–1940.* Caldwell, Idaho: Caxton Press, 2002.

Gates, Robert A. *American Literary Humor during the Depression.* New York: Greenwood Press, 1992.

Gelderman, Carol. *Mary McCarthy: A Life.* New York: St. Martin's Press, 1988.

Ghirardo, Diane. *Building New Communities: New Deal America and Fascist Italy.* Princeton, N.J.: Princeton University Press, 1989.

Gody, Lou, et al. *The WPA Guide to New York City.* Introduction by William H. White. New York: New Press, 1992.

Goldman, Emma. *My Disillusionment in Russia.* 1925. Reprint, Gloucester, Mass.: Peter Smith, 1974.

Gunther, John. *Inside USA.* New York: Harper and Brothers, 1947.

Hackett, Alice Payne. *Seventy Years of Bestsellers.* New York: R. R. Bowker, 1967.

Hall, Thomas E., and J. David Ferguson. *The Great Depression: An International Disaster of Perverse Economic Policies.* Ann Arbor: University of Michigan Press, 1998.

Hamby, Alonzo L. *For the Survival of Democracy.* New York: Free Press, 2004.

Hannaford, Peter. *The Quotable Coolidge: Sensible Words for a New Century.* Bennington, Vt.: Images from the Past, 2001.

Harris, Sara Drucker, and Harriet Crittenden. *Father Divine, Holy Husband.* Garden City, N.Y.: Doubleday, 1953.

Hartigan, Francis. *Bill W.: A Biography of Alcoholics Anonymous Cofounder Bill Wilson.* New York: St. Martin's Press, 2000.

Haynes, John Earl, and Harvey Klehr. *Venona: Decoding Soviet Espionage in America.* New Haven, Conn.: Yale University Press, 1999.

Higgs, Robert. *Against Leviathan: Government Power and a Free Society.* Oakland, Calif.: Independent Institute, 2004.

Hillis, Marjorie. *Orchids on Your Budget.* New York: Bobbs-Merrill, 1937.

Hirsch, H. N. *The Enigma of Felix Frankfurter.* New York: Basic Books, 1981.

Hofstadter, Richard. *The Age of Reform.* New York: Vintage Books/Random House, 1955.

Holbrook, Stewart H. *The Age of Moguls.* Garden City, N.Y.: Doubleday, 1954.

Hollander, Paul. *Political Pilgrims: Travels of Western Intellectuals to the Soviet Union, China and Cuba.* New York: HarperCollins, 1983.

Holzer, Henry Mark. *The Gold Clause.* New York: Books in Focus, 1980.

Hoover, Herbert. *American Individualism.* Garden City, N.Y.: Doubleday, 1922.

———. *The Memoirs of Herbert Hoover, 1929–1941: The Great Depression.* New York: Macmillan, 1952.

Hoshor, John. *God in a Rolls Royce: The Rise of Father Divine, Madman, Menace, or Messiah.* New York: Hillman-Curl, 1936.

Houseman, John. *Arena: The Story of the Federal Theater.* New York: Limelight, 1985.

Hubbard, Preston J. *Origins of the TVA: The Muscle Shoals Controversy, 1920–1932.* New York: Norton Library, 1961.

Hutchinson, Dennis J., and David J. Garrow, eds. *The Forgotten Memoir of John Knox: A Year in the Life of a Supreme Court Justice in FDR's Washington.* Chicago: University of Chicago Press, 2002.

Hyman, Sidney. *Marriner S. Eccles: Private Entrepreneur and Public Servant.* Stanford, Calif.: Graduate School of Business, 1976.

Ickes, Harold L. *The Autobiography of a Curmudgeon.* New York: Quadrangle, 1943.

———. *The Secret Diary of Harold L. Ickes.* 2 vols. New York: Simon & Schuster, 1954.

Irey, Elmer L., with William J. Slocum. *The Tax Dodgers.* New York: Greenberg Press, 1948.

Irwin, Douglas. "From Smoot Hawley to Reciprocal Trade Agreements." National Bureau of Economic Research Working Paper 5895, 1997.

———. "The Smoot Hawley Tariff: A Quantitative Assessment." National Bureau of Economic Research Working Paper 5509, 1996.

Irwin, Will. *Herbert Hoover: A Reminiscent Biography.* New York: The Century Company, 1928.

Jackson, Robert H. *That Man: An Insider's Portrait of Franklin D. Roosevelt.* Edited by John Q. Barrett. New York: Oxford University Press, 2003.

Jeansonne, Glen. *Gerald L. K. Smith: Minister of Hate.* Baton Rouge: Louisiana State University Press, 1997.

Joslin, Theodore G. *Hoover off the Record, 1934.* Garden City, N.Y.: Doubleday, Doran, 1937.

Kanigel, Robert. *The One Best Way: Frederick Winslow Taylor and the Enigma of Efficiency.* New York: Penguin, 1997.

Kindleberger, Charles P. *The World in Depression: 1929–1939*, revised edition. Berkeley: University of California Press, 1986.

Kirk, Russell, ed. *The Portable Conservative Reader.* New York: Penguin, 1996.

Klein, Maury. *Rainbow's End: The Crash of 1929.* Oxford, England: Oxford University Press, 2001.

Knepper, Cathy D. *Greenbelt, Maryland: A Living Legacy of the New Deal.* Baltimore: Johns Hopkins University Press, 2001.

Komarovsky, Mirra. *The Unemployed Man and His Family: The Effect of Unemployment on the Status of the Man in 59 Families.* New York: Farrar, Straus and Giroux, 1971.

Kopper, Philip. *America's National Gallery of Art: A Gift to the Nation.* New York: Harry N. Abrams, 1991.

Koskoff, David E. *The Mellons: The Chronicle of America's Richest Family.* New York: Thomas Y. Crowell, 1978.

Kurland, Philip B., and Gerhard Casper, eds. *Landmark Briefs and Arguments of the Supreme Court of the United States: Constitutional Law.* Vol. 28. Bethesda, Md.: University Publications of America, 1997.

Kurth, Peter. *American Cassandra: The Life of Dorothy Thompson.* Boston: Little, Brown, 1990.

Lampman, Robert J., ed. *Social Security Perspectives: Essays of Edwin Witte.* Madison: University of Wisconsin Press, 1962.

Lamson, Peggy. *Roger Baldwin: Founder of the American Civil Liberties Union.* Boston: Houghton Mifflin, 1976.

Lash, Joseph. *From the Diaries of Felix Frankfurter.* New York: W. W. Norton, 1975.

Lastrapes, William D., and George Selgin. "The Check Tax: Fiscal Folly and the Great Monetary Contraction." *Journal of Economic History* 57, no. 4 (December 1997).

Lathem, Edward Connery, ed. *Meet Calvin Coolidge: The Man behind the Myth.* Brattleboro, Vt.: Stephen Greene Press, 1960.

Lebergott, Stanley. *Manpower in Economic Growth. The American Record Since 1800.* New York: McGraw-Hill, 1964.

Leff, Mark H. *The Limits of Symbolic Reform: Taxation and the New Deal.* Cambridge, England: Cambridge University Press, 1984.

———. *Dealers and Dreamers: A New Look at the New Deal.* New York, Doubleday, 1988.

Lender, Mark Edward, and James Kirby Martin. *Drinking in America: A History.* New York: Free Press, 1982.

Leuchtenburg, William E. *In the Shadow of FDR.* Ithaca, N.Y.: Cornell University Press, 1983.

———, ed. *The New Deal: A Documentary History.* New York: Harper and Row, 1968.

Levine, Lawrence W., and Cornelia R. Levine. *The People and the President: America's Conversation with FDR.* Boston: Beacon Press, 2002.

Lilienthal, David Eli. *Democracy on the March.* New York: Harper and Brothers, 1944.

———. *The Journals of David Lilienthal.* Vol. 1, *The TVA Years.* New York: Harper and Row, 1964.

Louchheim, Katie, ed. *The Making of the New Deal.* Cambridge: Harvard University Press, 1983.

Lowry, Edward G. *Washington Closeups.* Boston: Houghton Mifflin, 1921.

McCarthy, Mary. *Intellectual Memoirs: New York, 1936–1938.* New York: Harvest, 1993.

McCormick, Anne O'Hare. *The Hammer and the Scythe: Soviet Russia Enters the Second Decade.* New York: Alfred A Knopf, 1929.

McCullough, C. Hax Jr. *One Hundred Years of Banking: The History of Mellon National Bank and Trust.* Pittsburgh, Pa.: Mellon National Bank, Herbick and Held, Printer, ca. 1969.

McDonald, Forrest. *Insull.* Chicago: University of Chicago Press, 1962.

McDonald, Michael J., and John Muldowny. *TVA and the Dispossessed.* Knoxville: University of Tennessee Press, 1982.

McElvaine, Robert S. *Down and Out in the Great Depression: Letters from the Forgotten Man.* Durham: University of North Carolina Press, 1983.

———. *The Great Depression.* New York: Crown/Three Rivers Press, 1985.

McElway, St. Clair, and A. J. Liebling. "Who Is This King of Glory?" (profile of Father Divine). *New Yorker,* June 13, 1936; June 20, 1936; June 27, 1936.

McGrattan, Ellen R., and Edward C. Prescott. "The 1929 Market: Irving Fisher Was Right." Research Department Staff Report 294, Federal Reserve Bank of Minneapolis, 2003.

McKenna, Marian. *Franklin Roosevelt and the Great Constitutional War.* New York: Fordham University Press, 2002.

Makey, Herman O. *Wendell Willkie of Elwood.* Elwood, Ind.: National Book Company, 1940.

Mangione, Jerre. *The Dream and the Deal: The Federal Writers' Project, 1935–1943.* Philadelphia: University of Pennsylvania Press, 1982.

Margulies, Sylvia. *The Pilgrimage to Russia: The Soviet Union and the Treatment of Foreigners, 1924–1937.* Madison: University of Wisconsin Press, 1968.

Martin, George. *Madam Secretary: Frances Perkins.* Boston: Houghton Mifflin, 1976.

Maurer, James Hudson. *It Can Be Done.* New York: Rand School Press, 1938.

Mellon, Andrew, *Taxation: The People's Business.* New York: Macmillan, 1924.

Mellon, Paul, with John Baskett. *Reflections in a Silver Spoon: A Memoir.* New York: William Morrow, 1992.

Meltzer, Allan. *A History of the Federal Reserve.* Vol. 1. Chicago: University of Chicago Press, 2002.

Meltzer, Milton. *Dorothea Lange: A Photographer's Life.* Syracuse, N.Y.: Syracuse University Press, 2000.

Meyers, William Starr, and Walter H. Newton. *The Hoover Administration: A Documented Narrative.* New York: Charles Scribners' Sons, 1936.

Mills, Ogden L. *The Seventeen Million.* New York: Macmillan, 1937.

———. *What of Tomorrow?* New York: Macmillan, 1935.

Moley, Raymond. *After Seven Years.* New York: Harper and Brothers, 1939.

———. *Twenty-seven Masters of Politics.* New York: Funk and Wagnalls, 1949.

Morgenthau, Henry III. *Mostly Morgenthaus: A Family History.* New York: Ticknor and Fields, 1991.

Moscow, Warren. *Roosevelt and Willkie.* Englewood Cliffs, N.J.: Prentice-Hall, 1968.

Mott, Frank Luther. *Golden Multitudes: The Story of Best Sellers in the United States.* New York: Macmillan, 1947.

Moulton, Harold G., et al. *The Recovery Problem in the United States.* Washington, D.C.: Brookings Institution, 1936.

Namorato, Michael V. *Rexford G. Tugwell: A Biography.* New York: Praeger, 1988.

Nash, George H. *The Life of Herbert Hoover.* Vols. I, II, III. New York: W.W. Norton, 1988, 1996.

Neal, Steve. *Dark Horse: A Biography of Wendell Willkie.* Garden City, N.Y.: Doubleday, 1984.

Neuse, Steven M. *David E. Lilienthal: The Journey of An American Liberal.* Knoxville: University of Tennessee Press, 1996.

Parker, Robert Allerton. *The Incredible Messiah: The Deification of Father Divine.* Boston: Atlantic Monthly Press, 1937.

Parrish, Michael E. *Felix Frankfurter and His Times: The Reform Years.* New York: Free Press, 1982.

Parrish, Wayne, and Wayne Weishaar. *Men without Money: The Challenge of Barter and Scrip.* New York: G. P. Putnam's Sons, 1933.

Paul, Randolph. *Taxation in the United States.* Boston: Little, Brown, 1954.

Patterson, James T. *America's Struggle against Poverty, 1900–1985.* Cambridge, Mass.: Harvard University Press, 1986.

Pearson, Drew. *Washington Merry-Go-Round.* New York: Liveright, 1931.

Pearson, Drew, and Robert S. Allen. *The Nine Old Men.* Garden City, N.Y.: Doubleday Doran, 1936.

Perkins, Frances. *The Roosevelt I Knew.* New York: Viking Press, 1946.

Peters, Charles. *Five Days in Philadelphia: The Amazing 'We Want Willkie!' Convention of 1940 and How It Freed FDR to Save the Western World.* New York: Public Affairs, 2005.

Phillips, Harlan B. *Felix Frankfurter Reminisces: An Intimate Portrait as Recorded in Talks with Dr. Harlan B. Phillips.* New York: Reynal, 1960.

Potter, David M. *People of Plenty: Economic Abundance and the American Character.* Chicago: University of Chicago Press, 1954.

Powell, Jim. *FDR's Folly: How Roosevelt and His New Deal Prolonged the Great Depression.* New York: Crown Forum, 2003.

Prestbo, John A. *The Market's Measure: An Illustrated History of America Told through the Dow Jones Industrial Average.* Princeton, N.J.: Dow Jones, 2000.

Pritchett, C. Herman. *The Tennessee Valley Authority: A Study in Public Administration.* Chapel Hill: University of North Carolina Press, 1943.

Reddekopp, G. Neil. "The Schechter Case and the Constitutionality of the NIRA." Unpublished paper, Department of History, University of Calgary, 1977.

Richberg, Donald R. *The Rainbow.* Garden City, N.Y.: Doubleday Doran, 1936.

Roosevelt, Eleanor. *My Day: The Best of Eleanor Roosevelt's Acclaimed Newspaper Columns, 1936–1962.* Edited by David Emblidge. New York: Da Capo Press, 2001.

Roosevelt, Elliott, ed. *The Roosevelt Letters.* Vol. 3. London: George G. Harrap, 1952.

Roosevelt, James, with Bill Libby. *My Parents: A Differing View.* Chicago: Playboy Press, 1976.

Root, Oren. *Persons and Persuasions.* New York: W. W. Norton, 1974.

Rosenman, Samuel. *Working with Roosevelt.* New York: Harper and Brothers, 1952.

Rothbard, Murray N. *America's Great Depression.* Auburn, Ala.: Mises Institute, 1962.

Saloutos, Theodore. *The American Farmer and the New Deal.* Ames: Iowa State University Press, 1982.

Schivelbusch, Wolfgang. *Three New Deals: Reflections on Roosevelt's America, Mussolini's Italy, and Hitler's Germany, 1933–1939.* Translated by Jefferson Chase. New York: Metropolitan Books, 2006.

Schlesinger, Arthur. *The Age of Roosevelt.* Boston: Houghton Mifflin, 2003.

Schwarz, Jordan A. *Interregnum of Despair: Hoover, Congress, and the Depression.* Champaign: University of Illinois Press, 1970.

———. *Liberal: Adolf A. Berle and the Vision of an American Era.* New York: Free Press, 1987.

———. *The New Dealers: Power Politics in the Age of Roosevelt.* New York: Vintage Books, Random House, 1994.

Schweikart, Larry. *The Entrepreneurial Adventure.* Albany, N.Y.: Thomson Learning, 2000.

Shannon, David A., ed. *The Great Depression.* New York: Prentice Hall, 1960.

Shoup, Carl, et al. *Facing the Tax Problem: A Survey of Taxation in the US and a Program for the Future.* New York: Twentieth Century Fund, 1937.

Sitkoff, Harvard. *A New Deal for Blacks: The Emergence of Civil Rights as a National Issue.* New York: Oxford University Press, 1978.

Skidelsky, Robert. *The Economist as Savior, 1920–1937.* New York: Penguin, 1992.

Smiley, Gene. *Rethinking the Great Depression: A New View of Its Causes and Consequences.* Chicago: Ivan R. Dee, 2002.

Sobel, Robert. *Coolidge: An American Enigma.* Chicago: Regnery, 1998.

Stalin, Joseph. *Interviews with Foreign Workers' Delegations.* New York: International Publishers (Glavlit), 1932.

Steel, Ronald. *Walter Lippmann and the American Century.* Boston: Little, Brown, 1980.

Stein, Herbert. *Fiscal Revolution in America: Policy in Pursuit of Reality.* Washington, D.C.: AEI Press, 1990.

———. *Presidential Economics: The Making of Economic Policy from Roosevelt to Reagan and Beyond.* New York: Simon & Schuster, 1984.

Sternsher, Bernard. *Rexford Tugwell and the New Deal.* New Brunswick, N.J.: Rutgers University Press, 1964.

Stevens, Joseph E. *Hoover Dam: An American Adventure.* Norman: University of Oklahoma Press, 1988.

Strum, Philippa. *Louis D. Brandeis: Justice for the People.* New York: Schocken Books, 1984.

Stryker, Roy Emerson, and Nancy Wood. *In This Proud Land: America, 1935–1943, as Seen in the FSA Photography.* New York: New York Graphic Society, 1975.

Sumner, William Graham. *The Forgotten Man and Other Essays.* New Haven, Conn.: Yale University Press, 1919.

———. *What Social Classes Owe Each Other.* Caldwell, Ohio: Caxton Press, 2003.

Talbert, Roy Jr. *FDR's Utopian: Arthur Morgan of the TVA.* Jackson: University Press of Mississippi, 1987.

Taft, Charles P. *You and I—and Roosevelt.* New York: Farrar and Rinehart, 1936.

Temin, Peter. *Lessons from the Great Depression.* Cambridge: Massachusetts Institute of Technology, 1989.

Thomsen, Robert. *Bill W.: The Absorbing and Deeply Moving Life Story of Bill Wilson, Co-Founder of Alcoholics Anonymous.* Center City, Minn.: Hazelden Press, 1975.

Tobin, James. *Great Projects: The Epic Story of the Building of America from the Taming of the Mississippi to the Invention of the Internet.* New York: Free Press, 2001.

Tugwell, Rexford G. *The Brains Trust.* New York: Viking Press, 1968.

———. *The Diary of Rexford G. Tugwell, 1932–1944.* Edited by Michael Vincent Namorato. New York: Greenwood Press, 1992.

———. *The Light of Other Days.* Garden City, N.Y.: Doubleday, 1962.

———. "Roosevelt and the Bonus Marchers of 1932." *Political Science Quarterly* 87, no. 3 (September 1972): 363–76.

———. *To the Lesser Heights of Morningside: A Memoir.* Philadelphia: University of Pennsylvania Press, 1982.

Tugwell, Rexford G., and Howard C. Hill. *Our Economic Society and Its Problems.* New York: Harcourt Brace, 1934.

Vangermeersch, Richard. *The Life and Writings of Stuart Chase (1888–1985).* Studies in the Development of Accounting Thought, vol. 8. New York: Elsevier, 2005.

Vedder, Richard K., and Lowell E. Gallaway. *Out of Work: Unemployment and Government in Twentieth-Century America.* New York: Holmes & Meier, 1993.

Villard, Oswald Garrison. *Russia from a Car Window.* New York: Nation Magazine, 1929.

Warburg, James P. *Hell Bent for Election.* Garden City, N.Y.: Doubleday, Doran, 1935.

Watts, Jill. *God, Harlem, USA: The Father Divine Story.* Berkeley and Los Angeles: University of California Press, 1992.

Weisbrot, Robert. *Father Divine: The Utopian Evangelist of the Depression Era Who Became an American Legend.* Boston: Beacon Books, 1983.

Wenger, Beth S. *New York Jews and the Great Depression: Uncertain Promise.* Syracuse, N.Y.: Syracuse University Press, 1999.

Whitman, Wilson. *David Lilienthal: Public Servant in a Power Age.* New York: Henry Holt, 1948.

Wickenden, Dorothy, ed. *The New Republic Reader: Eighty Years of Opinion and Debate.* New York: Basic Books, 1994.

Wicker, Elmus. *The Banking Panics of the Great Depression: Studies in Macroeconomic History.* Cambridge: Cambridge University Press, 2000.

Willkie, Wendell. *This Is Wendell Willkie.* New York: Dodd, Mead, 1940.

Wilson, Joan Hoff. *Herbert Hoover: Forgotten Progressive.* New York: HarperCollins, 1992.

Wilson, Lois. *Lois Remembers.* New York: Al-Anon Family Group HQ, 1979.

Zinn, Howard, ed. *New Deal Thought.* Indianapolis: Bobbs-Merrill, 1966.

Films

Our Daily Bread and Other Great Films of the Great Depression (DVD). Image Entertainment, 1999.

Index